Lecture Notes in Computer Scier

Edited by G. Goos, J. Hartmanis and J. van L

T0250744

Springer
Berlin
Heidelberg
New York
Barcelona
Hong Kong
London
Milan
Paris
Tokyo

Yiannis Cotronis Jack Dongarra (Eds.)

Recent Advances in Parallel Virtual Machine and Message Passing Interface

8th European PVM/MPI Users' Group Meeting
Santorini/Thera, Greece, September 23-26, 2001
Proceedings

 Springer

Series Editors

Gerhard Goos, Karlsruhe University, Germany
Juris Hartmanis, Cornell University, NY, USA
Jan van Leeuwen, Utrecht University, The Netherlands

Volume Editors

Yiannis Cotronis
University of Athens
Department of Informatics and Telecommunications
Panepistimiopolis, 157 84, Greece
E-mail: cotronis@di.uoa.gr

Jack Dongarra
University of Tennessee
Innovative Computing Lab., Computer Science Department
1122 Volunteer Blvd, Knoxville, TN, 37996-3450, USA
E-mail: dongarra@cs.utk.edu

Cataloging-in-Publication Data applied for

Die Deutsche Bibliothek - CIP-Einheitsaufnahme

Recent advances in parallel virtual machine and message passing interface :
proceedings / 8th European PVM MPI Users' Group Meeting, Santorini/Thera,
Greece, September 23 - 26, 2001. Yiannis Cotronis ; Jack Dongarra (ed.). -
Berlin ; Heidelberg ; New York ; Barcelona ; Hong Kong ; London ; Milan ;
Paris ; Tokyo : Springer, 2001
    (Lecture notes in computer science ; Vol. 2131)
    ISBN 3-540-42609-4

CR Subject Classification (1998): D.1.3, D.3.2, F.1.2, G.1.0, B.2.1, C.1.2

ISSN 0302-9743
ISBN 3-540-42609-4 Springer-Verlag Berlin Heidelberg New York

Springer-Verlag Berlin Heidelberg New York
a member of BertelsmannSpringer Science+Business Media GmbH

http://www.springer.de

© Springer-Verlag Berlin Heidelberg 2001
Printed in Germany

Typesetting: Camera-ready by author, data conversion by Olgun Computergrafik
Printed on acid-free paper   SPIN 10840070   06/3142   5 4 3 2 1 0

Preface

Parallel Virtual Machine (PVM) and Message Passing Interface (MPI) are the most frequently used tools for programming according to the message passing paradigm, which is considered one of the best ways to develop parallel applications.

This volume comprises 50 revised contributions presented at the Eighth European PVM/MPI Users' Group Meeting, which was held on Santorini (Thera), Greece, 23–26 September 2001. The conference was organized by the Department of Informatics and Telecommunications, University of Athens, Greece.

This conference has been previously held in Balatofüred, Hungary (2000), Barcelona, Spain (1999), Liverpool, UK (1998), and Krakow, Poland (1997). The first three conferences were devoted to PVM and were held at the TU Munich, Germany (1996), the ENS Lyon, France (1995), and the University of Rome (1994).

This conference has become a forum for users and developers of PVM, MPI, and other message passing environments. Interaction between these groups has proved to be very useful for developing new ideas in parallel computing and for applying some of those already existent to new practical fields. The main topics of the meeting were evaluation and performance of PVM and MPI, extensions and improvements to PVM and MPI, algorithms using the message passing paradigm, and applications in science and engineering based on message passing. The conference included one tutorial on MPI and 9 invited talks on advances in MPI, cluster computing, network computing, Grid computing, and parallel programming and programming systems. These proceedings contain papers on the 46 oral presentations together with 4 poster presentations.

Invited speakers of Euro PVM/MPI were Frederica Darema, Al Geist, Bill Gropp, Domenico Laforenza, Phil Papadopoulos, Alexander Reinefeld, Thomas Sterling, and Vaidy Sunderam.

We would like to express our gratitude for the kind support of Compaq, HP-Germany, IBM, Microsoft, MPI Software Technology, M-Data, Silicon Computers Ltd, SUN Microsystems, TurboLinux, the Ministry of Education and Religious Affairs of Greece, and the University of Athens. Also, we would like to thank the members of the Program Committee and the other reviewers for their work in refereeing the submitted papers and ensuring the high quality of Euro PVM/MPI.

September 2001

Yiannis Cotronis
Jack Dongarra

Program Committee

Vassil Alexandrov   University of Reading, UK
Ranieri Baraglia   CNUCE, Pisa, Italy
Arndt Bode   LRR – Technische Universität München, Germany
Marian Bubak   Institute of Computer Science and ACC CYFRONET, AGH, Krakow, Poland

Jacques Chassin-
de-Kergommeaux   ID IMAG – Grenoble, France
Yiannis Cotronis   University of Athens, Greece
José Cunha   Universidade Nova de Lisboa, Portugal
Erik D'Hollander   University of Gent, Belgium
Frederic Desprez   LIP – ENS Lyon and INRIA France
Jack Dongarra   University of Tennessee and ORNL, USA
Graham Fagg   University of Tennessee, USA
Al Geist   Oak Ridge National Labs, USA
Michael Gerndt   LRR – Technische Universität München, Germany
Andrzej Goscinski   Deakin University, Australia
Rolf Hempel   C&C Research Labs, NEC Europe Ltd., Germany
Ladislav Hluchý   Slovak Academy of Science – Bratislava, Slovakia
Peter Kacsuk   SZTAKI, Hungary
Jan Kwiatkowski   Wroclaw University of Technology, Poland
Domenico Laforeza   CNUCE-Inst. of the Italian National Res. Council, Italy
Miron Livny   University of Wisconsin – Madison, USA
Thomas Ludwig   LRR – Technische Universität München, Germany
Emilio Luque   Universitat Autònoma de Barcelona, Spain
Tomàs Margalef   Universitat Autònoma de Barcelona, Spain
Hermann Mierendorff   GMD, Germany
Shirley Moore   University of Tennessee, USA
Benno Overeinder   University of Amsterdam, The Netherlands
Andrew Rau-Chaplin   Dalhousie University – Halifax, Canada
Jeff Reeve   University of Southampton, UK
Yves Robert   LIP – Ecole Normale Superieure de Lyon, France
Casiano Rodríguez   Universidad de La Laguna, SPAIN
Wolfgang Schreiner   RISC-Linz, – Johannes Kepler Univeristy, Linz, Austria
Miquel A. Senar   Universitat Autònoma de Barcelona, Spain
João Gabriel Silva   Universidade de Coimbra, Portugal
Vaidy Sunderam   Emory University – Atlanta, USA
Francisco Tirado   Universidad Computense de Madrid, Spain
Bernard Tourancheau   RESAM, UCB-Lyon and INRIA, France
Pavel Tvrdík   Czech Technical University, Czeck Republic
Jerzy Wásniewski   The Danish Computing Centre for Research and Education, Lyngby, Denmark
Roland Wismüller   LRR – Technische Universität München, Germany

Additional Reviewers

Astaloš Jan	Slovak Academy of Science – Bratislava, Slovakia
Balis Bartosz	Institute of Computer Science AGH, Cracow, Poland
Balogh Zoltan	Slovak Academy of Science – Bratislava, Slovakia
Birra Fernando	Universidade Nova de Lisboa, Portugal
Buhmann Martin	Universität Giessen, Germany
Caron Eddy	LIP ENS Lyon, France
Dobrucky Miroslav	Slovak Academy of Science – Bratislava, Slovakia
Eavis Todd	Dalhousie University, Canada
Ferrini Renato	CNUCE, Pisa, Italy
Funika Wlodzimierz	Institute of Computer Science AGH, Cracow, Poland
Hernandez Mario	Universidad de las Palmas de Gran Canaria, Spain
Leon Coromoto	Universidad de La Laguna, Spain
Lindermeier Markus	LRR – Technische Universität München, Germany
Luksch Peter	LRR – Technische Universität München, Germany
Martakos Drakoulis	University of Athens, Greece
Miranda Javier	Universidad de las Palmas de Gran Canaria, Spain
Namyst Raymond	LIP ENS Lyon, France
Palmerini Paolo	CNUCE, Pisa, Italy
Perego Raffaele	CNUCE, Pisa, Italy
Perez Juan C.	Universidad de La Laguna, Spain
Quinson Martin	LIP ENS Lyon, France
Rackl Gunther	LRR – Technische Universität München, Germany
Sayas Francisco-Javier	Universidad de Zaragoza, Spain
Suter Frederic	LIP ENS Lyon, France
Tran D.V.	Slovak Academy of Science – Bratislava, Slovakia
Zajac Katarzyna	Institute of Computer Science AGH, Cracow, Poland

Table of Contents

Invited Speakers

Implementation, Evaluation and Performence of PVM/MPI

Extensions and Improvements on PVM/MPI

Tools for PVM and MPI

Algorithms Using Message Passing

Algorithms in Science and Engineering

The SPMD Model: **Past, Present and Future**

Frederica Darema

National Science Foundation

I proposed the *SPMD* (Single Program Multiple Data) model, in January 1984[1], as a means for enabling parallel execution of applications on multiprocessors, and in particular for highly parallel machines like the RP3 (the IBM Research Parallel Processor Prototype[2]). This talk will provide a review of the origins of the SPMD, it's early use in enabling parallel execution of scientific applications and it's implementation in one of the first[3] parallel programming environments. In fact [3]was the first programming environment that implemented the *SPMD* model; other environments in the 1985 timeframe were based on the fork-and-join (*F&J*) model.

The *SPMD* model allowed processes cooperating in the execution of a program to self-schedule themselves, and dynamically get assigned the work to be done in this co-operative execution. The model also allowed the processes to act on different sets of data, and the designation of data as private to each process as well as data shared among the processes. The implementation also introduced the notion of "*replicate*" execution where appropriate as more efficient. From its inception the *SPMD* model was more general than the *vector*, the *SIMD*, and the *data-parallel* model. While a number of people have identified and characterized the *SPMD* model as a *data-parallel* model, *SPMD* has been from the outset more general than that, as it has allowed multiple (and distinct) instruction streams to be executed concurrently and allow these streams to act on different data (not limited only to *data-parallel*). The *SPMD* model is different from the *F&J* model, and the generality of the *SPMD* was challenged by some who had adopted F&J for parallel execution of end-user applications. However the *SPMD* model allowed efficient parallel execution by providing control at the application level rather the costly OS level control; F&J entailed more complex, costly, and laborious process control. The *SPMD* model demonstrated that it was easy and efficient to map applications on parallel machines, and has served adequately the efforts to exploit parallel/distributed computing for scientific/engineering (*S/E*) applications. Many environments have implemented *SPMD* for *S/E* computing, the most popular today being *PVM* and *MPI*. *SPMD* has also been implemented in *COBOL* environments to exploit parallel processing of commercial applications. The model is also used on today's computational grid platforms (*GRIDS*), which employ a number of SPMD as well as F&J programming environments (*MPI, multithreading, CORBA*). Given today's dynamic computing platforms and applications, the question arises of whether *SPMD* is sufficient and what the model(s) of the future will be. The talk will elaborate more on these issues.

---

[1] IBM internal memo January 1984, and a *VM Parallel Environment* by F. Darema-Rogers, D. George, V. A. Norton and G. Pfister, at the Proceedings of the IBM Kingston Parallel Processing Symposium, Nov. 27-29, 1984 – IBM Confidential

[2] G. Pfister , et al, Proceedings of the ICPP, August 1985

[3] F. Darema-Rogers, et al in "VM/EPEX, A VM-based Environment for Parallel Execution", IBM Res.Rep. RC11407(1985); and Parallel Computing 7(1988)11-24, North Holland

Y. Cotronis and J. Dongarra (Eds.): Euro PVM/MPI 2001, LNCS 2131, p. 1, 2001.
© Springer-Verlag Berlin Heidelberg 2001

Building a Foundation for the Next PVM: Petascale Virtual Machines

G. Al Geist

Oak Ridge National Laboratory,
PO Box 2008,
Oak Ridge, TN 37831-6367
gst@ornl.gov
http://www.csm.ornl.gov/~geist

Abstract. High performance computing is on the verge of another major jump in scale and now is the time for us to begin creating super-scalable applications and systems software if we are to exploit the coming machines. In the early 1990s there was a major jump in scalability from a few hundred to a few thousand processors. Throughout the 1990s the size of the largest machines remained around this scale and PVM and MPI were key to making these machines useable.

Now we are seeing changes such as SETI@home and IBM's Blue Gene systems that suddenly jump the scale of machines to 100,000 processors or more. Our software environment and applications are unprepared for this jump. This talk looks at the common software needs between these very different ways to get to this scale. The talk will argue that software may need to fundamentally change to allow us to exploit machines that approach a petascale. Interspersed with this discussion will be results from research going on today to increase the scalability, fault tolerance, and adaptability of MPI and systems software.

1 Background

Advanced scientific computing is an essential part of all 21st Century scientific research programs. Extraordinary advances in computing technology in the past decade have set the stage for major advances in scientific computing. In the 1994 ORNL installed a 3000 processor Intel Paragon. At the time it was the most powerful and largest scale computing system in the world. Although other computers soon surpassed the Paragon in power, their scale remained in the range of a few thousand processors. Even today the largest computer systems have only 9000 processors. PVM and MPI are the key software packages that allowed these large-scale machines to be usable and will continue to be the programming paradigm of choice. But this level of scalablity is about to dramatically change. Much like the early 1990s, we are on the verge of a two orders of magnitude jump in scalability. IBM has recently announced the development of BlueGene. This computer, designed to model a folding protein, will have a million processors and a petaflop computing capacity. In a completely different approach

Y. Cotronis and J. Dongarra (Eds.): Euro PVM/MPI 2001, LNCS 2131, pp. 2–6, 2001.

to large-scale computing, companies like SETI@home are utilizing hundreds of thousands of PCs scattered across the Internet to solve business and research problems. SETI@home is primarily focused on radio signal processing to detect signs of intelligence outside earth. Efforts by such companies are creating huge, planet-wide virtual machines.

Both approaches share common needs in fault tolerant, adaptable systems software and highly scalable applications. This paper explores the need for super scalable algorithms and describes some efforts that have started in the U.S.A. under the DOE Scientific Discovery through Advanced Computing (SciDAC) initiative to create a standard set of scalable systems software interfaces. The goal of this effort is to do for system software what MPI did for message-passing. The paper will discuss new approaches that may be required for applications to survive on systems of such large scale. Message-passing may no longer be enough. Fault tolerance and adaptability may become as important as latency and bandwidth. Research into fault tolerant MPI, adaptable collective communication, and distributed control will be described to show where we are today in developing these new approaches.

2 Need for Super-scalable Algorithms

Within the next five to ten years, computers 1,000 times faster than today's computers comprised of 100,000 processors will become available. There is little understanding of how to manage or run such large systems, nor how applications can exploit such architectures. To utilize such Petascale Virtual Machines, new computational tools and codes are needed to enable the creation of realistic simulations of physical situations and to provide new insights into a host of scientific problems. Research, development, and deployment of mathematical models, computational methods, numerical libraries, and scientific codes are needed to take full advantage of the capabilities of such computers for strategic or critical problems in biology, materials sciences, chemistry, high-energy physics, climate, and others. Today's system software only works up to several hundred processors. Current research in scalable software looks to push this up to a few thousand. These solutions are likely to be inadequate for computers with 20,000-100,000 processors. Radically new approaches are needed in the development of "Super Scalable Software" both system software and application software, but first the community must build a foundation and establish standard interfaces on which to work together on this software.

High-end systems software is a key area of need on the large machines. The systems software problems for teraop class computers with thousands of processors are significantly more difficult than for small-scale systems with respect to fault-tolerance, reliability, manageability, and ease of use for systems administrators and users. Layered on top of these are issues of security, heterogeneity and scalability found in today's large computer centers. The computer industry is not going to solve these problems because business trends push them towards smaller systems aimed at web serving, database farms, and departmental sized

systems. In the longer term, the operating system issues faced by next generation petaop class computers will require research into innovative approaches to
systems software that must be started today in order to be ready when these
systems arrive.

In August 2001, the DOE SciDAC initiative awarded funding for the creation
of a Scalable Systems Software Center. The Center is not a physical place but
rather a virtual community of experts working to make systems software more
scalable. In addition to solving problems for the largest systems, the open source
software produced by the Center will provide many benefits for the myriad of
smaller scale systems. The standard interfaces established by the Center will
provide a basis for the eventual adoption and deployment of these technologies
in the commercial marketplace.

There are several concepts that are important to the long-term success of any
effort to improve system reliability and management. First is modularity, since a
modular system is much easier to adapt, update, and maintain, in order to keep
up with improvements in hardware and other software. Publicly documented
interfaces are a requirement because it is unlikely that any package can provide
the flexibility to meet the needs of every site. So, a well-defined API will allow a
site to replace or augment individual components as needed. Defining the API's
between components across the entire system software architecture, provides an
integrating force between the system components as a whole and improves the
long-term usability and manageability of terascale systems around the world.

Center will use the MPI approach to the development of standardized interfaces. A team of experts (including industry participation) collectively agree on
and specify standardized interfaces between system components through a series
of regular meetings of the group. The interfaces must be platform independent
wherever possible. Before defining the interfaces the team needs to converge
on an architecture for system software, including tools for hierarchical booting,
system monitoring, and resource management. As part of specifying interfaces,
the homegrown tools at various supercomputer sites as well as packages like
PBS will be modified to conform to the architecture, and thus help validate the
completeness of the interfaces before they are finalized.

Organizations interested in participating in this scalable systems effort should
contact the project coordinator, Al Geist, at gst@ornl.gov.

3   New Approaches

Exploiting Petascale Virtual Machines is going to take more than changes to
systems software. The need to solve ever harder scientific problems in biology and
chemistry are this decade's driving force to petascale computing. Super scalable
applications are a key need and fundamentally new approaches to application
development will be required to allow programs to run across 100,000 processor
systems. Today scientists are satisfied if they can get a fast MPI package to
link to their application. Issues like the mean time to failure for the machine
being less than the length of the job are rarely considered. With very large scale

systems like BlueGene and SETI@home failure of some portion of the machine is constant. In addition to message passing, applications will need some form of fault tolerance built in and supported by the systems software.

As part of the Harness project, UTK is developing a fault tolerant MPI called FT-MPI. This package of the 30 most used functions in MPI allows applications to be developed that can dynamically recover from failures of tasks within the MPI job. FT-MPI allows the application to recreate a new MPI_COMM_WORLD communicator on the fly and continue with the computation. FT-MPI also supplies the means to notify the application that it has had a failure and needs to recover. MPI_COMM_SPAWN is included in FT-MPI to provide a means to replace failed tasks where appropriate. The FT-MPI spawn function has several recovery options such as allowing the new task to replace the rank of the failed task in the same communicator.

Failure inside a MPI collective communication call is a particularly tricky case since some tasks may have already received their data and returned before the failure occurs. FT-MPI has defined the semantics of all its collective communication calls consistently. Every task in a failed collective communication call will either return with the data it would have received in a non-failure case or return with an error message about the failure. More fault tolerant MPI implementations and the use of them in super scalable algorithms is expected to increase significantly in the next decade.

Adaptability of applications is another area that needs significant work in order to be able to exploit Petascale Virtual Machines. There are many aspects of adaptability. One reason for adaptability is performance. This can include selecting versions of math libraries that are tuned to a particular node or machine. Work along these lines includes the ATLAS work by Jack Dongarra. It can also include selecting optimal collective communication operations based on the number of tasks participating and the underlying architecture of the machine. This is an ongoing effort in the Harness project, which will be described in more detail in the talk.

A second reason for adaptability is failure. The approach today is based on checkpoint/restart, but this becomes less attractive as the number of tasks grows very large. For example, you don't want to restart 40,000 jobs because 1 failed. And if the restart takes too long then another failure is likely before the restart is even complete, leading to a cycle where no progress is made on the application. Moving to packages like a fault tolerant MPI provide the opportunity to dynamically adapt to problems.

Underlying any fault tolerant or adaptability package is a runtime environment that is responsible for monitoring, notification, and management of all the tasks running on the machine. This runtime environment itself can be viewed as a distributed application that must be fault tolerant. Harness can be viewed as an experimental version of such a distributed runtime environment. Research in the past year has significantly improved the scalability and fault tolerance of the distributed control algorithms in Harness. An update on this work will be given in the talk.

A third advantage of adaptability is predicting failure and migrating to avoid it in the first place. While little work has been done in this area for parallel computers, it could be an important approach for super scalable algorithms running for extended periods on petascale machines. Such adaptability requires the integration of several existing areas: system monitoring, statistical analysis of normal and abnormal conditions, applications listening for notification, and tasks able to be migrated to other parts of the machine. Research has been done in each of these areas over the past decade. The next step is to integrate all this work into a predictive failure system for petascale machines.

4   Conclusion

High performance computing is on the verge of another major jump in scale and now is the time for us to begin creating super-scalable applications and systems software if we are to exploit the coming machines. Within the next ten years, computers 1,000 times faster than today's computers comprised of 100,000 processors will be available. There is little understanding of how to manage or run such large systems, nor how applications can exploit such architectures. To utilize such Petascale Virtual Machines, we will require new approaches to developing application software that incorporate fault tolerance and adaptability. Scalable system software for such machines is non-existent. In this paper we described the need for new approaches to application software and how ongoing research in fault tolerant MPI and Harness are just beginning to address these needs. We also describe a newly formed Scalable Systems Software Center to begin to build an integrated suite of software as a foundation for coming Petascale Virtual Machines.

Challenges and Successes
in Achieving the Potential of MPI*

William D. Gropp

Mathematics and Computer Science Division
Argonne National Laboratory
Argonne, Illinois 60439

Abstract. The first MPI standard specified a powerful and general message-passing model, including both point-to-point and collective communications. MPI-2 took MPI beyond simple message-passing, adding support for remote memory operations and parallel I/O. Implementations of MPI-1 appeared with the MPI standard; implementations of MPI-2 are continuing to appear. But many implementations build on top of a point-to-point communication base, leading to inefficiencies in the performance of the MPI implementation. Even for MPI-1, many MPI implementations base their collective operations on relatively simple algorithms, built on top of MPI point-to-point (or a simple lower-level communication layer). These implementations achieve the functionality but not the scalable performance that is possible in MPI. In MPI-2, providing a high-performance implementation of the remote-memory operations requires great care and attention to the opportunities for performance that are contained in the MPI standard.

One of the goals of the MPICH2 project is to provide an easily extended example of an implementation of MPI that goes beyond a simple point-to-point communication model. This talk will discuss some of the challenges in implementing collective, remote-memory, and I/O operations in MPI. For example, many of the best algorithms for collective operations involve the use of message subdivision (possibly involving less than one instance of a MPI derived datatype) and multisend or store-and-forward operations. As another example, the remote memory operations in MPI-2 specify semantics that are designed to specify precise behavior, excluding ambiguities or race conditions. These clean (if somewhat complex) semantics are sometimes seen as a barrier to performance. This talk will discuss some of the methods that can be used to exploit the RMA semantics to provide higher performance for typical application codes. The approaches taken in MPICH2, along with current results from the MPICH2 project, will be discussed.

* This work was supported by the Mathematical, Information, and Computational Sciences Division subprogram of the Office of Advanced Scientific Computing, U.S. Department of Energy, under Contract W-31-109-Eng-38.

Y. Cotronis and J. Dongarra (Eds.): Euro PVM/MPI 2001, LNCS 2131, p. 7, 2001.

Programming High Performance Applications in Grid Environments

Domenico Laforenza

Advanced Computing Department
CNUCE-Institute of the Italian National Research Council
via Vittorio Alfieri, 1, I-56010 Ghezzano, Pisa, Italy
Domenico.Laforenza@cnuce.cnr.it

Abstract. The need for realistic simulations of complex systems relevant to the modeling of several modern technologies and environmental phenomena increasingly stimulates the development of advanced computing approaches. Nowadays it is possible to cluster or couple a wide variety of resources including supercomputers, storage systems, data sources, and special classes of devices distributed geographically and use them as a single unified resource, thus forming what is popularly known as a "computational grid" [1,2].

Grid Computing enables the development of large scientific applications on an unprecedented scale. Grid-aware applications (meta-applications, multidisciplinary applications) make use of coupled computational resources that cannot be replicated at a single site. In this light, grids let scientists solve larger or new problems by pooling together resources that could not be coupled easily before. Designing and implementing grid-aware applications often require interdisciplinary collaborations involving aspects of scientific computing, visualization, and data management [3]. Multi-disciplinary applications are typically composed of large and complex components, and some of them are characterized by huge high performance requirements [4,5,6,7]. In order to get better performance, the challenge is to map each component onto the best candidate computational resource having a high degree of affinity with the software component. This kind of mapping is a non-trivial task. Moreover, it is well known that, in general, the programmer's productivity in designing and implementing efficient parallel applications on high performance computers remains a very time-consuming task. Grid computing makes the situation worse as heterogeneous computing environments are combined so that the programmer must manage an enormous amount of details. Consequently, the development of grid programming environments that would enable programmers to efficiently exploit this technology is an important and hot research issue. A grid programming environment should include interfaces, APIs, utilities and tools so as to provide a rich development environment. Common scientific languages such as C, C++, Java and Fortran should be available, as should application-level interfaces like MPI and PVM. A range of programming paradigms should be supported, such as message passing and distributed shared memory. In addition, a suite of numerical and other commonly used libraries should be available.

Today, an interesting discussion is opened about the need to think at new abstract programming models and develop novel programming techniques address-

Y. Cotronis and J. Dongarra (Eds.): Euro PVM/MPI 2001, LNCS 2131, pp. 8–9, 2001.

ing specifically the grid, which would deal with the heterogeneity and distributed computing aspects of grid programming [8].

In this talk, after an introduction on the main grid programming issues, an overview of the most important approaches/projects conducted in this field worldwide will be presented. In particular, the speaker's contribution in designing some grid extension for a new programming environment will be shown. This work constitutes a joint effort conducted by some academic and industrial Italian partners, in particular the Department of Computer Science of the Pisa University and CNUCE-CNR, in the framework of the ASI-PQE2000 National Project aimed at building ASSIST (A Software development System based on Integrated Skeleton Technology) [9,10,11]. The main target for the ASSIST Team is to build of a new programming environment for the development of high performance applications, based on the integration of the structured parallel programming model and the objects (components) model. In this way, ASSIST should be available for a wide range of hardware platforms from the homogeneous parallel computers (MPP, SMP, CoWs) to the heterogeneous ones (Grids).

References

1. I. Foster, C. Kesselman (eds), *The Grid: Blueprint for a future computing infrastructure*, Morgan Kaufmann, 1999.
2. I. M.A. Baker, R. Buyya, and D. Laforenza, *The Grid: International Efforts in Global Computing*, Proceedings of the International Conference on Computer and eBusiness, Scuola Superiore Reiss Romoli, L'Aquila, Italy, July 31.2000 - August 6, 2000.
3. K. Keahey, P. Beckman, and J. Ahrens, *Ligature: Component Architecture for High Performance Applications*, The International Journal of High Performance Computing Applications, Volume 14, No 4, Winter 2000, pp. 347-356.
4. R. Armstrong, D. Gannon, A. Geist, K. Keahey, S. Kohn, L. McInnes, S. Parker, and B. Smolinski, *Toward a common component architecture for high performance scientific computing*, In Proceedings of the 8th High Performance Distributed Computing (HPDC'99), 1999.
5. C. René, T. Priol, *MPI Code Encapsulation using Parallel CORBA Object*, Proceedings of the Eighth IEEE International Symposium on High Performance Distributed Computing, IEEE, pages 3–10, August 1999.
6. J. Darlington et al, *Parallel programming using skeleton functions*, LNCS 694, 1995, Springer-Verlag.
7. S. Newhouse, A. Mayer, and John Darlington, *A Software Architecture for HPC Grid Applications*, A. Bode et al. (Eds.): Euro-Par 2000, LNCS 1900, pp. 686-689, 2000, Springer-Verlag.
8. Global Grid Forum, *Advanced Programming Models Working Group*, www.gridforum.org and www.eece.unm.edu/~apm.
9. M. Vanneschi, *PQE2000: HPC tools for industrial applications*, IEEE Concurrency, Oct.-Dec. 1998.
10. B. Bacci, M. Danelutto, S. Pelagatti, and M. Vanneschi, *SkIE: a heterogeneous environment for HPC applications*, Parallel Computing, Volume 25, pages 1827–1852, December 1999.
11. P. Ciullo, M. Danelutto, L. Vaglini, M. Vanneschi, D. Guerri, M. Lettere, *Ambiente ASSIST: Modello di programmazione e linguaggio di coordinamento ASSIST-CL (Versione 1.0)*, Maggio 2001 (Progetto ASI-PQE2000, Deliverable no. 1, in Italian).

NPACI Rocks Clusters:
Tools for Easily Deploying and Maintaining Manageable High-Performance Linux Clusters

Philip M. Papadopoulos, Mason J. Katz, and Greg Bruno

San Diego Supercomputer Center, La Jolla, CA, USA
{phil,mjk,bruno}@sdsc.edu
http://rocks.npaci.edu

Abstract. High-performance computing clusters (commodity hardware with low-latency, high-bandwidth interconnects) based on Linux, are rapidly becoming the dominant computing platform for a wide range of scientific disciplines. Yet, straightforward software installation, maintenance, and health monitoring for large-scale clusters has been a consistent and nagging problem for non-cluster experts. The complexity of managing hardware heterogeneity, tracking security and bug fixes, insuring consistency of software across nodes, and orchestrating wholesale (or forklift) upgrades of Linux OS releases (every 6 months) often discourages would-be cluster users.

The NPACI Rocks toolkit takes a fresh perspective on management and installation of clusters to dramatically simplify this software tracking. The basic notion is that complete (re)installation of OS images on every node is an easy function and the preferred mode of software management. The NPACI Rocks toolkit builds on this simple notion by leveraging existing single-node installation software (Red Hat's Kickstart), scalable services (e.g., NIS, HTTP), automation, and database-driven configuration management (MySQL) to make clusters approachable and maintainable by non-experts. The benefits include straightforward methods to derive user-defined distributions that facilitate testing and system development and methods to easily include the latest upgrades and security enhancements for production environments. Installation performance has good scaling properties with a complete reinstallation (from a single server 100 Mbit http server) of a 96-node cluster taking only 28 minutes. This figure is only 3 times longer than reinstalling just a single node.

The toolkit incorporates the latest Red Hat distribution (including security patches) with additional cluster-specific software. Using the identical software tools that are used to create the base distribution, users can customize and localize Rocks for their site. This flexibility means that the software structure is dynamic enough to meet the needs of cluster-software developers, yet simple enough to allow non-experts to effectively manage clusters. Rocks is a solid infrastructure and is extensible so that the community can adapt the software toolset to incorporate the latest functionality that defines a modern computing cluster. Strong adherence to widely-used (*de facto*) tools allows Rocks to move with the rapid pace of Linux development.

Y. Cotronis and J. Dongarra (Eds.): Euro PVM/MPI 2001, LNCS 2131, pp. 10–11, 2001.

Rocks is designed to build HPC clusters and has direct support for Myrinet, but can support other high-speed networking technologies as well. Our techniques greatly simplify the deployment of Myrinet-connected clusters and these methods will be described in detail during the talk, including how we manage device driver/kernel compatibility, network topology/routing files, and port reservation. We will also give "untuned" Linpack performance numbers using the University of Tennessee HPL suite to illustrate the performance that a generic Rocks cluster would expect to attain with no specialized effort.

Version 2.1 (corresponding to Redhat Version 7.1) of the toolkit is available for download and installation.

Clusters for Data-Intensive Applications in the Grid

Alexander Reinefeld

Konrad-Zuse-Zentrum für Informationstechnik Berlin
Humboldt-Universität zu Berlin
ar@zib.de

1 Background

Large-scale data-intensive applications like the high energy physics codes developed at Cern rely to a great extent on the ability to utilize geographically distributed computing and data-storage resources. Consequently, much effort has been spent on the design of Grid middleware that organizes the coordinated use of distributed resources. Software packages like Globus, Cactus, Legion, WebFlow, NetSolve, Ninf, Nimrod-G, Condor-G, and many others have already been successfully deployed, and some of them are in daily use in the various science and research laboratories throughout the world.

Once the idea of Grid Computing [1] grows beyond that of an academic experiment, the demand for computing power will outgrow the current supply. It will then be no longer feasible to use costly supercomputers as compute nodes. Commodity clusters will become cost-effective alternatives—not only for high-throughput computing, but also for running parallel high-performance applications in Grid environments.

Our work focuses on the design of grid-aware high-performance compute clusters that can be used as nodes in the Grid. The impetus for this work came from the Datagrid project, which is the first project that completely relies on the utilization of worldwide distributed commodity clusters. The clusters are used for the analysis and simulation of the particle collision data that will be generated by the four experiments planned for the Large Hadron Collider (LHC). The economic impact is enormous: It is expected that the computing power of approximately 50,000 PCs (organized in distributed clusters) will be necessary for the analysis of the LHC data.

But LHC computing is only one example. Also the processing of satellite data and the distributed screening of potential drug targets in rational drug design rely on the efficient utilization of distributed computing and storage resources. The applications used in Bioinformatics do not only read/write a huge amount of data, but the codes themselves are often highly parallel, requiring fast interconnects. In this application domain, simple throughput computing schemes are insufficient. More sophisticated methods for program and data mapping and also for communication must be used.

Y. Cotronis and J. Dongarra (Eds.): Euro PVM/MPI 2001, LNCS 2131, pp. 12–13, 2001.

2   Framework for a Grid-Aware Cluster

In our talk, we present a framework for a high-performance compute cluster that can be used as a scalable, reliable, and manageable node in the Grid. Our design was guided by three goals: A *low total cost of ownership* by providing improved fault tolerance and resilience as compared to conventional Beowulf clusters, a *high degree of scalability* allowing clusters with 10^1 to 10^3 nodes, and *maximum performance* for parallel applications by using improved protocols on commodity interconnects. In particular, we present

- criteria for the selection of CPUs and chipsets for a given class of application,
- considerations on the choice of a fast system area network,
- performance results of our zero-copy implementation of the Scheduled Transfer Protocol (ST) on Gigabit Ethernet,
- a scheme for utilizing cluster IDE disks as a network RAID.

Building on the cluster as a compute node, we present middleware for the proper integration of clusters in Grid environments. Here, the focus is on the specification and management of distributed resources.

More detailed information on this ongoing research project can be found in a recent report [2].

Acknowledgements

This is joint work with Volker Lindenstruth, Kirchhoff Institut für Physik, Universität Heidelberg. Part of the work has been funded by the EU project Datagrid.

References

1. I. Foster and C. Kesselman (eds.): *The Grid: Blueprint for a New Computing Infrastructure.* Morgan Kaufman Publishers, 1999.
2. A. Reinefeld, V. Lindenstruth: *How to Build a High-Performance Compute Cluster for the Grid.* Konrad-Zuse-Zentrum für Informationstechnik Berlin, ZIB Report 01-11, May 2001.

A Comparative Analysis of PVM/MPI and Computational Grids

Vaidy Sunderam[1] and Zsolt Németh[2]

[1] Emory University, Atlanta, GA 30322, USA
[2] MTA SZTAKI, Budapest, H-1518, Hungary

Concurrent computing on clusters or heterogeneous networks may be accomplished via traditional environments such as PVM [2] and MPI[3], or with emerging software frameworks termed computational grids[1]. In both scenarios, parallel distributed applications are comprised of a number of cooperating processes that exploit the resources of loosely coupled computer systems. However, there are some important differences, semantic, rather than technical in nature, that distinguish distributed computing systems from grids. This paper qualitatively analyzes the differences between these modes of concurrent computing and characterizes them in an attempt to assist developers and end users in selecting appropriate platforms for their needs.

The essential difference between distributed systems and grids is the way in which they establish a virtual hypothetical concurrent machine from the available components. A conventional distributed application assumes a pool of computational nodes (see left side in Figure 1) that form a virtual parallel machine. The pool of nodes consists of PCs, workstations, and possibly supercomputers, provided that the user has personal access (a valid login name and password) on all of them. From these candidate nodes an actual virtual machine is configured according to the needs of the given application. Access to the virtual machine is realized by login (or equivalent authentication) on all constituent nodes. In general, once access has been obtained to a node on the virtual machine, all resources belonging to or attached to that node may be used without further authorization. Since personal accounts are used, the user is explicitly aware of the specifics of the local node (architecture, computational power and capacities, operating system, security policies, etc). Furthermore, the virtual pool of nodes that can be included in any virtual machine is static since the set of nodes on which the user has login access changes very rarely. Typical number of nodes in the pool is of the order of 10-100.

As opposed to conventional distributed systems that harness resources that users have personal access to, computational grids are aimed at large-scale resource sharing. Grids assume a virtual pool of resources rather than computational nodes (right side of Figure 1). Although, computational resources obviously form a subset of resources, grid systems are expected to operate on a wider range like storage, network, data, software, graphical and audio input/output devices, manipulators, sensors, and so on, although resources typically exist within nodes. Nodes are geographically distributed, and span multiple administrative domains. The virtual machine is constituted by a set of resources taken from the pool. Access to the virtual machine is controlled by credentials that are validated by the owners of resources; credentials strictly specify the type and level

Y. Cotronis and J. Dongarra (Eds.): Euro PVM/MPI 2001, LNCS 2131, pp. 14–15, 2001.

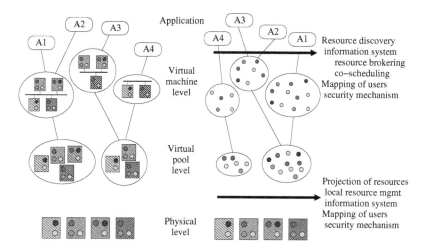

Fig. 1. Conventional distributed systems vs. computational grids

of access. The virtual pool of resources is dynamic and diverse. Resources can be added and withdrawn at any time, and their performance or load can change frequently over time. The user has very little or no *a priori* knowledge about the actual type, state and features of the resources constituting the pool. The typical number of resources in the pool is of the order of 1000 or more.

In conventional distributed systems abstract levels (virtual pools and virtual machines) are just different views of the same computational nodes. Access changes, reconfiguration of the abstract machine, and mapping processes are all possible at the application level based on the user's awareness of the system. Grid systems operate on real abstract levels. Applications may require resources without knowing their location, present status and availability. Conversely, resource owners may permit resource use by users that do not have local accounts. These facilities cannot be availed of at the application level, additional services are necessary. The most fundamental and distinguishing features of grids are services that provide a proper mapping between users (and/or their applications) and local accounts, and mapping the abstract resource needs to existing physical resources. These are the two primary characteristics of grids. On top of these, more sophisticated and efficient services can be built, such as resource brokering, co-allocation, and staging. Other services, such as monitoring and fault detection are also supported by conventional systems and are not unique to grids.

References

1. I. Foster, C. Kesselman (eds.): The Grid: Blueprint for a New Computing Infrastructure. Morgan Kaufmann Publishers, 1999.
2. A. Geist, A. Beguelin, J. Dongarra, W. Jiang, R. Manchek, V. Sunderam: PVM: Parallel Virtual Machine A Users' Guide and Tutorial for Networked Parallel Computing. MIT Press, 1994
3. The Message Passing Interface Standard. http://www-unix.mcs.anl.gov/mpi/

MPI-2 One-Sided Communications
on a Giganet SMP Cluster

Maciej Gołębiewski and Jesper Larsson Träff

C&C Research Laboratories, NEC Europe Ltd.
Rathausallee 10, D-53757 Sankt Augustin, Germany
{maciej,traff}@ccrl-nece.de

Abstract. We describe and evaluate an implementation of the MPI-2
one-sided communications on a Giganet SMP Cluster. The cluster runs
under Linux with our own port of MPICH 1.2.1, a well-known, portable,
non-threaded MPI implementation, to the Virtual Interface Provider Library (VIPL). We call this implementation MPI/VIA. The one-sided
communications part was adapted from MPI/SX, a full MPI-2 implementation for the NEC SX-5 vector-parallel supercomputer. We evaluate the performance of the one-sided communications by comparing
to point-to-point communication. For applications that can be implemented naturally in both the one-sided and the point-to-point model,
the MPI/VIA implementation is such that the user is not penalized by
choosing one model over the other.

1 Introduction

Remote memory access is a natural complement to the point-to-point message-
passing programming model, appropriate for applications with irregular communication patterns where some processes either produce data to, or are in need of
data from other processes without their partners being aware of this. A model
for remote memory access termed *one-sided communication* was introduced into
the Message Passing Library (MPI) with the MPI-2 standard [4]. The main characteristics of MPI one-sided communication setting it apart from point-to-point
communication are that

- communication between a pair of processes is initiated by one process only,
  the *origin*, which supplies *all* parameters necessary for communication with
  the other process, the *target*, and
- communication and synchronization are always explicitly separated.

We describe an implementation of the MPI-2 one-sided communications for a
Giganet SMP cluster. The cluster runs under Linux with our own port of MPICH
1.2.1 to the Virtual Interface Provider Library (VIPL). We call this MPI imple-
mentation MPI/VIA. The one-sided communication part was adapted from the
MPI/SX implementation for the NEC SX-5 vector-parallel supercomputer [6].
This choice came naturally since the NEC SX-5 is itself a clustered, hierarchical

Y. Cotronis and J. Dongarra (Eds.): Euro PVM/MPI 2001, LNCS 2131, pp. 16–23, 2001.

Table 1. MPI/VIA point-to-point performance; communication buffers *not* in cache.

interface	latency [μs]	bandwidth [MB/s]
intra-node: ch_proc	1.7	178
inter-node: cLAN	16.6	105

system, since the MPI/SX one-sided communication implementation fits well with the MPICH design, and since MPI/SX performs very well on the NEC SX-5. Fitting the MPI/SX one-sided communications into MPI/VIA on the other hand prompted further refinements and optimizations of MPI/SX. We evaluate the MPI/VIA one-sided communications by a simple benchmark which compares the performance of one-sided and point-to-point communication.

The last year or so has seen several implementations of the one-sided communication part of the MPI-2. The Windows NT and SUN implementations described in [5,2] are both multi-threaded, and differ in their completion mechanisms and their handling of MPI derived datatypes from MPI/VIA and MPI/SX. The Fujitsu implementation [1] is based on the Fujitsu MPILib interface that already supports one-sided communication. At least partial implementations of the one-sided communications exist for systems from Cray, Hitachi, HP, IBM and others. The open-source MPI implementation LAM (http://www.lam-mpi.org/) also implements the one-sided communications.

2   The SMP Cluster

Our Giganet SMP cluster is a heterogeneous cluster consisting of 4-way Pentium III Xeon SMP nodes with varying CPU clock (500 and 550 MHz) and L2 cache (512 KB, one node with 1 MB). The front-side-bus is only 66 MHz. Each node is equipped with two 100 MBit Ethernet cards for NFS, remote login and management functions. The MPI applications use a separate interconnect system based on the Virtual Interface Architecture (VIA) [3]. We use the VIA implementation from Giganet (now Emulex). All nodes connect to a 30 port cLAN 5300 switch (http://wwwip.emulex.com/ip/products/clan5000.html) via cLAN 1000 NICs (http://wwwip.emulex.com/ip/products/clan1000.html) inserted into a 64 bit, 33 MHz PCI slot. Link bandwidth is 1.25 Gbit/s, full-duplex.

For the MPI library we use our own port of MPICH 1.2.1 to VIPL. The implementation optimizes the intra-node communication by a separate set of protocols, the ch_proc device, when both sender and receiver are on the same SMP node. The ch_proc device uses a small shared memory segment for exchanging control packets and short messages. Long messages are read directly by the receiver from the sender's memory, using Linux's /proc pseudo-filesystem. To allow reading from the /proc/pid/mem files, a minor kernel patch is required (see Section 4.2). Unlike the shared-memory devices ch_shmem and ch_lfshmem present in MPICH, ch_proc supports the MPMD programming model natively and without any additional overhead. The basic MPI point-to-point communication performance of MPI/VIA on our cluster is summarized in Table 1.

3   Memory Allocation and Window Creation

In MPI the one-sided communication operations can access segments of other processes' memory, which are made available in a *window object*. A window object is set up by a collective call

```
MPI_Win_create(base,size,disp_unit,...,comm,&win)
```

in which each process in the communicator `comm` contributes a segment of memory starting from local address `base`. Locally, information on the window object is stored in the `win` data structure. In our implementation the function broadcasts base address, size and displacement unit among all processes by an `MPI_Allgather` operation. Range checking for one-sided communication operations can then be performed locally. The addresses of the local window data structures are also broadcast, and used by other processes to get access to `win`.

Window memory can be allocated by either `malloc` or the special MPI-2 memory allocator `MPI_Alloc_mem`. In MPI/SX `MPI_Alloc_mem` is used to allocate shared memory outside of the MPI process, which allows one-sided communication by direct memory copy. On the SMP cluster access to the memory of other processes is possible via Linux's `/proc` pseudo-filesystem, and hence there is no need for special allocation in MPI/VIA.

4   Communication

MPI-2 has three one-sided communication operations, namely

- `MPI_Put`, which moves data from *origin* to the *target* window,
- `MPI_Get`, which moves data from the *target* window to *origin*, and
- `MPI_Accumulate`, which combines data from *origin* with data in *target* window using an MPI built-in operation.

The MPI-2 standard allows for various types of blocking behavior of the one-sided communication calls. A most liberal possibility, now implemented in both MPI/VIA and MPI/SX, is to queue communication requests when the target window is not ready. Only the closing synchronization call (ie. `MPI_Win_complete` and `MPI_Win_unlock`) may block until the target becomes ready, at which point queued communications are processed.

4.1   Put and Accumulate

MPI/SX and MPI/VIA are non-threaded implementations. MPI processes check, at each communication call, for incoming (one-sided) communication requests. We have introduced into our Abstract Device Interface (ADI) a function

```
int MPID_Put(void *origin_addr, MPI_Aint origin_count,
             struct MPIR_DATATYPE *origin_type,
             int target_rank,
             MPI_Aint target_off, MPI_Aint target_count,
             struct MPIR_DATATYPE *target_type, int cache,
             MPI_Win win, int *posted, int *completion)
```

which transmits origin data to a window of target rank. If the target window is in shared memory, transfer is by memory copy. Otherwise, the call is non-blocking, and completion in the sense that the data is out of the origin buffer is signaled by incrementing the completion counter. Data transfer is initiated by sending a control message containing target window offset, data size, datatype description, and, for accumulate, the operation to be performed. Short data are tagged onto the control message; in MPI/SX "piggy-backing" of short data on the control message gave the expected factor 2 gain for messages up to 200 KBytes. Longer data are sent separately by a special, non-blocking send operation, which signals completion by incrementing the completion counter. On the target the control message is used to set up a corresponding non-blocking receive call. Also this receive signals completion by incrementing a counter. The argument posted returns the number of messages actually sent as a result of the operation. Local completion of one-sided operations can be asserted by waiting for the completion count to reach the total number of posted messages. On the target, access to the local window data structure is facilitated by sending as part of the control message the target local win address. The cache argument controls caching of the target datatype. Analogous ADI functions exist for one-sided get and accumulate.

In MPI/VIA inter-node put is done by the direct memory access mechanism. The Remote DMA Write operation is a required feature of the VIA specification. The origin specifies both the local and remote address and the network hardware transfers the data directly into the target window without intervention of the target process. The VIA standard requires that target memory is properly set up before the origin executes an RDMA operation. This involves locking the buffer into physical memory pages and obtaining a special memory handle that must match on both origin and target. This memory handle serves to prevent the origin from corrupting target memory, and is set up at window creation. As the maximum allowed size for a single RDMA operation varies between different VIA implementations, the put function breaks large transfers into smaller chunks that can be handled by the hardware. This introduces extra start-up overheads.

4.2   Get

In MPI/SX get on a target window in global shared memory is done by direct memory copy. Otherwise, a control message which includes a description of the target datatype is sent, and a non-blocking receive is posted. On the target, a non-blocking send operation with the requested data is set up. Completion at origin is signaled by incrementing the completion counter by the number of internally posted messages.

In MPI/VIA intra-node get is handled by reading directly from the target window using the target process' virtual memory representation in the Linux /proc pseudo-filesystem. The /proc/pid/mem files represent the memory as a file. We use lseek to locate the desired address and access it by calling read. As the Linux kernel has a relatively low overhead for calling system functions, we are able to achieve good performance even for transfers of moderate size. To use the /proc interface this way, we had to slightly modify the Linux kernel by relaxing the limitations on access to the /proc/pid/mem files: reading the file is allowed for any process with the same user and group id as the owner. Also normal point-to-point operations within a node are optimized this way. The advantage is that data is transferred by a single copy operation, and that long messages only rarely have to be broken into shorter chunks.

4.3 Handling of Derived Datatypes

Derived datatypes pose a problem since the target datatype is provided by the origin, and does in general not exist at the target. Hence a datatype description has to be transmitted together with the data. This is part of the one-sided request message. A light-weight representation of each derived datatype is constructed incrementally by the type constructors as the type is created. The light-weight type is a linear representation of the type stored in an array; it contains only minimal information for the internal pack-unpack functions to work properly. Caching at the target is used to avoid having to send data type information over and over again. Note that for each derived datatype the whole type is cached; there is no implicit caching of subtypes. This seems to be different from the SUN implementation [2], and could be a drawback for very complex datatypes. On the other hand it allows for simple replacement (no need to worry about subtypes disappearing from the cache etc.).

5   Synchronization

One-sided communication takes place within communication (either *access* or *exposure* or both) *epochs*, which are opened and closed by synchronization calls. When a process closes an access epoch all pending one-sided communication requests must have completed locally. When an exposure epoch is closed, all one-sided communication requests with this process as target must have completed.

In MPI/SX and MPI/VIA completion of pending requests is enforced by counting messages. Local completion (at the origin) is enforced by waiting for the completion counters to reach the total number of posted requests. At the target the number of requests from different origins have to be computed before doing the completion counting.

5.1   Fence

An MPI_Allreduce call, in which each process contributes the number of communication requests posted to all other processes, is used to compute, for

Table 2. MPI-2 one-sided synchronization overheads for $n = 1$ and $n = p$ neighbors as a function of number of processes. Synchronization times in microseconds.

p	Fence	Ded-1	Ded-p	Lock-1	Lock-p
2	25.04	13.96	13.95	20.99	22.01
4	44.99	15.00	39.98	19.01	59.97
8	111.04	40.96	210.02	73.03	513.98
16	170.03	42.96	618.98	75.95	1384.95
20	194.03	44.97	800.00	77.96	1851.02
24	218.05	44.98	1019.04	77.06	2316.96

each process, the number of requests for which it is a target. It then waits for these requests to complete. If the fence does not open a new epoch (assertion MPI_MODE_NOSUCCEED) a subsequent barrier synchronization is not necessary. If the epoch is not preceded by an open epoch (assertion MPI_MODE_NOPRECEDE) the MPI_Allreduce is not necessary; a barrier suffices to ensure that the window is ready for communication. This more careful implementation of fence reduced the overhead by about two thirds.

5.2   Dedicated

MPI_Win_start opens an epoch for access to a group of processes by posting a non-blocking receive for a ready message from the targets. MPI_Win_complete signalizes the end of the access epoch by sending to each target the number of one-sided communication requests posted. MPI_Win_post opens an epoch for exposure to a group of processes, and sends a ready message to each of these. MPI_Win_wait waits for the completion message from each process of the access group, computes the total number of target requests, and waits for completion of these. With this synchronization model (and the passive target model) we expect it to be of advantage that one-sided communication calls do *not* block even when the target is not ready.

5.3   Passive Target

Opening an exposure epoch is done by sending a non-blocking lock request to the target. The target eventually responds by sending a ready message to the origin when the lock is granted. Locking messages can be saved if the user asserts that no conflicting locks will be placed (assertion MPI_MODE_NOCHECK). In MPI/SX this lead to significant performance gains for locks. As explained, one-sided communication calls do not block, but the unlock call will block until the target has granted the lock, and all action at both origin and target has completed. Note that this mechanism may not work well in the case where the locked target window performs no MPI calls. In that case the origin may block until, say, the window is freed.

Table 3. The exchange test with 24 processors and varying message sizes (in bytes). Each message is sent in $c = 1$ chunk. Running times are in microseconds, and are maxima over all processes of the minimum time over 25 epochs.

Data size	Put-Fence	Get-Fence	Put-ded	Get-ded	Put-Lock	Get-Lock	Sendrecv
			$n = 1$ neighbor				
1024	389.00	300.94	165.97	124.99	179.97	161.04	58.97
4096	390.03	421.05	228.98	247.03	190.03	272.98	134.95
16384	585.95	704.01	358.05	522.00	435.97	498.05	418.98
65536	1568.95	1996.98	1241.02	1918.06	1529.99	1585.05	1541.04
262144	8630.03	9988.98	7436.02	9000.97	8685.99	8078.04	7724.94
			$n = 12$ neighbors				
1024	1403.96	2701.94	1360.00	2291.96	2217.98	3578.01	901.05
4096	3301.02	5839.96	3207.00	6046.97	4053.03	7739.96	3303.00
16384	11267.02	12646.05	10379.01	12275.99	11626.00	15293.95	11329.94
65536	45553.97	44042.94	39646.98	43864.00	43437.00	52119.00	42714.01
262144	192873.03	174341.02	160303.99	173470.02	180551.95	190830.95	170299.03
			$n = 24$ neighbors				
1024	2421.03	5275.00	2593.98	5220.02	4297.01	7164.04	1911.01
4096	6057.95	11899.03	6175.03	11939.99	7853.99	16080.95	7396.00
16384	21466.03	26701.05	20156.01	24301.04	22862.96	29915.01	21718.98
65536	85219.04	89412.02	77918.94	87021.04	87702.99	99621.00	82285.96
262144	345777.96	358598.04	303756.04	335580.01	342668.03	379980.01	333018.98

6   Experimental Evaluation

In the MPI-2 one-sided communication model the time spent in an isolated communication operation is not well-defined since it cannot be known when communication actually takes place within an epoch. We use a simple *exchange benchmark* to compare the one-sided communication performance to that of point-to-point. Let p be the number of processes. Each process transfers data to n neighbors, $p + i \bmod p$ for $i = 1, \ldots, n$. For each neighbor the same amount of data is communicated in $c \geq 1$ chunks. This scheme can be implemented with either MPI_Put or MPI_Get and any of the three synchronization mechanisms, as well as with MPI_Sendrecv. Each process records the minimum time over a number of epochs, and the maximum over these minima reported.

Table 2 shows the overheads of the three MPI synchronization mechanisms for MPI/VIA, measured by opening and closing empty epochs. Overheads are very tolerable. Fence synchronization grows logarithmically in p (independent of n), whereas both dedicated and lock synchronization is linear in n (independent of p). Already for $p = 4$ it is better to use fence for full synchronization among all processes. The jump in the times from 4 to 8 processors is because more than one SMP node is starting to be used.

Table 3 shows the communication performance of MPI/VIA for $p = 24$ processes (the number of processors in our cluster), and $n = 1$, $n = p/2$, and $n = p$ neighbors. Communication times increase linearly with message volume. Subtracting the synchronization overhead, the performance of put compares well to the performance of `MPI_Sendrecv`, independently of message length, although the synchronization overhead apparently cannot always be amortized even with $n > 1$. For messages up to about 16K Bytes the get performance is only within a factor 2 of point-to-point. This should be improved upon.

7 Concluding Remarks

We have described streamlined implementations of the MPI-2 one-sided communications in MPI/SX and MPI/VIA. For MPI/SX the streamlining led to worthwhile performance improvements. On the Giganet SMP cluster the MPI/SX implementation shows reasonable to good performance of one-sided communications compared to point-to-point as measured by the exchange benchmark. We therefore encourage users to actually use the MPI-2 one-sided communications for applications that lend themselves naturally to the one-sided programming paradigm.

Acknowledgments

Thanks to Michael Gerndt for a discussion on the semantics of `MPI_Win_fence`, which eventually lead to certain performance improvements.

References

1. N. Asai, T. Kentemich, and P. Lagier. MPI-2 implementation on Fujitsu generic message passing kernel. In *Supercomputing*, 1999. http://www.sc99.org/proceedings/papers/lagier.pdf.
2. S. Booth and E. Mourão. Single sided MPI implementations for SUN MPI. In *Supercomputing*, 2000. http://www.sc2000.org/proceedings/techpapr/index.htm\#01.
3. Compaq, Intel, Microsoft. *Virtual Interface Specification*, 1997.
4. W. Gropp, S. Huss-Lederman, A. Lumsdaine, E. Lusk, B. Nitzberg, W. Saphir, and M. Snir. *MPI – The Complete Reference*, volume 2, The MPI Extensions. MIT Press, 1998.
5. F. E. Mourão and J. G. Silva. Implementing MPI's one-sided communications in WMPI. In *Recent Advances in Parallel Virtual Machine and Message Passing Interface. 6th European PVM/MPI Users' Group Meeting*, volume 1697 of *Lecture Notes in Computer Science*, pages 231–238, 1999.
6. J. L. Träff, H. Ritzdorf, and R. Hempel. The implementation of MPI-2 one-sided communication for the NEC SX-5. In *Supercomputing*, 2000. http://www.sc2000.org/proceedings/techpapr/index.htm\#01.

Effective Communication
and File-I/O Bandwidth Benchmarks

Rolf Rabenseifner[1] and Alice E. Koniges[2]

[1] High-Performance Computing-Center (HLRS), University of Stuttgart
Allmandring 30, D-70550 Stuttgart, Germany
`rabenseifner@hlrs.de`
`www.hlrs.de/people/rabenseifner/`
[2] Lawrence Livermore National Laboratory, Livermore, CA 94550
`koniges@llnl.gov`
`www.rzg.mpg.de/~ack`

Abstract. We describe the design and MPI implementation of two benchmarks created to characterize the balanced system performance of high-performance clusters and supercomputers: b_eff, the communication-specific benchmark examines the parallel message passing performance of a system, and b_eff_io, which characterizes the effective I/O bandwidth. Both benchmarks have two goals: a) to get a detailed insight into the performance strengths and weaknesses of different parallel communication and I/O patterns, and based on this, b) to obtain a *single* bandwidth number that characterizes the *average* performance of the system namely communication and I/O bandwidth. Both benchmarks use a time-driven approach and loop over a variety of communication and access patterns to characterize a system in an automated fashion. Results of the two benchmarks are given for several systems including IBM SPs, Cray T3E, NEC SX-5, and Hitachi SR 8000. After a redesign of b_eff_io, I/O bandwidth results for several compute partition sizes are achieved in an appropriate time for rapid benchmarking.

1 Introduction and Design Criteria

Characterization of a system's usable performance requires more than vendor-supplied tables such as peak performance or memory size. On the other hand, a simple number characterizing the computational speed (as detailed by the TOP500 figures [8]) has much appeal in giving both the user of a system and those procuring a new system a basis for quick comparison. Such application performance statistics are vital and most often quoted in press releases, yet do not tell the whole story. Usable high-performance systems require a balance between this computational speed and other aspects in particular communication scalability and I/O performance. We focus on these latter areas.

There are several communication test suites that serve to characterize relative communication performance and I/O performance. The key concept that differentiates the effective bandwidth benchmarks described here from these other test suites is the use of sampling techniques to automatically scan a subset of

Y. Cotronis and J. Dongarra (Eds.): Euro PVM/MPI 2001, LNCS 2131, pp. 24–35, 2001.

the parameter space and pick out key features, followed by averaging and use of maxima to combine the results into a single numerical value. But this single value is only half of our goal. The **detailed insight** given by the numerous results for each measured pattern is the second salient feature. Additionally, both benchmarks are optimized in their execution time; b_eff needs about 3-5 minutes to examine its communication patterns, and b_eff_io, adjusted appropriately for the slower I/O communication, needs about 30 minutes. To get detailed insight, it is important to choose a set of patterns that reflects typical application kernels.

Effective Bandwidth Benchmark: The effective bandwidth benchmark (b_eff) measures the accumulated bandwidth of the communication network of parallel and/or distributed computing systems. Several message sizes, communication patterns, and methods are used. A fundamental difference between the classical ping-pong benchmarks and this effective bandwidth benchmark is that **all** processes are sending messages to neighbors in parallel, i.e., at the same time. The algorithm uses an average to take into account that short and long messages are transferred with different bandwidth values in real application scenarios. The result of this benchmark is a single number, called the *effective bandwidth*.

Effective I/O Bandwidth Benchmark: Most parallel I/O benchmarks and benchmarking studies characterize the hardware and file system performance limits [1,5]. Often, they focus on determining conditions that maximize file system performance. To formulate b_eff_io, we first consider the likely I/O requests of parallel applications using the MPI-I/O interface [7]. This interface serves both to express the user's needs in a concise fashion and to allow for optimized implementations based on the underlying file system characteristics [2,9,11]. Based on our benchmarking goals, note that the effective I/O bandwidth benchmark (b_eff_io) should measure different access patterns, report the detailed results, and calculate an average I/O bandwidth value that characterizes the whole system. Notably, I/O benchmark measures the bandwidth of data transfers between memory and disk. Such measurements are (1) highly influenced by buffering mechanisms of the underlying I/O middleware and filesystem details, and (2) high I/O bandwidth on disk requires, especially on striped filesystems, that a large amount of data must be transferred between these buffers and disk. On well-balanced systems an **I/O** bandwidth should be sufficient to write or read the total memory in approximately 10 **minutes**. Based on this rule, an I/O benchmark should be able to examine several patterns in 30 minutes accounting for buffer effects.

2 Multidimensional Benchmarking Space

Often, benchmark calculations sample only a small subspace of a multidimensional parameter space. One extreme example is to measure only one point. Our goal here is to sample a reasonable amount of the relevant space.

Effective Bandwidth Benchmark: For communication benchmarks, the major parameters are message size, communication patterns, (how many processes are communicating in parallel, how many messages are sent in parallel and which communication graph is used), and at least the communication method (MPI_Sendrecv, nonblocking or collective communication, e.g., MPI_Alltoallv). For b_eff, 21 different message sizes are used, 13 fixed sizes (1 byte to 4 kb) and 8 variable sizes (from 4 kb to the 1/128 of the memory of each processor). The communication graphs are defined in two groups, (a) as rings of different sizes and (b) by a random polygon. Details are discussed later in the definition of the b_eff benchmark. A first approach [16,17] was based on the bi-section bandwidth, but it has violated some of the benchmarking rules defined in [3,4]. Therefore a redesign was necessary.

Effective I/O Bandwidth Benchmark: For I/O benchmarking, a huge number of parameters exist. We divide the parameters into 6 general categories. At the end of each category in the following list, a first hint about handling these aspects in b_eff_io is noted. The detailed definition of b_eff_io is given in Sec. 4.

1. Application parameters are (a) the size of contiguous chunks in the memory, (b) the size of contiguous chunks on disk, which may be different in the case of scatter/gather access patterns, (c) the number of such contiguous chunks that are accessed with each call to a read or write routine, (d) the file size, (e) the distribution scheme, e.g., segmented or long strides, short strides, random or regular, or separate files for each node, and (f) whether or not the chunk size and alignment are wellformed, e.g., a power of two or a multiple of the striping unit. For b_eff_io, 36 different patterns are used to cover most of these aspects.
2. Usage parameters are (a) how many processes are used and (b) how many parallel processors and threads are used for each process. To keep these parameters outside of the benchmark, b_eff_io is defined as a maximum over these parameters and one must report the usage parameters used to achieve this maximum. Filesystem parameters are also outside the scope of b_eff_io.
3. The major programming interface parameter is specification of which I/O interface is used: Posix I/O buffered or raw, special filesystem I/O of the vendor's filesystem, or MPI-I/O, which is a standard designed for high performance parallel I/O [12] and therefore used in b_eff_io.
4. MPI-I/O defines the following orthogonal parameters: (a) access methods, i.e., first writing of a file, rewriting or reading, (b) positioning method, (c) collective or noncollective coordination, (d) synchronism, i.e., blocking or not. For b_eff_io there is no overlap of I/O and computation, therefore only blocking calls are used. Because explicit offsets are semantically identical to individual file pointers, only the individual and shared file pointers are benchmarked. All three access methods and five different pattern types implement a major subset of this parameter space.

For the design of b_eff_io, it is important to choose the grid points based more on general application needs than on optimal system behavior. These needs were a major design goal in the standardization of MPI-2 [7]. Therefore the b_eff_io pattern types were chosen according to the key features of MPI-2. The exact definition of the pattern types are given in Sec. 4 and Fig. 1.

3   The Effective Bandwidth: Definition and Results

The effective bandwidth is defined as (a) a logarithmic average over the ring patterns and the random patterns, (b) using the average over all message sizes, (c) and the maximum over all the three communication methods (d) of the bandwidth achieved for the given pattern, message size and communication method. As formula, the total definition can be expressed as:

$$b\_eff = logavg$$
$$(\, logavg_{\mathrm{ringpat.s}} \quad (avg_{\mathrm{L}} \; (max_{\mathrm{mthd}} \; (max_{\mathrm{rep}}(b_{\mathrm{ringpat.,L,mthd,rep}} \qquad))) \,)$$
$$, logavg_{\mathrm{randompat.s}} \; (avg_{\mathrm{L}} \; (max_{\mathrm{mthd}} \; (max_{\mathrm{rep}}(b_{\mathrm{randompat.,L,mthd,rep}} \,))) \,) \,)$$

with $b_{\mathrm{pat,L,mthd,rep}}$ = L * (total number of messages of a pattern "pat") * looplength / (maximum time on each process for executing the communication pattern looplength times)

Additional rules are: Each measurement is repeated 3 times (rep=1..3). The maximum bandwidth of all repetitions is used (see max_{mthd} in the formula above). Each pattern is programmed with three methods. The maximum bandwidth of all methods is used (max_{mthd}). The measurement is done for different sizes of a message. The message length L has the following 21 values: L = 1B, 2B, 4B, ... 2kB, 4kB, 4kB*(a**1), 4kB*(a**2), ... 4kB*(a**8) with and 4kB*(a**8) = L_{max} and L_{max} = (memory per processor) / 128 and looplength = 300 for the shortest message. The looplength is dynamically reduced to achieve an execution time for each loop between 2.5 and 5 msec. The minimum looplength is 1. The average of the bandwidth of all messages sizes is computed ($sum_{\mathrm{L}}(...)/21$). A set of ring patterns and random patterns is used (see details section below). The average for all ring patterns and the average of all random patterns is computed on the logarithmic scale (geometric average): $logavg_{\mathrm{ringpatterns}}$ and $logavg_{\mathrm{randompatterns}}$. Finally the effective bandwidth is the logarithmic average of these two values: $logavg(logavg_{\mathrm{ringpatterns}}, logavg_{\mathrm{randompatterns}})$.

Only for the detailed analysis of the communication behavior, the following additional patterns are measured: a worst case cycle, a best and a worst bi-section, the communication of a two dimensional Cartesian partitioning in the both directions separately and together, the same for a three dimensional Cartesian partitioning, a simple ping-pong between the first two MPI processes.

On communication methods: The communication is programmed with several methods. This allows the measurement of the effective bandwidth independent of which MPI methods are optimized on a given platform. The maximum bandwidth of the following methods is used: (a) MPI_Sendrecv, (b) MPI_Alltoallv, and (c) nonblocking with MPI_Irecv and MPI_Isend and MPI_Waitall.

On communication patterns: To produce a balanced measurement on any network topology, different communication patterns are used: Each node sends, in each measurement, a messages to its left neighbor in a ring and receives such a message from its right neighbor. Afterwards it sends a message back to its right neighbor and receives such a message from its left neighbor. Using the method

Table 1. Effective Benchmark Results

System	number of processors	b_eff MByte/s	b_eff per proc. MByte/s	L_{max}	ping-pong bandw. MB/s	b_eff at L_{max} MB/s	b_eff per proc. at L_{max} MByte/s	b_eff per proc. at L_{max} ring pat.
Distributed memory systems								
Cray T3E/900-512	512	19919	39	1 MB	330	50018	98	193
	256	10056	39	1 MB	330	22738	89	190
	128	5620	44	1 MB	330	12664	99	195
	64	3159	49	1 MB	330	7044	110	192
	24	1522	63	1 MB	330	3407	142	205
	2	183	91	1 MB	330	421	210	210
Hitachi SR 8000 round-robin	128	3695	29	8 MB	776	11609	90	105
	24	915	38	8 MB	741	2764	115	110
Hitachi SR 8000 sequential	24	1806	75	8 MB	954	5415	226	400
Hitachi SR 2201	16	528	33	2 MB		1451	91	96
Shared memory systems								
NEC SX-5/8B	4	5439	1360	2 MB		35047	8762	8758
NEC SX-4/32	16	9670	604	2 MB		50250	3141	3242
	8	5766	641	2 MB		28439	3555	3552
	4	2622	656	2 MB		14254	3564	3552
HP-V 9000	7	435	62	8 MB		1135	162	162
SGI Cray SV1-B/16-8	15	1445	96	4 MB	994	5591	373	375

MPI_Sendrecv, the two messages are sent one after the other in each node, if a ring has more than 2 processes. In all other cases, the two messages may be sent in parallel by the MPI implementation. Six ring patterns are used based on a one dimensional cyclic topology on MPI_COMM_WORLD: In the first ring pattern, all rings have the size 2 (except the last ring which may have the size 2 or three). In the 2nd and 3rd ring pattern, the size of each ring is 4 and 8 (except last rings, see [14]). In the 4th and 5th ring pattern the standard ring size is max(16, size/4) and max(32, size/2). And in the 6th ring pattern, one ring includes all processes. For the random patterns, one ring with all processes is used, but the processes are sorted by random ranks. The average is computed in two steps to guarantee that the ring patterns and random patterns are weighted the same.

On maximum message size L_{max}**:** On systems with sizeof(int)<64, L_{max} must be less or equal 128 MB, i.e., $L_{max} = \min(128\text{ MB}, (\text{memory per processor})/128)$; on all other systems L_{max} is equal to the 128th of the memory per processor.

Averaging method: The effective bandwidth value should represent the accumulated communication capability of the total system usable for large-scale applications. The geometric mean, i.e., the average on the logarithmic values, is chosen because it takes all network components into account. Thus the fastest connection and the slowest one are considered. The arithmetic mean is not used because the average would never be less than $50\,\%^1$ of the fastest bandwidth value and thus the lower bound of the average is independent of the magnitude of the lowest measured value, i.e., the speed of the weakest network component.

Latency: For small message sizes, the b_eff value is dominated by latency effects, starting with a message size of one byte. The value *b_eff per process* is

---

[1] 50 % in the case of averaging two values. (One Hundred%/n in the case of n values.)

determined mainly by the asymptotic bandwidth (b_∞) and by the message size that is necessary to achieve a significant part of b_∞. If the transfer time t is modeled by $latency + messagesize/b_\infty$, then $b_\infty/2$ is achieved if the message size is larger than $latency * b_\infty$. This product can be viewed as a characteristic value for the balance of latency and bandwidth. The first 13 message sizes are fixed, i.e., independent of the memory size of the system, to reduce the influence of the memory size on calculating *b_eff per process* as function of *latency* and b_∞. The upper 9 message sizes depend on the memory size and reflect that the size of a large application's message typically scales with the application's data size, which in turn scales with the size of the available memory.

3.1 Effective Benchmark Results

Table 1 shows some results on distributed and shared memory platforms. On some platforms, either the total system was not available for the measurements or the system was not configured to be used by one dedicated MPI application. But the *b_eff per processor* column extrapolates to the network performance if all processors are communicating to a neighbor. On shared memory platforms, the results generally reflect half of the memory-to-memory copy bandwidth because most MPI implementations have to buffer the message in a shared memory section. To compare these results with the traditional asymptotic ping-pong bandwidth for large message sizes, one should remember that b_eff is defined as an average over several message sizes. In the last three columns, the result is based only on the maximum message size L_{max}. In the last column, only the ring patterns are used. Comparing the last two columns, we see the negative effect of random neighbor locations. Comparing the last column with ping-pong results from the vendor we see the impact of communicating in parallel on each processor. For example, on a T3E the asymptotic ping-pong bandwidth is about 300 MByte/s for 2 processors. In contrast, b_eff per processor is 210 MByte/s. For ring patterns, there is virtually no degradation for larger number of processes. The measurement protocols can be found in [10]. The Hitachi results depend on the numbering of the MPI processes on the cluster of SMP nodes: *round-robin* means, that the numbering starts with the first processor on each SMP node, *sequential* means, that first all processors of the first SMP node are used, and so on. The numbering has a heavy impact on the communication bandwidth of the ring patterns and therefore of the b_eff result.

4   The I/O Benchmark: Definition and Results

The effective I/O bandwidth benchmark measures the following aspects:
- a *set of partitions*: a partition is defined by the number of nodes used for the b_eff_io benchmark and – if a node is a multiprocessor node – by the number of MPI processes on each node,
- the access methods *initial write*, *rewrite*, and *read*,
- the *pattern types* (see Fig. 1): (0) strided collective access, scattering large chunks in memory with size L each with one MPI-I/O call to/from disk chunks

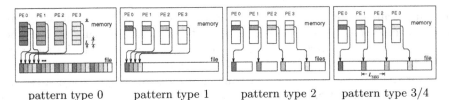

pattern type 0   pattern type 1   pattern type 2   pattern type 3/4

Fig. 1. Data transfer patterns used in b_eff_io. Each diagram shows the data transferred by **one** MPI-I/O write call.

Table 2. The pattern details used in b_eff_io

Pattern Type	No.	l	L	U
0: scatter, collect.	0	1 MB	1 MB	0
	1	M_{PART}	:=l	4
	2	1 MB	2 MB	4
	3	1 MB	1 MB	4
	4	32 kB	1 MB	2
	5	1 kB	1 MB	2
	6	32 kB +8B	1 MB + 256B	2
	7	1 kB +8B	1 MB + 8kB	2
	8	1 MB +8B	1 MB + 8B	2
1: shared, collect.	9	1 MB	:=l	0
	10	M_{PART}	:=l	4
	11	1 MB	:=l	2
	12	32 kB	:=l	1
	13	1 kB	:=l	1
	14	32 kB +8B	:=l	1
	15	1 kB +8B	:=l	1
	16	1 MB +8B	:=l	2

Pattern Type	No.	l	L	U
2: separated files, non-coll.	17	1 MB	:=l	0
	18	M_{PART}	:=l	2
	19	1 MB	:=l	2
	20	32 kB	:=l	1
	21	1 kB	:=l	1
	22	32 kB +8B	:=l	1
	23	1 kB +8B	:=l	1
	24	1 MB +8B	:=l	2
3: segmented, non-coll.	25f	same as patterns 17–24		
	33	fill up segments	:=l	0
4: segmented, collective	34f	same as patterns 25–33		
			$\Sigma U = 64$	

with size l; (1) strided collective access, but one read or write call per disk chunk; (2) noncollective access to one file per MPI process, i.e., on separated files; (3) is the same as (2), but the individual files are assembled to one segmented file; (4) is the same as (3), but the access to the segmented file is done with collective routines. For each pattern type, an individual file is used.

- the contiguous chunk size is chosen *wellformed*, i.e., as a power of 2, and *non-wellformed* by adding 8 bytes to the wellformed size,
- different chunk sizes, mainly 1 kB, 32 kB, 1 MB, and the maximum of 2 MB and 1/128 of the memory size of a node executing one MPI process.

The total list of patterns is shown in Table 2. A pattern is a pattern type combined with a fixed chunk size and alignment of the first byte[2]. The column "l" defines the contiguous chunks that are written from memory to disk and vice versa. The value M_{PART} is defined as *max(2 MB, memory of one node / 128)*. The column "L" defines the contiguous chunk in the memory. In case of pattern type (0), non-contiguous fileviews are used. If l is less than L,, then in each MPI-I/O read/write call, the L bytes in memory are scattered/gathered to/from the portions of l bytes at the different locations on disk, see the leftmost scenario in Fig. 1. In all other cases, the contiguous chunk handled by each

---

[2] The alignment is implicitly defined by the data written by all previous patterns in the same pattern type

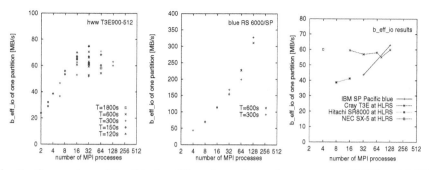

Fig. 2. Comparison of b_eff_io for different numbers of processes at HLRS and LLNL, measured partially without pattern type 3. Here T is in seconds, b_eff_io releases 0.x (left pictures and NEC on right picture) and release 1.x (right picture).

call to MPI_Write or MPI_Read is equivalent in memory and on disk. This is denoted by ":=l" in the L column. U is a time unit.

Each pattern is benchmarked by repeating the pattern for a given amount of time. For write access, this loop is finished with a call to MPI_File_sync. This time is given by the allowed time for a whole partition, e.g., $T = 15$ minutes, multiplied by $U/\Sigma U/3$, as given in the table. This time-driven approach allows one to limit the total execution time. For the pattern types (3) and (4) a fixed segment size must be computed before starting the pattern of these types. Therefore, the time-driven approach is substituted by a size-driven approach, and the repeating factors are initialized based on the measurements for types (0) to (2).

The b_eff_io value **of one pattern type** is defined as the total number of transferred bytes divided by the total amount of time from opening till closing the file. The b_eff_io value **of one access method** is defined as the average of all pattern types with double weighting of the scattering type. The b_eff_io value **of one partition** is defined as the average of the access methods with the weights 25 % for *initial write*, 25 % for *rewrite*, and 50 % for *read*. **The b_eff_io of a system** is defined as the maximum over any b_eff_io of a single partition of the system, measured with a scheduled execution time T of at least 15 minutes. This definition permits the user of the benchmark to freely choose the usage aspects and enlarge the total filesize as desired. The minimum filesize is given by the bandwidth for an initial write multiplied by 300 sec (= 15 minutes / 3 access methods). For using this benchmark to compare systems as in the TOP 500 list, more restrictive rules are under development.

4.1 Comparing Systems Using b_eff_io

First, we test b_eff_io on two systems, the Cray T3E900-512 at HLRS/RUS in Stuttgart and an RS 6000/SP system at LLNL called "blue Pacific." Figure 2 shows the b_eff_io values for different partition sizes and different values of T, the time scheduled for benchmarking one partition. All measurements were taken in a non-dedicated mode.

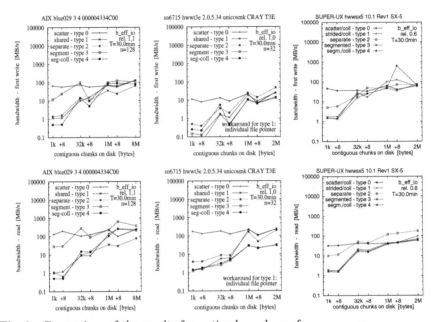

Fig. 3. Comparison of the results for optimal numbers of processes on
- IBM RS 6000/SP *blue Pacific* at LLNL, 128 nodes used, b_eff_io = 63 MB/s,
- Cray T3E-900/512 at HLRS, 32 PEs used, b_eff_io = 57 MB/s [13],
- NEC SX5-5Be/32M2 at HLRS, 4 CPUs used, b_eff_io = 60 MB/s.

Besides the different absolute values that correlate to the amount of memory
in each system, one can see very different behavior. For the T3E, the maximum
is reached at 32 application processes, with little variation from 8 to 128 processors, i.e., the I/O bandwidth is a global resource. In contrast, on the IBM
SP the I/O bandwidth tracks the number of compute nodes until it saturates.
In general, an application only makes I/O requests for a small fraction of the
compute time. On large systems, such as those at the High-Performance Computing Center at Stuttgart and the Computing Center at Lawrence Livermore
National Laboratory, several applications are sharing the I/O nodes, especially
during prime time usage. In this situation, I/O capabilities would not be requested by a significant proportion of the CPU's at the same time. "Hero" runs,
where one application ties up the entire machine for a single calculation are rarer
and generally run during non-prime time. Such hero runs can require the full
I/O performance by all processors at the same time. The middle diagram shows
that the RS 6000/SP fits more to this latter usage model.

The b_eff_io benchmark gives also a detailed insight into the I/O bandwidth
for several chunk sizes and patterns. The bandwidth is reported in a table that
can be plotted as in the pictures shown in each column in Fig. 3. The two diagrams in each column show the bandwidth achieved for *writing* and *reading* with
different patterns and chunk sizes. The *rewriting*-diagrams are omitted because
they show similar values as the *writing*-diagrams on these platforms. On each

diagram, the bandwidth is plotted on a logarithmic scale, separately for each pattern type and as a function of the chunk size. The chunk size on disk is shown on a pseudo-logarithmic scale. The points labeled "+8" are the non-wellformed counterparts of the power of two values. The maximum chunk size is different on the systems because the maximum chunk size was chosen proportional to the usable memory size per node to reflect the scaling up of applications on larger systems. Further topics on b_eff_io results are discussed in [6].

In general, our results show that the b_eff_io benchmark is a very fast method to analyze the parallel I/O capabilities available for applications using the standardized MPI-I/O programming interface. The resulting b_eff_io value summarizes I/O capabilities of a system in one significant I/O bandwidth value.

5   The Time-Driven Approach

Figure 2 shows interesting results. There is a difference between the maximum I/O bandwidth and the sampled bandwidth for several partition sizes. In the redesign from release 0.x to 1.x we have incorporated that the averaging for each pattern type can not be done by using the average of the bandwidth values for all chunk sizes. The bandwidth of one pattern must be computed as the total amount of transfered data divided by the total amount of time used for all chunk sizes. With this approach, it is possible to reduce caching effects and to allow a total scheduled time of 30 minutes for measuring all five patterns with the three access directions (write, rewrite, read) for **one** compute partition size.

Both benchmarks are proposed for the *Top Clusters* list [18]. For this, the I/O benchmark can be done automatically in 30 minutes for **three** compute partition sizes. This is implemented by reorganizing the sequence of the experiments: First, all files are written with the three different compute partition sizes, followed by rewriting, and then by all reading. Additionally, the rewriting experiments only use pattern type 0. Of course, if one wants to achieve very specific results, one can run this b_eff_io release 2.0 benchmark for the longer time period and with all rewriting patterns included.

6   Summary and Future Work

In this paper we have described in detail two benchmarks, the effective bandwidth and its I/O counterpart. We use these two benchmarks to characterize the performance of common computing platforms. We have shown how these benchmarks can provide both detailed insight into the performance of high-performance platforms and how they can reduce these data to a single number averaging important information about that system's performance. We give suggestions for interpreting and improving the benchmarks, and for testing the benchmarks on one's own system.

We plan to use this benchmark to compare several additional systems. Both benchmarks will also be enhanced to write an additional output that can be used in the SKaMPI *comparison page* [15].

Acknowledgments

The authors would like to acknowledge their colleagues and all the people that supported these projects with suggestions and helpful discussions. They would especially like to thank Karl Solchenbach and Rolf Hempel for productive discussions for the redesign of b_eff. We also gratefully acknowledge discussions with Jean-Pierre Prost and Richard Treumann of IBM. Work at LLNL was performed under the auspices of the U.S. Department of Energy by University of California Lawrence Livermore National Laboratory under contract No. W-7405-Eng-48.

References

1. Ulrich Detert, *High-Performance I/O on Cray T3E*; Peter W. Haas, *Scalability and Performance of Distributed I/O on Massively Parallel Processors*; Kent Koeninger, *Performance Tips for GigaRing Disk I/O*; 40th Cray User Group Conf., June 1998.
2. Philip M. Dickens, *A Performance Study of Two-Phase I/O*, in D. Pritchard, J. Reeve (eds.), Proceedings of the 4th Internatinal Euro-Par Conference, Euro-Par'98, Parallel Processing, LNCS–1470, pages 959–965, Southampton, UK, 1998.
3. William Gropp and Ewing Lusk, *Reproducible Measurement of MPI Performance Characteristics*, in J. Dongarra et al. (eds.), proceedings of the 6th European PVM/MPI Users' Group Meeting, EuroPVM/MPI'99, Barcelona, Spain, Sept. 26-29, 1999, LNCS 1697, pp 11–18. (Summary on the web: www.mcs.anl.gov/mpi/mpptest/hownot.html).
4. Rolf Hempel, *Basic Message Passing Benchmarks, Methodology and Pitfalls*, SPEC Workshop on Benchmarking Parallel and High-Performance Computing Systems, Wuppertal, Germany, Sept. 13, 1999, www.hlrs.de/mpi/b_eff/hempel_wuppertal.ppt.
5. Terry Jones, Alice Koniges, R. Kim Yates, *Performance of the IBM General Parallel File System*, to be published in Proceedings of the International Parallel and Distributed Processing Symposium, May 2000. Also available as UCRL JC135828.
6. Alice E. Koniges, Rolf Rabenseifner, Karl Solchenbach, *Benchmark Design for Characterization of Balanced High-Performance Architectures*, in proceedings, 15th International Parallel and Distributed Processing Symposium (IPDPS'01), Workshop on Massively Parallel Processing, April 23-27, 2001, San Francisco, USA.
7. Message Passing Interface Forum. *MPI-2: Extensions to the Message-Passing Interface*, July 1997, www.mpi-forum.org.
8. Hans Meuer, Erich Strohmaier, Jack Dongarra, Horst D. Simon, *TOP500 Supercomputer Sites*, www.top500.org.
9. J.P. Prost, R. Treumann, R. Blackmore, C. Harman, R. Hedges, B. Jia, A. Koniges, A. White, *Towards a High-Performance and Robust Implementation of MPI-IO on top of GPFS*, EuroPar2000, Munich, August 2000,in A. Bode et al. (Eds.): Euro-Par 2000, LNCS 1900, pp. 1253–1262, 2000. (Springer-Verlag: Berlin).
10. Rolf Rabenseifner, *Effective Bandwidth (b_eff) and I/O Bandwidth (b_eff_io) Benchmark*, www.hlrs.de/mpi/b_eff/ and www.hlrs.de/mpi/b_eff_io/.
11. Rajeev Thakur, William Gropp, and Ewing Lusk, *On Implementing MPI-IO Portably and with High Performance*, in Proc. of the Sixth Workshop on I/O in Parallel and Distributed Systems, pp 23–32, May 1999. www.mcs.anl.gov/romio/.
12. Rajeev Thakur, William Gropp, and Ewing Lusk, *I/O in parallel applications: The weakest link*, in The International Journal of High Performance Computing Applications, Vol. 12, No. 4, Winter 1998, pp. 389–395.

13. Rolf Rabenseifner, *Striped MPI-I/O*, `www.hlrs.de/mpi/mpi_t3e.html#StripedIO`.
14. Rolf Rabenseifner, *Ring Pattern List*, Nov. 1999.
    `www.hlrs.de/mpi/b_eff/ring_pattern_list` &
    `www.hlrs.de/mpi/b_eff/ring_numbers.c`
15. Ralf Reussner, Peter Sanders, Lutz Prechelt and Matthias Müller, *SKaMPI: A detailed, accurate MPI benchmark*, in proceedings, 5th European PVM/MPI Users' Group Meeting, LNCS 1497, pages 52-59, 1998. `wwwipd.ira.uka.de/~skampi/`
16. Karl Solchenbach, *Benchmarking the Balance of Parallel Computers*, SPEC Workshop on Benchmarking Parallel and High-Performance Computing Systems, Wuppertal, Germany, Sept. 13, 1999.
17. Karl Solchenbach, Hans-Joachim Plum and Gero Ritzenhoefer, *Pallas Effective Bandwidth Benchmark – source code and sample results*,
    `ftp://ftp.pallas.de/pub/PALLAS/PMB/EFF_BW.tar.gz`.
18. TFCC – IEEE `www.ieeetfcc.org`, and Top Clusters `www.TopClusters.org`.

Performance of PENTRAN™ 3-D Parallel Particle Transport Code on the IBM SP2 and PCTRAN Cluster

Vefa Kucukboyaci, Alireza Haghighat, and Glenn E. Sjoden

Mechanical and Nuclear Engineering Department
228 Reber Building
The Pennsylvania State University University Park, PA 16802
{haghighat,vefa}@psu.edu, joedean@spreintmail.com

Abstract. This paper discusses the algorithm and performance of a 3-D parallel particle transport code system, PENTRAN™ (Parallel Environment Neutral-particle TRANsport). This code has been developed in F90 using the MPI library. Performance of the code is measured on an IBM SP2 and a PC cluster. Detailed analysis is performed for a sample problem, and the code is used for determination of radiation field in a real-life BWR (Boiling Water Reactor). Using 48 IBM-SP2 processors with 256 Mbytes memory each, we have solved this large problem in ~12 hours, obtaining a detailed energy-dependent flux distribution.

1   Introduction

Neutral particle transport simulation in nuclear systems requires the use of linear Boltzmann equation. One of the most widely used techniques to solve this equation is the discrete ordinates (Sn) technique. This technique, because it discretizes all of the independent variables including space, energy, and direction, requires large memory (10-100's GB) and computation time.

In the past several years, different domain decomposition algorithms have been tested for the Sn method in different geometries and for different parallel architectures. [1,2,3,4] Most of these investigations are performed on CRAY and iPSC hypercube machines, and do not use standard multitasking libraries such as MPI or PVM.

Between 1988 and 1995, we developed several angular and spatial domain decomposition algorithms for the 1-D and 2-D curvilinear Sn method [5,6] on IBM and CRAY shared-memory machines.

Based on experience gained from above studies, in 1995, we initiated development of a new parallel 3-D discrete ordinates code for distributed-memory and distributed computing environments. The code (PENTRAN: Parallel Environment Neutral-particle TRANsport) [7] is designed based on full phase-space decomposition, parallel I/O, and partitioned memory. These design features in addition to new differencing and acceleration formulations make the code capable of modeling large real-life radiation transport problems.

Different features of the code have been benchmarked against standard production codes. Moreover, the code has been benchmarked based on the Kobayashi benchmark problems [8] and the VENUS-3 benchmark experiment [9].

Y. Cotronis and J. Dongarra (Eds.): Euro PVM/MPI 2001, LNCS 2131, pp. 36–43, 2001.

PENTRAN has been used for determination of radiation fields in real-life problems including the VENUS-3 facility, a BWR core shroud [10], a PGNAA assaying device [11], an X-ray room [12], a CT scan [13], and a time-of-flight experiment [14]. These problems are complex and require gigabytes of memory and hours/days of computation time.

This paper discusses the numerical features and algorithms of PENTRAN, and its use of MPI [15]. Further, performance of the code is measured on an IBM SP2 and a PC cluster. Detailed performance analysis is performed on SP2 for a sample problem, and both machines are used to solve a real-life Boiling Water Reactor (BWR).

The remaining of the paper is organized as follows: i) A discussion on the Sn formulation; ii) Features of PENTRAN™; iii) MPI implementation in PENTRAN; iv) Specifications of SP2 and PC-cluster; iv) Performance of PENTRAN on the IBM SP2 for a sample problem; and vi) Performance of PENTRAN™ on PCTRAN and IBM-SP2 for a real-life BWR problem.

2   Discussion on Sn Formulation

The time-independent Linear Boltzmann equation is given by

$$\hat{\Omega} \cdot \nabla \Psi(\vec{r}, E, \hat{\Omega}) + \sigma(\vec{r}, E) \Psi(\vec{r}, E, \hat{\Omega}) = \int_0^\infty dE' \int_{4\pi} d\Omega' \sigma_s(\vec{r}, E' \to E, \hat{\Omega}' \cdot \hat{\Omega}) \Psi(\vec{r}, E', \hat{\Omega}) +$$

$$\frac{\chi(E)}{4\pi} \int_0^\infty dE' \int_{4\pi} d\Omega' \upsilon \sigma_f(\vec{r}, E') \Psi(\vec{r}, E, \hat{\Omega}') + q_{ext}(\vec{r}, E, \hat{\Omega}) \tag{1}$$

Here, Ψ is the expected flux of particles in phase space ($dVdEd\Omega$), q_{ext} is the expected number of source particles emitted in the same phase space. The remaining terms are standard [16].

In the Sn method, all the independent variables (energy, space, and angle) are discretized. For the angular variable, a discrete set of directions (angles) is chosen and the Boltzmann equation is solved for these directions only. The directions are selected such that physical symmetries and particles are conserved. For the energy variable, the energy is divided into a number of sub-intervals. For the spatial variable, the Boltzmann equation is integrated over a mesh cell to obtain a system of equations in terms of cell boundary angular fluxes and the cell averaged angular flux.

The Cartesian discrete ordinates form of Eq. 1, used in PENTRAN, is given by

$$\frac{|\mu_m|}{\Delta x}(\Psi_{out,g}^m - \Psi_{in,g}^m) + \frac{|\eta_m|}{\Delta y}(\Psi_{left,g}^m - \Psi_{right,g}^m) + \frac{|\xi_m|}{\Delta z_k}(\Psi_{top,g}^m - \Psi_{bottom,g}^m)$$

$$+ \sigma_{avg}\Psi_{avg,g}^m = Q_{avg,g}^m \tag{2}$$

where indices m and g refer to direction and energy group, respectively, and *in/out*, *left/right*, and *top/bottom* refer to different sides of Cartesian mesh cell along x, y, and z axes, respectively. Note that Q includes scattering, fission, and external sources. In order to solve for the average angular flux (Ψ_{avg}), we need to know the six boundary angular fluxes. Three of these fluxes are boundary values, while the remaining three are obtained from three auxiliary equations, referred to as "differencing" equations. Commonly, on a serial machine, the above system of linear equations is solved via the Gauss-Seidel iterative technique.

3   Parallel Algorithms and Features of PENTRAN™

PENTRAN™ is a 3-D parallel Sn code with complete, automatic phase space (angle, energy and space) decomposition for distributed memory and computing environments. It has been developed in Fortran-77 with the MPI for distributed computing, and FORTRAN 90 constructs for dynamic memory allocation. PENTRAN™ has been implemented on different platforms such as IBM-SP2, SUN multi-processors, and PC clusters. Unique features and algorithms of PENTRAN™ include: Complete phase space decomposition; block-Jacobi and red-black with Alternating Direction Sweep (ADS) iterative techniques; a new adaptive differencing strategy [8] including linear diamond differencing (DZ), directional Θ-weighted (DTW) and an Exponential Directional Weighted (EDW); *Variable meshing with TPMC* (Taylor Projection Mesh Coupling) [17]; and Pre- and Post-processing Utilities.

4   MPI Implementation in PENTRAN

In this section, we discuss MPI implementation for task distribution, parallel I/O, message passing, and memory partitioning in PENTRAN. The PENTRAN code is ~30,000 lines, and highly modular. In all, the code is composed of 120 separate subroutines and functions.

All input to PENTRAN is performed using an ASCII input file read and processed by each independent processor. The filename is always input by the user on processor 1, and processor 1 broadcasts this input filename to all other processors. This approach is very effective, because no major message passing is needed for input processing. To accomplish parallel data input on each processor that is completely consistent with F77 character /numeric data restrictions, a small (typically 5-50 kb) scratch file is created. To avoid file I/O conflicts in a distributed file system, a uniquely named scratch file is generated and used independently by each processor; the scratch file name is based on the processor number. Similarly, another set of independent scratch files is created for material specifications for each fine mesh contained within each coarse mesh. This is performed to retain scalability, since only the local phase space is stored on each processor, and we do not want to read and store the fine mesh material data for the whole problem.

Ultimately, the auto scheduling in PENTRAN leads to a 3-D Cartesian, virtual processor array topology, with angular, energy group, and spatial decomposition axes.

Efficient communications over an arbitrary domain decomposition are obtained from the use of *communicators*. These *communicators* are generated on each process during the distribution of problem data over the virtual processor array. The methodology by which communicators are set up on PENTRAN is one of the most complex algorithms for parallelization in the code. It is because of these communicators that PENTRAN accomplishes complete phase space decomposition with a very high parallel fraction, close to 98%. PENTRAN builds *communicators* to selectively communicate with processors containing: all angles for a specific coarse cell and energy group; all energy groups and angles for a specific coarse cell; all coarse cells and angles for a specific energy group. All *communicators* are constructed by the BLDCOM routine, which is called by the CPMAP routine. CPMAP first sets up global variable mapping assignments over the virtual processor array. Once the virtual topology and work assignments are set up, they are tracked by a global process-mapping array (*kpmap*). Using *kpmap*, processor lists are generated by PENTRAN that contain the member process numbers belonging to communicator type. Once processors are identified, MPI communicator build commands are executed in the BLDCOM routine.

Following the assignment of the processors and communicators in the virtual processor array, phase space variables assigned to particular processor are tracked and computed independently. Local variable dimensions and process mapping arrays are used on each processor. This allows for partitioning the memory among processors.

5   Specifications of the IBM SP2 and the PCTRAN PC-Cluster

The IBM SP2 is located at the San Diego Supercomputing Center (SDSC). It includes 128 nodes, each containing 8 133 MHz CPUs with 4 GB of memory.

PCTRAN cluster has four 600 MHz Intel Pentium-III processors, each with 1024 Mbytes of memory. Processors are interconnected with a 100 Mbps fast Ethernet network switch and interface cards. Each machine runs the RedHat Linux 6.2 operating system and the MPICH implementation of MPI.

6   Parallel Performance of PENTRAN™ for a Sample Problem

Here, we evaluate the parallel performance of PENTRAN for a simple test problem, referred to as "box-in-a-box" problem. This is a 3-D, 3-group, 2-region symmetric, fixed source problem with upscattering between groups 2 and 3, and placed in a vacuum. Problem size is 24x24x24 cm^3, which contains 27 coarse meshes, each containing 64 uniformly spaced fine meshes, for a total of 1728 meshes, and an S_8 angular quadrature set (80 directions).

For the "box-in-a-box" problem, we tested different domain decomposition algorithms for increasing number of processors. Table 1 gives the speedup, efficiency and the communication overhead for each decomposition scheme. As the degree of decomposition increases, the efficiency decreases because the grain size reduces while the communication time increases. To examine the performance of PENTRAN, we compared the speedups given in Table 1 with maximum theoretical values predicted by the Amdahl's law:

$$speedup(\max.) = \frac{1}{(1 - f_p) + \dfrac{f_p}{p} + \dfrac{T_c}{T_s}}$$ (3)

where T_s is serial wall-clock time, T_c is the parallel communications time, f_p is the parallelizable fraction of the code, and p is the number of processors. Comparing the results of Table 1 to the Amdahl's law (assuming a ratio of T_c/T_s equal to .072 based on 24 processors, and a parallel fraction of 98%) show that that estimated speedups follow very closely the predicted values by the Amdahl's law. These results demonstrate that PENTRAN is highly parallelizable.

Table 1. Parallel performance of PENTRAN for the "box-in-a-box" problem with a constant problem size

Decomposition Strategy	Processor	Speedup	Efficiency (speedup/proc.)	Communication (%) wall-clock
Angular	2	1.90	95	5
Group	3	2.70	90	9
Spatial	3	2.59	86	12
Angular	4	3.26	82	16
Angular-Group (2,3)	6	4.78	80	22
Spatial	9	5.49	61	21
Angular-Group (4,3)	12	6.72	56	42
Angular-Spatial (4,3)	12	5.76	48	43
Angular-Spatial (2,9)	18	8.74	49	32
Angular-Group-Spatial (2,3,3)	18	8.74	49	34
Angular-Group (8,3)	24	7.96	33	64
Spatial	27	8.74	32	37

7   Parallel Performance of PENTRAN
for a Real-Life Problem-BWR Core Shroud Model

Here, we evaluate the parallel performance of PENTRAN for a large, real-life problem. A 3-D PENTRAN™ model is developed for a BWR core shroud. The model extends 290 cm and 346.5 cm in radial and axial directions, respectively. A total of 432 coarse and 265,264 fine meshes are used in the model. Coarse meshes are distributed over 12 axial zones with 36 coarse meshes per axial zone. For solving this problem we have used both an IBM SP2 system and a PC-Cluster, and compared the performance.

Forty-seven (47) neutron groups and 20 gamma groups are used for this simulation. For parallel processing, we have partitioned the problem into eight angular and six spatial sub-domains and processed them on 48 processors. The problem requires ~235 MB of memory per processor.

The wall-clock time required to render a PENTRAN™ solution for the BWR core shroud problem (265264 meshes, 80 directions, 67 groups) is ~12 hours on 48 nodes of the SDSC IBM-SP2 system.

Table 3 presents scalability tests using different quadrature orders (i.e., number of angles), number of processors and space-angle domain decomposition algorithms. To avoid loss of granularity (ratio of computation to communication), we increase the quadrature order as the number of processors is doubled. Cases 2-3 give scalability of 91% and 85%, respectively, while case 4 gives a scalability of 73%. The decrease in efficiency, especially in the last case can be attributed to the increase in message passing overhead with the increasing decomposition. These results demonstrate that PENTRAN™ is quite efficient for rendering a solution in a minimum amount of wall-clock time with a high parallel efficiency and scalability even for a large model such as the BWR problem.

We have performed a similar performance test on the PCTRAN cluster. For this, we have partitioned the problem into four angular subdomains and processed them on 4 processors of the PCTRAN cluster. Table 4 presents tests using different quadrature orders. In this test, we keep the number of processors constant and increase the quadrature order. Columns 2 and 4 show the work and the time ratios, respectively relative to case 1. We observe that, as we increase the number of directions, we get better performance. For example, comparing cases 1 and 4, we increase the workload by a factor of 5; however, wall-clock time/iteration increases only by a factor of 2.31. This behavior can be attributed to the granularity, i.e., case 1 has the lowest computation (with 24 directions) to communication ratio.

These preliminary results indicate that, due to slower network and limited bandwidth, PCTRAN performance is sensitive to granularity. However, for coarse grain problems, performance of PCTRAN is very competitive to that of IBM-SP2. Note that, IBM-SP2 has RISC6000 processors running at 133 MHz, while PCTRAN has Intel Pentium processors running at 600 MHz.

Table 2. Scalability of PENTRAN™ on IBM-SP2

Case	No. of Directions	No. of Processors	Decomposition (A/G/S)[1]	Wall-clock/iteration (s)
1	24	6	1/1/6	30.12
2	48	12	1/1/12	33.28
3	80	24	4/1/6	29.52
4	168	48	8/1/6	36.12

[1] (A/G/S) refers to the number of Angular, Group, and Spatial subdomains.

Table 3. Scalability of PENTRAN™ on PCTRAN.

Case	No. of Directions	Work Ratio	Wall-clock/iteration (s)	Time Ratio	Work Ratio/Time Ratio
1	24	1.00	87.04	1.00	1.00
2	48	2.00	108.87	1.25	1.60
3	80	3.33	142.08	1.63	2.04
4	120	5.00	188.07	2.16	2.31

8   Summary and Conclusions

This paper introduces PENTRAN™, a 3-D particle transport theory code, which has been developed in F90 with MPI library. It provides a discussion on PENTRAN features and its MPI implementation. It analyzes the parallel performance of the code using the SDSC IBM SP2 and the PCTRAN cluster. For a sample problem, it is concluded that PENTRAN can achieve a parallel fraction of ~98% on SP2. Further, the PENTRAN performance is measured for solving a real-life BWR reactor problem on both SP2 and PCTRAN. Using 48 IBM-SP2 processors with 256 Mbytes memory each, this large problem is solved in ~12 hours. Timing tests have demonstrated that PENTRAN™ is highly scalable (~90%) on IBM-SP2. Our preliminary tests indicate that the low-cost, high-performance PCTRAN cluster offers a cost effective computing environment for performing large-scale, 3-D, real-life simulations.

References

1. Wienke and Hiromoto,(1985) Wienke, B. and R. Hiromoto, "Parallel Sn Iteration Schemes", *Nucl. Sci. Eng.*, **90,** 116-123 (1985)

2. Yavuz, M. and E. Larsen, "Iterative methods for Solving x-y Geometry Sn Problems on Parallel Architecture Computers", *Nucl. Sci. Eng*, **112,** 32-42 (1992)

3. Azmy, Y., "On the Adequacy of Message-Passing Parallel Supercomputers for Solving Neutron Transport Problems," *Proc. IEEE Supercomputing '90*, (1990)

4. Dorr, M, and C. Still, "Concurrent Source Iteration in the Solution of Three-Dimensional, Multigroup Discrete Ordinates Neutron Transport Equations", *Nucl. Sci. Eng*, **122,** 287-308 (1996)

5. Haghighat, A. "Spatial and Angular Domain Decomposition Algorithms for the Curvilinear Sn Transport Theory Method," *Transport Theory and Statistical Physics*, **22,** 391-417 (1993).

6. Haghighat A, M. Hunter and R. Mattis, "Iterative Schemes for Parallel Sn Algorithms in a Shared-Memory Computing Environment", *Nucl. Sci. Eng*, **121,** 103-113 (1995b)

7. Sjoden, G. and Haghighat, A., "PENTRAN™: Parallel Environment Neutral-particle TRANsport in 3-D Cartesian Geometry", *Proc. Int. Conf. on Mathematical Methods and Supercomputing for Nuclear Applications*, pp. 232-234. Saratoga Springs, NY (1997).

8. Haghighat, A., G. E. Sjoden, and V. Kucukboyaci, "Effectiveness of PENTRAN™'s Unique Numerics for Simulation of the Kobayashi Benchmarks," to be published in *Progress in Nuclear Energy* (2001).

9. Haghighat, A., H. A. Abderrahim, and G.E. Sjoden "Accuracy and Parallel Performance of PENTRAN Using the VENUS-3 Benchmark Experiment", *Reactor Dosimetry*, ASTM STP 1398, John G. Williams, et al., Eds., ASTM, West Conshohocken, PA (2000).

10. Kucukboyaci, V., Haghighat, A., Sjoden, G. E., and Petrovic B., "Modeling of BWR for Neutron and Gamma Fields Using PENTRAN™," *Reactor Dosimetry*, ASTM STP 1398, John G. Williams, et al., Eds., ASTM, West Conshohocken, PA, (2000).

11. Petrovic, B., A. Haghighat, T. Congedo, and A. Dulloo "Hybrid Forward Monte Carlo-Adjoint Sn Methodology for Simulation of PGNAA Systems," *Proc. Int. Conf. M&C'99*, Madrid, Spain, pp. 1016-1025 (1999).

12. Sjoden, G. E., R. N. Gilchrist, D. L. Hall and C. A. Nusser "Modeling a Radiographic X-Ray Imaging Facility with the PENTRAN Parallel Sn Code," *Proceedings of the PHYSOR 2000*, Pittsburgh, PA (2000).

13. Brown, J. F. and A. Haghighat, "A PENTRAN Model for a Medical Computed Tomography Scanner," *Proc. Radiation Protection and Shielding 2000*, Spokane, WA (2000).

14. Kucukboyaci, V., A. Haghighat, J. Adams, A. Carlson, S. Grimes, and T. Massey, "PENTRAN Modeling for Design and Optimization of the Spherical-Shell Transmission Experiments," to be presented at the 2001 ANS Summer meeting, Milwaukee, Wisconsin (2001).

15. Gropp, W., Lusk, E., and Skjellum, A., USING MPI- Portable Parallel Programming with Message Passing Interface, MIT Press, Boston, 1995.

16. Lewis, E. E. and W. F. Miller, *Computational Methods of Neutron Transport*, LaGrange Park, Illinois: American Nuclear Society (1993)

17. Sjoden, G. E. and A. Haghighat, "Taylor Projection Mesh Coupling between 3-D Discontinuous Grids for Sn," *Trans. Am. Nucl. Soc.*, 74, 178 (1996).

Layering SHMEM on Top of MPI

Lars Paul Huse

Scali AS, Olaf Helsets vei 6, P.O. Box 70, Bogerud, N-0621 Oslo, Norway
lph@scali.com
http://www.scali.com

Abstract. In this paper we present the Scali ShMem - ScaShMem - library, a compatibility library for Cray SHMEM. SHMEM is a popular application programmers interface for MPP (Massive Parallel Processor) programming, but has previously been limited to MPPs from Cray Inc. and SGI Inc. ScaShMem is layered on top of Scali's MPI implementation; ScaMPI, using multiple threads enabled by ScaMPI's thread-hot & -safe features. Since ScaShMem is layered on top of MPI, one-sided communication of SHMEM can be mixed with MPI message passing. Near native Scali MPI performance is demonstrated for ScaShMem, justifying porting SHMEM applications to a price favorable cluster environment.

1 Introduction

The Shared Memory Access Library - *SHMEM* [5] - is now an integral part of *MPT* (Message Passing Toolkit) [6] from Cray Inc. and SGI Inc. SHMEM is an *API* (Application Programmers Interface) for *RMA* (Remote Memory Access) and collective operations, containing approximately 140 Fortran and 150 C functions.

SCI (Scalable Coherent Interface) [24] is a standardized high-speed interconnect based on shared memory, with the *SMP*s (Shared Memory Processor) connected in closed rings. SCI's built-in error-checking mechanisms in hardware enables very low latency communication. Dolphin ICS implementation of SCI [8], attaching to a standard PCI bus, has built-in hardware support traffic routing between multiple SCI rings connected to the same adapter. This enables building clusters with multi-dimensional meshes as network topology, with scalable network performance to large configurations [3].

MPI (Message Passing Interface) [16,17] is a well-established communication standard for message passing, one-sided communication and parallel IO. *ScaMPI* [11] is Scali's high performance thread-hot & -safe implementation of MPI. ScaMPI currently runs over local and SCI shared memory on Solaris and Linux for x86-, IA64-, Alpha- and SPARC-based workstations. Scali ShMem, *ScaShMem*, is a compatibility library for SHMEM, layered on top of ScaMPI. Care has been taken to allow application programmers to mix MPI communication and SHMEM exchange, similar to the MPT environment. ScaMPI's built-in timing and tracing facility [23] has been extended to include all SHMEM calls.

In the rest of the paper we will first introduce the details of SHMEM and the ScaShMem implementation, then present performance measurements and finally make conclusions.

Y. Cotronis and J. Dongarra (Eds.): Euro PVM/MPI 2001, LNCS 2131, pp. 44–51, 2001.

2   ScaShMem Implementation

The SHMEM library was initially made to provide very efficient communication for the Cray T3D, but can also run on Cray T3E, Cray SV1 and SGI Origin machines, and more recent Compaq AlphaServer SC series. SHMEM includes functionality for remote read (get) and write (put), remote atomic operations (e.g. compare-and-swap and fetch-and-add) and collective operations (e.g. barrier, broadcast, gather, reduce and all-reduce). SHMEM does not use globally shared memory, but instead all references to remote data are done by giving the address of the corresponding local data and the process rank. Memory can either be allocated implicitly (statically) at startup or explicitly (dynamically) at runtime (C and Fortran 90) trough special memory allocation calls. This is in contrast to MPI-2 one-sided communication, where all the communication memory first is allocated by the application and then presented to the MPI library (MPI_Win_create()). In SHMEM dynamic memory allocation is a collective operation, and all processes have to allocate equal sized memory blocks symmetrically. The default SHMEM data types are all 64 bit, but 32 bit data types are also available and in the C interface some operations are even defined for 16 bit integers. Unlike MPI, SHMEM has no explicit start/initialization and stop/termination calls.

Scali ShMem is a compatibility library for SHMEM, layered on top of ScaMPI [11]. In message passing the data source process initiates a data transfer by issuing a send request, and the destination process issues a corresponding receive request - cooperating to make the data transfer. In one-sided communication the application process submitted to an RMA request makes no function calls to hint the communication library of forthcoming communication. The shmem_put() and shmem_get() operations can therefore be viewed as active messages [9], i.e. messages containing data and a pointer to the associated operation (with parameters), are sent to a remote node for execution. All the memory in SHMEM applications must therefore, directly or indirectly, be globally visible. A Cray T3D and T3E process operates directly on physical memory, and SHMEM can therefore directly access remote process memory, through special E-registers, in a one-to-one memory mapping [7]. Applications running on a workstation operate in virtual address space, with dynamic mapping to physical memory (or swap-space on disc). In a cluster environment with inter-node shared memory capability, e.g. SCI based systems, a direct access approach is possible. Since communication adapters connected to an IO bus usually do not have virtual memory support (operating directly on physical memory), a direct access approach would involve pinning down all application memory and exporting a complete mapping to all other processes. As can be imagined this approach has serious scaling limitations on 32 bit systems (only 4 Gbyte address space), and has high probability of security holes and is very resource demanding (e.g. no process memory can be swapped out). ScaShMem therefore uses an indirect mapping approach not to be exposed to these limitations, raising the next issue; How to access the applications' memory from within its own context.

In ScaShMem an RMA is initiated when the process sends an active message to its communication partner (using MPI_Send()). A remote write (put) request contains the data to write and size and destination address information. A remote read (get) request contains size and source address information, and the requesting process then has to wait for the requested data (using MPI_Recv()). The RMA requests have to be handled within the application context or by a separate process/thread for each process. Most MPI implementations handle immediate communication within the application context; immediately when called or when other MPI calls are made i.e. no separate handler as in e.g ScaMPI (default mode), MPICH [19] and LAM/MPI [15]. While this usually provides a resource optimum solution for message passing, this approach may lead to deadlock for SHMEM applications[1]. ScaShMem therefore uses a separate server thread (pthreads [21]), having full access to the application's virtual address space. Concurrent MPI calls are made possible due to ScaMPI's thread-hot & -safe features. Prosess local put and get operations are translated to standard OS memory copy (memcpy()).

To clarify the communication sequence, here is a detailed description on how process A (application thread) get's data from process B (server thread):

- B lies ready to service the request - in MPI_Recv().
- A generates a request and sends it to B - using MPI_Send.
- B receiving the request from A, decodes it and replies back to A with the requested data - using MPI_Send().
- A receives the requested data - using MPI_Recv() - and then continues executing the application.
- B prepares for the next request - by calling MPI_Recv().

One observation that can be made from this is that all MPI_Send()'s have a matching MPI_Recv(), so no buffering in MPI is needed and no communication deadlock can occur.

Since the SHMEM API does not contain a start call, a.k.a. MPI_Init(), the server thread is started when the first SHMEM call is made. ScaShMem has however a non-standard function shmem_start() to hasten initialization. Since all MPI applications are required to call MPI_Init() as their first task [16], detecting if the application mix SHMEM and MPI can be done at ScaShMem startup (by calling MPI_Initialized()). If the application is running in a SHMEM only environment ScaShMem uses MPI_COMM_WORLD for communication. Otherwise, in mixed SHMEM and MPI environment, a new communicator, based on MPI_COMM_WORLD, is generated for ScaShMem communication in order not to interfere with direct MPI communication.

2.1 Dynamic Memory Handling

For a workstation application, memory (statically) allocated at startup is always assigned the same virtual address, hence a remote or local static data item have

---

[1] Deadlock example: a process waiting for data updates from a remote process without using a SHMEM call, e.g. by spin-reading (polling) private memory waiting for special data value.

the same virtual address in all concurrent processes. Allocating dynamic memory, by e.g. shmalloc() or shmemallign(), is in SHMEM programming considered a collective operation, where all processes have to allocate equal sized memory chunks. Even though these memory chunks are allocated equal in size and in the same order, equal virtual addresses for all processes are not guaranteed. Drivers mapping memory directly into user-space, e.g. the SCI driver, may contribute to this. ScaShMem therefore handles memory references by keeping track of all allocated memory for SHMEM communication, transforming all addresses for out-bound requests to its allocation sequence number and offset and transforming back to the local virtual address for incoming requests. For completeness, the static/initial data are in this context viewed as the first allocation performed, starting at NULL. Releasing dynamic memory is similar to allocation.

2.2   Atomic Operations

SHMEM contains a set of atomic operations to local and remote memory. The operations include compare-and-swap, masked-swap, swap, fetch-and-add and add. In addition, SHMEM has calls for data dependent waiting e.g. shmem_-wait_until(). The low-level building blocks for these operations are an integral part of the Scali low level toolbox [14].

2.3   Collective Operations and Communication Groups

All SHMEM collective calls can with minor adaptations be mapped to collective calls in MPI. shmem_broadcast() and shmem_collect() have a one-to-one mapping to MPI_Bcast() and MPI_Gather(), while shmem_{*datatype*}-{*operation*}_to_all() (e.g. shmem_double_sum_to_all()) maps to MPI_Allreduce() with appropriate parameters.

The shmem_barrier(), in addition to being an synchronization (MPI_Barrier()), has the side-effect that all previous remote SHMEM requests have to be completed before returning from the call - similar to doing an MPI_Win_Fence() for all MPI-2 shared memory windows. ScaShMem has solved this by using counters for all sent and all handled requests in each process for all SHMEM processes. The synchronization of these counters are piggy-backed to the message envelope of all requests or explicitly sent when executing a shmem_barrier().

SHMEM does not have MPI's explicit group abstractions, but operates in the global communicator (MPI_COMM_WORLD) and supplies each collective call with three parameters; *PE_start* it the rank of the first process in the group. *logPE_stride* is the log_2 rank spacing between group members. *PE_size* is the total number of members in the group.

Since MPI requires all members of a group to participate in collective operations, explicit MPI groups have to be generated to match SHMEM implicit ones[2]. To avoid race conditions, one of the server threads is appointed master of group creation, and all requests for new groups have to be sent through the master. All

---

[2] The alternative would have been to reimplement all MPI collective calls to match SHMEM ad hoc groups.

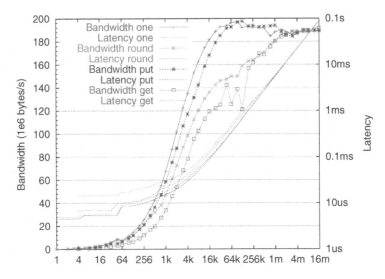

Fig. 1. SCI ping-ping communication performance over 66 MHz/32 bit PCI bus

ScaShMem groups are based on `MPI_COMM_WORLD`, and the group master therefore sends a group creation request to all other processes. All servers then collectivly execute `MPI_Comm_Splitt()`, using `MPI_UNDEFINED` for non members of the new group. Since groups are expected to be used more than once, ScaShMem keeps all generated groups in a lookup table for easy retrieval and reuse.

3   ScaShMem Performance

In communication tests two parameters are usually evaluated; short message latency and long message bandwidth (throughput). Communication performance was measured between two Serverworks HE-LE workstations (100 MHz memory bus) with dual 800 MHz PIII's interconnected with Dolphin D330 PCI-SCI interfaces over a 66 MHz/32 bit PCI bus. ScaMPI over SCI has a latency of 4.2 μs (ping-pong-half) and a peak one-way bandwidth of 197 MByte/s (96 KByte messages). ScaMPI SMP internal performance is 1.6 μs latency and 431 MByte/s (48 KByte messages).

Figure 1 shows the point-to-point communication performance of **shmem_-put()** and **shmem_get()** for a ping-ping communication test, i.e. one way communication, compared to ScaMPI `MPI_Send()` & `MPI_Recv()` for **one way** (ping-ping) and **roundtrip** (ping-pong) communication. For 16 MByte payload the numbers are 189 MByte/s for put (194 MByte/s at 1 MByte), 191 MByte/s for get and 190 MByte/s for MPI_Send, i.e. long message bandwidth is little affected by the ScaShMem protocol. For 4 bytes ScaShMem put uses 6.7 μs, while get uses 14.1 μs (get is actually a blocking ping-pong communication). This is close to the ScaMPI 4 byte latency of 4.6 μs. Using a ping-pong test, 4 byte latency

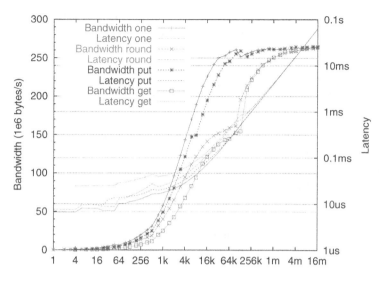

Fig. 2. SCI ping-ping communication performance over 66 MHz/64 bit PCI bus

is measured to 8.6 μs (ping-pong-half) for ScaShMem put, while ScaMPI uses 5.0 μs, i.e. latency is slightly increased when using ScaShMem.

Figure 2 shows preliminary performance results, for the same communication test as in figure 1, between two IA-64 based workstations (dual 667 MHz Itanium) with 66 MHz/64 bit PCI bus using SCI as interconnect. The ScaMPI performance is 5.8 μs & 265 MByte/s for one way and 6.2 μs & 263 MByte/s for roundtrip communication (0.8 μs latency is added for 4 byte messages), while put performance is 9.6 μs & 265 MByte/s and get 24.2 μs & 264 MByte/s.

The communication performance of Cray T3E-900 is reported to 1 μs latency & 350 MByte/s for SHMEM and 14 μs latency & 260 MByte/s for MPI [1]. We expect the new Cray T3E-1350 to have similar latency, but higher bandwidth.

Since the collective operations are directly mapped to MPI collective operations, performance and scaling for ScaShMem and ScaMPI [12,13] are equal.

Unfortunately, most SHMEM applications we have tested are proprietary and under NDA by Scali AS, so no scalability figures can be published at this point. Porting SHMEM applications from true 64 architectures, e.g. Cray T3E, to ScaShMem on a workstation cluster, is a straightforward task and involves only the expected changes when porting applications between ILP64 and ILP32/LP64 data model [20] i.e. the size of integers and pointers may differ.

4   Related and Further Work

There are a lot of shared memory libraries available, but few are intended for writing applications directly. Most of them are used as a basis for high level communications libraries e.g. MPI [16,17] and PVM [22]. The novelty of ScaSh-

Mem is that we layer the complete (except stack manipulation) shared memory SHMEM API on top of a high level message passing interface.

The Illinois HPMV communication library [10] has partially implemented SHMEM on top of Fast Messages, missing support for e.g. point-to-point strided and gather/scatter operations. The performance on top of Myrinet and Windows NT 4.0 is given to 8 byte latency 20.9 μs (put) and 22.8 μs (get), and long message bandwidth of 93.7 MByte/s (put) and 86.4 MByte/s (get). Compaq Computer Corporation and Quadrics Supercomputers World Ltd have within the US ASCI Pathforward Program ported SHMEM to QsNet (also called Elan) on Alpha clusters. SHMEM is now available on Compaq AlphaServer SC series with latency of 3 μs [4].

MPI-2 one-sided communication is available from several vendors e.g. SGI (MPT 1.4 for IRIX), DEC [25] and SUN [2] and free implementations [15,18].

There is still potential for improving ScaShMem performance e.g. message handling of put and get can be made an integral part of the receive handling of ScaMPI and the lost CPU cycles, due to the server thread, can be saved.

5   Conclusions

SHMEM has been a de-facto standard for RMA on Cray and SGI MPPs for some years, while implementations of one-sided communication in MPI-2 has just recently become available. Therefore there are still relatively few application using MPI-2 one-sided communication, while SHMEM on the other hand is used by numerous applications, but have up until now been restricted to run on large MPPs or have limited performance and functionality.

Application programmers tend to have an acceptance for some reduced performance compared to adapting to new API's, hence all the dusty-decks in the world. Although ScaShMem can't deliver full MPP performance, it is not too far behind and its performance should be acceptable for a lot of applications. Compared to MPPs, the price per memory bit and MIPS for workstations is favorable. ScaShMem will therefore enable solving large problems on workstation clusters at a reasonable price.

In the paper we have described how SHMEM is mapped on top of ScaMPI. All major SHMEM features (except stack manipulation) are implemented. Layering ScaShMem on top of ScaMPI have enabled rapid development and the thin software layer preserves most of the performance of ScaMPI.

Acknowledgements

Special thanks to the SHMEM users who let us test their applications (which helped in the debugging and performance tuning), and to Øystein Gran Larsen who helped in porting and running them.

References

1. E. Anderson, J. Brooks, C. Grassl, S. Scott: Performance of the CRAY T3E Multiprocessor. In proceedings of Supercomputing 1997.
2. S. Booth, F.E. Mourão: Single sided MPI implementations for SUN MPI. In proceedings of Supercomputing 2000.
3. H. Bugge: Affordable Scalability using Multicubes. Proceedings of the 1st International Conference on SCI-based Technology and Research - SCI Europe (1998).
4. The Compaq AlphaServer SC series specification at http://www.compaq.com/-alphaserver/sc.
5. Cray, Inc.: Application Programmer's Library Reference Manual. (004-2231-002).
6. Cray / SGI Inc.: Message Passing Toolkit for IRIX (2001) http://www.cray.com/-products/software/mpt.html and http://www.sgi.com/software/mpt/index.html
7. Cray Inc.: Cray T3E white paper (2000). Available from http://www.cray.com/
8. Dolphin ICS Inc.: PCI-SCI Adapter Card D320/D321 Functional Overview Version 1.01, (1999) Available from http://www.dolphinics.com/whitepapers.html
9. T. von Eicken: Active Messages: An Efficient Communication Architecture for Multiprocessors, Ph.D. thesis at University of California at Berkeley (1993)
10. A.A. Chien, D. Reed, D. Padua et.al.: High-Performance Virtual Machines - HPVM (1999). Available from http://www-csag.ucsd.edu/projects/hpvm.html.
11. L.P. Huse, K. Omang, H. Bugge, H.W. Ry, A.T. Haugsdal, E. Rustad: ScaMPI - Design and Implementation. LNCS 11734; SCI: Scalable Coherent Interface. Architecture & Software for High-Performance Compute Clusters (1999)
12. L.P. Huse: Collective Communication on Dedicated Clusters of Workstations. Proceedings of 6th PVM/MPI European Users Meeting - EuroPVM/MPI (1999)
13. L.P. Huse: MPI Optimization for SMP Based Clusters Interconnected with SCI. Proceedings of 7th PVM/MPI European Users Meeting - EuroPVM/MPI (2000)
14. L.P. Huse, K. Omang, G. Krawezik, H. Bugge: Architectural Issues of Creating Portable SCI Cluster Middleware. Proceedings of the 3rd International Conference on SCI-based Technology and Research - SCI Europe (2000).
15. LAM/MPI (Local Area Multicomputer) Parallel Computing Environment - Version 6.5.1 (2000) Available from http://www.lam-mpi.org.
16. MPI Forum: MPI: A Message-Passing Interface Standard. Version 1.1 (1995)
17. MPI Forum: MPI-2: Extensions to the Message-Passing Interface (1997)
18. F.E. Mourão, J.G. Silva: Implementing MPI-2 One-Sided Communications for WMPI. Proceedings of 6th PVM/MPI European Users Meeting - EuroPVM/MPI (1999)
19. MPICH: Portable MPI Model Implementation. Version 1.2.1 (2000) Available from http://www.mcs.anl.gov/mpi/mpich.
20. The Open Group: Data Size Neutrality and 64-bit Support (2000) Available from http://www.unix-systems.org/whitepapers.
21. POSIX System Application Program Interface IEEE Std 1003.1c-1995 (1995)
22. PVM (Parallel Virtual Machine) Library version 3.4 (2001) Available from http://www.epm.ornl.gov/pvm/pvm_home.html
23. Scali AS: ScaMPI User's Guide v 1.10 (2001) Available from http://www.scali.com.
24. IEEE standard for Scalable Coherent Interface IEEE Std 1596-1992 (1993)
25. J.L. Träff, H. Ritzdorf, R. Hempel: The Implementation of MPI-2 One-Sided Communication for the NEC SX-5. In proceedings of Supercomputing 2000.

Support for MPI at the Network Interface Level

Bernard Tourancheau[1] and Roland Westrelin[2]

[1] Sun Labs Europe
Grenoble, France
Bernard.Tourancheau@sun.com
[2] ReSAM laboratory/Université de Lyon
RESO project/INRIA
ENS Lyon, Lyon, France
Roland.Westrelin@ens-lyon.fr

Abstract. Commodity components and high speed interconnects allow to build fast and cheap parallel machines. Several research works demonstrated that the raw performance of MPI ports on top of user level network interfaces are close to the hardware specifications. However, most of these implementations offer poor possibilities of communication/computation overlapping because the application thread on the host processor is responsible of a large part of the protocol. Thus, application performances are limited. One solution to this problem is to move the critical part of the protocol at the network interface card level. We implemented such a communication system that relies on the embedded processor of the Myrinet host interface. We present the problems moving part of the protocol at the network interface level introduces, our design choices and we also give performance measurements.

1 Introduction

Using commodity components, it is possible to build clusters of workstations with Gflop CPUs and Gb/s interconnects offering an unbeatable price/performance ratio. The main bottleneck in such a parallel machine is most of the time the communication software. Legacy protocol stacks such as TCP/IP rely heavily on system calls and memory accesses because they assume an unreliable, slow network and the need for internetworking capabilities. User Level Network Interfaces (ULNI) were designed to provide performance close to the hardware maximum at the application level. In such a design, the operating system is removed from the critical path and zero-copy communications are used.

BIP (Basic Interface for Parallelism) [10] is a ULNI dedicated to Myrinet developed by our team. It achieves 3 μs of latency and 250 MB/s of bandwidth with the latest Myrinet hardware. MPI-BIP [11] is the port of MPI on top of BIP. It adds a few μs of overhead. The BIP firmware evolved into an extensible design that uses the latest Myrinet hardware features and remains efficient. It is called BIPng [13]. Adding extensions to this firmware is easier because it offers a thin layer that interfaces with the hardware and relies on an event loop which acts as a simple scheduler.

Y. Cotronis and J. Dongarra (Eds.): Euro PVM/MPI 2001, LNCS 2131, pp. 52–60, 2001.

To achieve low latency, ULNI systems rely on polling to detect the completion of communication operations. Applications are then limited to one thread per processor. In the case of an MPI implementation, the application thread is also responsible of the MPI protocol. Because MPI ports on top of ULNI often use a rendez-vous scheme, overlapping of communication by computation is not always possible and MPI libraries, with raw performance close to the hardware specifications, often perform very poorly in application benchmarks. One solution to this problem, addressed in this paper, is to transfer the MPI communication library main functionalities into a programmable NIC (Network Interface Card) running a cluster-optimized firmware. Note that, even if our implementation was performed in the Myrinet framework, it could have been done with any programmable NIC.

Section 2 addresses the issues related to the MPI protocol and to the overlapping of computation and communications. Section 3 presents our design choices. Performance measurements of their implementation in the BIP firmware are given in section 4. Section 5 concludes.

2   Problem Statement

The problem introduced in this section was already discussed in [12] and demonstrated with the original MPI-BIP implementation in [3].

An MPI implementation typically uses two kinds of messages: control messages used internally for communication management and data messages which contain only application data. The underlying communication system whether it is a legacy protocol stack such as TCP-UDP/IP or a light weight communication layer is used to transfer both kinds of messages.

For flow control and buffer management reasons, the use of a rendez-vous protocol is an efficient scheme for large message communications. The sender sends a Request To Send (RTS) to the receiver. The RTS contains the communicator and tag. The receiver when it gets the RTS performs the matching phase according to the rules defined in the MPI standard [5] to find a matching posted receive. If one is found, it sends back a Clear To Send (CTS). Otherwise, it enqueues the send request to try to match it with the receives that will be posted subsequently.

Figure 1 illustrates the problem. Process 1 in node 1 posts a MPI send request. The MPI implementation sends a RTS and the process goes back to computation. Process 2 in node 2 posts the matching receive request and goes back to computation for a long time. The RTS reaches node 2 while process 2 is in a computation phase and thus is not processed. When process 2 reenters the MPI library, it gets the RTS, find the matching receive and sends the CTS. But, again, the CTS message reaches node 1 while process 1 is computing. The treatment of the message is postponed until process 1 reenters the MPI library, the data are then finally sent.

The first solutions rely on blocking communication primitives and asynchronous message delivery (through interrupts) at the host level. It ensures that

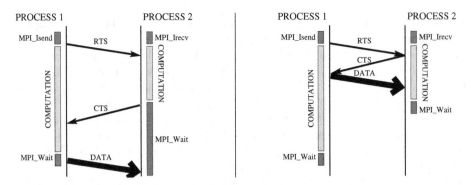

Fig. 1. The problem with communication/computation overlapping: on the left what happens in most implementations and, on the right the behavior of an efficient MPI implementation.

the communications progress even when the application is computing. For instance, in the implementation of MPI on top of VIA from MPI Software Technology [4], a pool of threads is dedicated to MPI communications. These mechanisms (threads, blocking calls, interrupts) results in expensive context switches and high impact on communication latency. A solution to minimize interrupt frequency is to rely on a watchdog mechanism comparable to the one described in [1]. The underlying communication system monitors the progressions of the communication at the network interface level. It triggers an interrupt to force the processing of pending messages by the host only when a message is waiting to be processed for a long time.

The second class of solutions uses NIC level support of MPI mechanisms. The RTS/CTS/data protocol (matching between send requests and posted receives and management of unexpected send requests) is handled at the NIC level, such as in MPI-NP [7]. Authors report a one-way latency of 22 μs and a maximum bandwidth of 92 MB/s for their custom Myrinet firmware [1]. There is no published result for application benchmarks. The Portals 3 [2] library, used to implement MPI, provides two communication primitives (put and get) that support a matching scheme. There are plans to implement the put and get primitives on top of Myrinet with a custom firmware but, as far as we know, it is not yet done.

3   Proposed Solution

The following is a description of a simple implementation of MPI protocols and matching mechanisms at the firmware level in BIPng.

---

[1] We use faster host interfaces in this paper so this numbers should not be used to compare with our measurements.

3.1 Basic Strategy

The firmware must keep the list of posted sends and receives to be able to match send and receive requests. When a send is posted on the send side, the firmware must first send the RTS. Once the send request matches a receive request on the receive side, the firmware sends the CTS. Our implementation provides zero-copy communications with dynamic registration of user buffers as it is done in BIP. So, for the data communications, the firmware needs DMA descriptors with the physical addresses of the pages of the buffers on the send and receive sides.

To have a simple design we chose to use processor writes (Programmed I/O) through the PCI bus in the LANai memory to pass requests from the MPI library to the firmware. To minimize the processing at the the firmware level, the send requests written by the host processor contain the RTS message ready to be sent. The receive requests contain the CTS message with only a few missing fields that are added by the firmware once the matching is done. For messages up to about 10 Kbytes, the requests written by the host processor also include the DMA descriptor for the data communications. For bigger messages, DMA descriptors are stored in main memory and fetched when the data transfer is ready using DMA by the firmware. Since the requests are written directly by the host in the network interface memory, an area in this memory is reserved for send and receive requests. It is managed by the host: it allocates and frees entries.

Underlying transfers (both RTS/CTS messages and data) use BIPng mechanisms. They provide asynchronous communications, notification mechanisms and efficient data transfers (with a fine grain pipeline for data messages). BIPng's firmware also includes a simple scheduler that allows multiple threads to execute in the firmware.

The firmware manages several queues: the send and receive posted queues and the unexpected send request queue. When a send is posted, the firmware immediately sends the RTS and then enqueues the request in the send posted queue. When a RTS is received, the firmware first checks for a matching receive in the receive posted queue. If none is found, the RTS is stored in the unexpected send request queue. When a receive is posted, the firmware first checks for a matching send request in the unexpected send request queue. If one is found, it fills the missing field of the CTS message passed by the host and sends the CTS. If none is found, it enqueues the receive in the posted receive queue.

The RTS contains a logical identifier of send request. On the send side, when the firmware sends back the CTS, it adds the identifier of the send request in the message. On the reception of the CTS, the firmware on the send side only has to retrieve this identifier and it can start the data phase.

In this first implementation, we chose to use simple FIFOs for all the queues. Using more complex data structure is also possible to ensure that the matching time is constant.

Events are passed from the host to the firmware using a FIFO in the network interface memory. The head of this FIFO is polled by the firmware when it is idle. The firmware notifies the host using a FIFO located in main memory and written through DMA by the firmware. This mechanism relies on cache snooping and cache coherency mechanisms to minimize the traffic over the memory bus when the processor is waiting for an event.

3.2   Small Messages and More Complex Strategy

Even if, for messages large enough, the time of the handshake between the network interfaces using the RTS/CTS protocol is negligible, for small messages these extra messages are very expensive. Since our goal is to provide a low latency MPI implementation, another protocol must be used for this range of sizes.

A Straight Forward Strategy. In MPI-BIP, small messages are composed and written by the host processor in a queue in the network interface memory. A request to send the content of such a buffer is then passed to the firmware which blindly sends the content of this buffer. On the receive side, the message just received is DMAed in a FIFO in main memory without any treatment. An extra memory copy is needed to put the data in the user buffer. A credit-based flow control is used to prevent overflow of the FIFO in main memory.

We chose to use this protocol. However, the strategy for small messages cannot be independent from the one for larger messages for a simple reason: an MPI message which uses the small message strategy can match a MPI receive which uses a large message strategy. Since matching is performed at the firmware level and a "small send" can match a "big receive", matching must also be performed at the firmware level for the small message strategy. It implies that receives for the small strategy must be posted a the firmware level.

It means that when a receive is posted for a small message, the MPI library must pass the request to the firmware. It has to wait until the firmware tells which send request matches the receive request to be able to complete the receive even if the message was already received and is ready to be copied into the user buffer.

An Optimized Strategy. The extra handshake between the host and the firmware is expensive especially if the firmware is busy. We chose to try to design a strategy in which the host could decide by itself which send request a small message received matches.

We use the following scheme. The host keeps a queue of every posted receive requests and a queue of every small messages received. It can then match send and receive requests by a simple lookup in these queues. All the receives are posted at the firmware level too because it is unavoidable as explained above.

When a small message is received by the firmware, it is directly passed to the MPI library at the host level to minimize the latency of the communication. Then while the message is being copied to main memory, the firmware analyzed it and if it does not match a receive request, enqueues the send request or, otherwise, removes a receive request from the queue of posted receive. When the host receives the messages it looks in the queue of posted receive for the matching receive. Small messages communication can be performed without any extra handshake with the firmware. The price paid is that both the firmware and the host match the received message against the send requests.

Since a small message can match a big receive, the host queue of receives posted must contain all the RTS. It means that when the firmware receives a RTS, it passes it to the host. Because the host has all the RTS and small messages in a queue, when a receive of any size is posted by the application, it can match this receive request with the corresponding RTS or small message already received.

Since the host and the firmware maintains their own queues, they can both perform the MPI matching. To avoid redundant processing at the firmware and host level and to further optimize, we split the processing of the requests between the host and the firmware: our implementation exploits the power of the host CPU as much as possible and relies on the firmware to ensure progression of the communications.

3.3   Support for Multiple Processes per Node

Our goal is to support SMP nodes with only one network interface per node. First, the data structures used by the firmware must be split among the processes so that the network interface can be shared by several processes.

It is also necessary to support communications within the SMP node among the different processes that share the network interface. Such communications cannot be handled only at the host level. Because of the MPI_ANY_SOURCE wild-card, matching must be performed at the NIC level even for the communications between processes local to an SMP node.

Our MPI implementation right now supports multiple processes per node but communications inside a machine do not rely on shared memory (such as in [6]). All the processes of the local machine are seen as independent processes. Messages are sent through the network whether they are for a local process or for a process on another machine.

Figure 2 presents a summary of data structures used in our implementation.

3.4   Integration with MPI

The mechanisms described above were introduced in BIPng. The MPICH [8] distribution was modified to plug BIPng at the ADI level.

The MPI standard is currently not fully implemented. However, enough is available so that we can experiment with our implementation of the MPI matching mechanisms in firmware.

4   Performance Measurements

The test-bed for the experiments presented in this section is composed of four dual PIII 600 Mhz and four LANai 7 Myrinet-1280 host interfaces. The network cards are plugged in 64 bit/66 Mhz PCI slots.

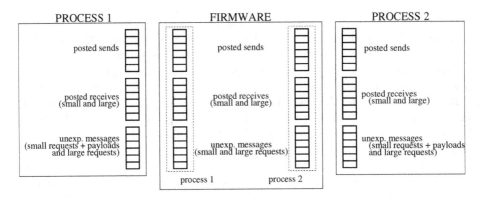

Fig. 2. The data structures at host and firmware level in the case of two processes on one machine that share the network interface.

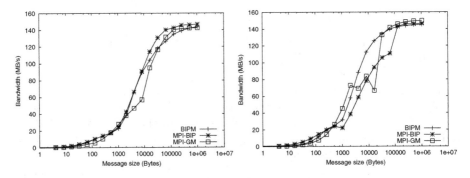

Fig. 3. Ping-pong and ping-ping tests.

We used three different implementations of MPI: the initial MPI-BIP, MPI-GM and our implementation with MPI support at the NIC level, named BIPM. For communications between process inside an SMP, MPI-GM relies on shared memory but, here, MPI-BIP and BIPM use the network.

Figure 3 gives the bandwidth of simple ping-pong and ping-ping tests. For large messages, the bandwidth of BIPM is not as good as the one of MPI-BIP mainly because BIPM uses network packets a lot smaller: 4 Kbytes for BIPM and 100 Kbytes for MPI-BIP. The latency (not represented on these graphs) is $6.5\mu s$ for MPI-BIP, $8.5\mu s$ for BIPM and $15\mu s$ for MPI-GM.

Table 1 gives the results for two application benchmarks from the NAS parallel benchmarks suite [9] for 4 processes (one process per machine) and 8 processes (2 processes per machine). The communication performance is very important for IS which is communication bound. This benchmark does not offer any possibilities of communication/computation overlapping as it only uses collective operations. As shown by the table, the execution times on 4 and 8 processors

Table 1. Some results with the NAS parallel benchmarks: execution time in seconds and speedups.

	sequential	4 processes			8 processes		
		MPI-BIP	MPI-GM	BIPM	MPI-BIP	MPI-GM	BIPM
IS	12.0 s	3.12 s (3.8)	3.38 s (3.5)	3.14 s (3.8)	2.52 s (4.8)	2.48 s (4.8)	2.62 s (4.6)
LU	1650 s	432 s (3.8)	786 s (2.1)	395 s (4.2)	237 s (7.0)	363 s (4.5)	241 s (6.8)

are close for all the MPI implementations. That shows that these MPI implementations are close in terms of communication performance.

The LU benchmark relies on non blocking MPI calls and thus can benefit from overlapping of communication by computation. It is also inherently unbalanced and results from that benchmark can be quite unstable. MPI-GM does not perform well because it does not allow communication/computation overlapping. The results for MPI-BIP are good not because of better communication performance, but because, by chance, the application could overlap communication by computation. Running this same benchmark with MPI-BIP on a different platform could give results as bad as the ones of MPI-GM. On the contrary, BIPM shows good performance because it ensures that communication can be overlapped by computation.

5   Open Issues and Conclusions

Our study shows that embedding MPI functionalities into the NIC is possible within today's cluster architectures. Preliminary experiments show that moving part of the protocol processing in the firmware does not affect significantly raw or application performances and that it can exploit communication/computation overlapping to provide dramatic improvements.

Several issues remain open. How does NIC level support of the MPI protocol compare in terms of performance with MPI implementations that rely on interrupts and operating system mechanisms to achieve communication progression? How to implement communications over shared memory in a node when multiple processes share the network interface? How to support multiple communication channels in a cluster with heterogenous network hardware? With a reliable network like Myrinet it is questionable whether recovery mechanisms should be implemented at the firmware level or at the host level. When the network interface is used for MPI protocol processing, is it possible to handle recovery mechanisms at the host level?

This is a first step toward embedding functionalities into NICs. There is a wide range of open studies for such network embedded computing; from the preceding communication functions to the networking operation such as filtering, multicasting, monitoring, and other services.

References

1. R.A.F. Bhoedjang, K. Verstoep, T. Ruhl, and H.E. Bal. Reducing data and control transfer overhead through network-interface support. In *First Myrinet User Group Conference (MUG)*, Lyon, France, September 2000.
2. Ron Brightwell, Tramm Hudson, Arthur B. Maccabe, and Rolf Riesen. The portals 3.0 message passing interface. Technical Report SAND99-2959, Sandia National Labs, November 1999.
3. Frédérique Chaussumier, Frédéric Desprez, and Loïc Prylli. Asynchronous communications in MPI – the BIP/Myrinet approach. In *Recent Advances in Parallel Virtual Machine and Message Passing Interface. Proc. 6th European PVM/MPI Users' Group (EuroPVM/MPI '99)*, volume 1697 of *Lect. Notes in Comp. Science*, pages 485–492. Springer-Verlag, September 1999.
4. R. Dimitrov and A. Skjellum. Efficient MPI for virtual interface (VI) architecture. Technical report, MPI Software Technology, 1998. Available from http://www.mpi-softtech.com/publications/default.asp.
5. Message Passing Interface Forum. MPI: A message passing standard, June 1995. Available from http://www.mpi-forum.org/docs/docs.html.
6. Patrick Geoffray, Loïc Prylli, and Bernard Tourancheau. BIP-SMP: High performance message passing over a cluster of commodity SMPs. In *Supercomputing (SC '99)*, Portland, OR, November 1999. Electronic proceedings only.
7. C. Keppitiyagama and A. Wagner. Asynchronous MPI Messaging on Myrinet. In *15th International Parallel and Distributed Processing Symposium*, page 50, San Francisco, California, April 2001.
8. MPICH-a portable implementation of MPI. http://www-unix.mcs.anl.gov/mpi/mpich/.
9. NAS parallel benchmarks. http://science.nas.nasa.gov/Software/NPB/.
10. Loïc Prylli and Bernard Tourancheau. BIP: a new protocol designed for high performance networking on Myrinet. In *1st Workshop on Personal Computer based Networks Of Workstations (PC-NOW '98)*, volume 1388 of *Lect. Notes in Comp. Science*, pages 472–485. Held in conjunction with IPPS/SPDP 1998. IEEE, Springer-Verlag, April 1998.
11. Loïc Prylli, Bernard Tourancheau, and Roland Westrelin. The design for a high performance MPI implementation on the Myrinet network. In *Recent Advances in Parallel Virtual Machine and Message Passing Interface. Proc. 6th European PVM/MPI Users' Group (EuroPVM/MPI '99)*, volume 1697 of *Lect. Notes in Comp. Science*, pages 223–230, Barcelona, Spain, September 1999. Springer Verlag.
12. Loïc Prylli and Roland Westrelin. Current issues in available implementations on Myrinet. In *Proc. First Myrinet User Group Conference*, pages 149–155, Lyon, France, September 2000.
13. Roland Westrelin. A new software architecture for the BIP/myrinet firmware. In IEEE Computer Society, editor, *International Symposium on Cluster Computing and the Grid (CCGRID)*, Brisbane, Australia, May 2001.

The Implementation of One-Sided Communications for WMPI II

Tiago Baptista[2], Hernani Pedroso[1], and João Gabriel Silva[2]

[1] Critical Software, S.A.,
Instituto Pedro Nunes, R. Pedro Nunes, 3000 Coimbra, Portugal
hpedroso@criticalsoftware.com
[2] CISUC/Dep. Engenharia Informática
Universidade de Coimbra – PoloII, 3030-097 Coimbra, Portugal
baptista@student.dei.uc.pt, jgabriel@dei.uc.pt

Abstract. This paper describes the implementation of MPI-2 one-sided communications (OSC) for the forthcoming WMPI II product, aimed at clusters of workstations (Windows workstations, for now). This implementation is layered directly on top of the WMPI Management Layer (WML), rather than being on top of the MPI layer and as such can draw more performance from the new features of the WMPI's WML. The major features of this implementation are presented, including the synchronization operations, the remote memory operations and the datatype handling mechanism. Performance benchmarks were taken, comparing the message passing and the one-sided communication models, as well as to compare this implementation with one layered on top of MPI.

1 Introduction

The second version of the MPI standard [1], MPI-2 [2], introduced several new important features. Among others, it introduced the ability to make message passing without the involvement of one of the processes – one-sided communication (OSC). This standard's chapter specifies functions for single sided communication, where a process specifies both origin and target parameters.

Although this functionality can easily be mapped over shared memory systems, its implementation for distributed memory systems (e.g. clusters) is not straightforward. Most of the OSC implementations use the traditional MPI point-to-point functions to do one-sided communication when processes don't share memory [3, 4]. Nevertheless this approach cannot use the internal specific capabilities of the implementation. This paper presents an implementation of the OSC functionality for NT clusters inside the WMPI [5] library. This implementation joins the experience gathered by a first OSC implementation on top of MPI point-to-point functions [3] with the new internal structure of the WMPI library, which allows the use of several communication devices, where some of them may have RMA support.

We begin by describing the background of this implementation and library it belongs to. Next, the specifics of the architecture are presented, the synchronization operations, the remote memory access operations and the datatype handling, followed

Y. Cotronis and J. Dongarra (Eds.): Euro PVM/MPI 2001, LNCS 2131, pp. 61–68, 2001.

by performance benchmarks results. Finally, conclusions are presented regarding current and future work.

2   Background

The implementation of One-Sided Communications presented here was created for WMPI commercial implementation. As part of this project, all MPI-2 [2] features are being implemented on WMPI 1.5 – the current MPI implementation – that will then evolve into the WMPI II product.

WMPI was first implemented in 1996 [6], based in the MPICH/p4 [7] implementation. It was the first MPI implementation for Windows™ systems. Since then the library has undergone several improvements and in the last year a complete new structure was created (WMPI 1.5) [5]. This implementation allows the use of several communication devices simultaneously, is thread-safe and was the base for the implementation of the dynamic process creation functionality introduced in the MPI-2 standard [8].

An OSC implementation had already been created on top of WMPI 1.1 [3]. That implementation had a somewhat different methodology as it was layered on top of the MPI point-to-point functions, whereas the new implementation is layered directly on top of WML (WMPI Management Layer) [5], the layer that translates the MPI actions into the library and operating system specific actions. WML is the responsible layer to manage the communications, choose the correct communication device, request to send the messages and verify when new data arrives.

Hence, we were able to design several enhancements to one-sided communications using the new features of WMPI 1.5. One of these new features is the ability to handle multiple devices simultaneously and to have those devices in a separate module. The one-sided communications were as such, implemented to allow direct RMAs for devices that support them. For now, the shared memory device is already implemented allowing these direct RMAs. In the future other devices may follow, like a device for VIA.

3 One-Sided Communications Architecture

This implementation of one-sided communications was designed always having in mind the new features of WMPI 1.5 and the best way to gain performance and stability using these features. The object-oriented methodology was chosen to design and implement OSC, as it allowed for better layering and encapsulation of functionality. The standard itself is not particularly pointed towards an object-oriented interpretation. However, if one reads it having in mind an object-oriented design, it all begins to look like it had been thought this way. It is rather straightforward to define the architecture in terms of classes, despite the lack of other implementations having taken this path. Take the case of windows and epochs for example. A window has both an access and an exposure epoch. These can be any kind of epoch. Thinking object-oriented we will have the class CWindow that will have two objects of the class CEpoch. This class is the base class for all epoch classes. This way, the class

CWindow won't require the knowledge of what type of epoch is open and the epoch classes can be implemented separately. That way they won't interfere with each other. As the object-oriented paradigm can be implemented in C++, the performance won't suffer.

One of the major architecture decisions taken regards the functions where to block: the epoch open, the remote memory access or the epoch close. The standard allows any of the stated. For this implementation the choice was to block on the RMA, if the target has not yet called the epoch's open function. This decision took primarily into account the fact that the RMA may be a direct access to the remote processes memory and as such it is required that the remote exposure epoch be opened prior to performing the operation.

3.1   The Synchronization Operations

There are three different synchronization methodologies, the fence, the start-post and the lock mechanisms.

Fence. The fence mechanism is the most basic form of synchronization in MPI-2 one-sided communications. Using a call to MPI_Win_fence, access and exposure epochs are opened for all the processes in the communicator of the window. This call is a collective call and as such involves all the processes that share the window. The same MPI_Win_fence function is used both to open and to close the epochs. As this mechanism involves all the processes in the communicator, the synchronization is achieved using a call to MPI_Barrier. As to ensure that all the remote memory accesses issued before this call are complete, every process checks and waits until all pending operations are complete.

Start-Post. For applications where processes communicate only with few other processes, there is no need to include all the processes of the window in each synchronization operation. Using the start-post mechanism, groups of communicating processes can synchronize themselves without involving the other processes of the window. The function MPI_Win_start opens an access epoch to a group of processes it receives as a parameter. This function will never block nor send any message. It will only initialize the access epoch and process any pending posts received by this process. The epoch created will only grant access to a process after having received its post message. Every process of the group passed to MPI_Win_start has a matching MPI_Win_post call. This function will create and initialize an exposure epoch allowing all processes of a group received as a parameter. It will send a post message to every process in the group allowing them to open an access epoch.

To close the access epoch, a process must call MPI_Win_complete. This function will send a complete message to all the processes in the group of this epoch (passed to MPI_Win_start) and checks that all the RMAs issued are complete at the origin. This function will also check if all posts were received prior to closing the epoch, thus ensuring that a subsequent call to MPI_Win_start won't receive a post that was meant for the previous epoch.

To close the exposure epoch at the target, a process must call MPI_Win_wait. No messages are sent. The function waits for the complete messages from all the processes in the group of this epoch (passed to MPI_Win_wait) and checks that all RMAs are complete at this process (the target). All RMAs are sure to be complete at the origin because the complete message was received.

Lock. For applications requiring a truly one-sided synchronization mechanism, the lock/unlock methods were introduced. These methods implement the so-called *passive target* synchronization. That name refers to the ability of one process to exclusively lock another process without the latter having an active part in the process.

By calling MPI_Win_lock, a process can open an access epoch to the process whose rank is passed as a parameter. No explicit exposure epoch is required to be opened at the target. However, an internal exposure epoch is opened to maintain consistency with the other two synchronization methods and to allow the RMAs to be generic, having no knowledge of the type of epoch in use. The lock acquired can be either exclusive or shared. Several processes can lock the same target simultaneously provided that all the calls to MPI_Win_lock are shared. However, if a process is exclusively locked, no other (shared or exclusive) lock can be acquired until the call to MPI_Win_unlock by the origin. The MPI_Win_lock function doesn't block. It only sends the request to the target process. A subsequent call to an RMA will, nevertheless, block until the lock is acquired.

To close the access epoch and release the lock, the origin process must call MPI_Win_unlock. As the MPI_Win_lock didn't block, this methos method must check if the lock is acquired before proceeding with the release. On exit of this function all RMAs must have completed both at the origin and at the target. To accomplish this, first the origin process checks if all RMAs are complete, next a message is sent to the target requesting the release of the lock. At the target, checks are made to ensure all RMAs are complete and sends to the origin the unlock reply message. The origin can now exit from the function.

3.2   The Remote Memory Access Operations

The remote memory access operations were implemented as generic, i.e., the type of epoch in use is not known by the function. This was easily accomplished due to the object-oriented paradigm followed.

There are three different situations that can occur when calling an RMA: the datatype is not contiguous, the datatype is contiguous and the device supports direct RMA or the datatype is contiguous and the device does not support direct RMA. To gain performance wherever possible, the implementation takes into account the kind of operation being performed and acts accordingly. Even so, there is some functionality that is common to all three. First the function checks the current epoch to ensure that the RMA can proceed (e.g. the matching post has been received), and then it verifies if the block to transfer fits the target location.

Now there are the three different options. If the datatype is not contiguous, the data must be packed and the datatype sent along with the packed data. If the datatype is contiguous and the device does not support RMA, there is no need to pack the data or send the datatype. A message is sent using the memory location of the RMA as the send buffer. If the datatype is contiguous and the device does support RMA, no message is sent. A direct call to the device is made to process the request.

The Put is implemented with none (the case of direct RMA) or one message exchange and the Get with none or two messages. The Accumulate is implemented the same way as Put but no direct RMA is possible. It would require two direct RMAs to perform the Accumulate and this would be less efficient than sending one message. The datatype is always needed to perform the operation at the target.

3.3   Datatype Handling

Datatype creation functions are local functions, which build local representations of datatypes, which may be unknown in other processes of the computation. Opposite to the point-to-point functions, in one-sided communication, the receive parameters are specified in the origin process. One of the parameters to specify is the receiving datatype, however there is no guarantee that the specified datatype is available at the remote process.

This new OSC implementation has some different conditions than the previous one, it is developed for homogeneous environments and may use communication devices that allow for remote memory access. These two new conditions simplify the situation when the receive datatype is contiguous and the communication device has support for RMAs. In this case the un-marshal and marshal of the data is done locally and the result buffer is copied directly to the target process's memory. If the send datatype is equal to the remote datatype (and is contiguous), then the user data buffer can be directly copied to the remote process without further computation (this is the best case).

Nevertheless if the receive datatype is non-contiguous it is preferable to pack the data and send it to the remote process, there the data should be unpacked and copied into the target memory region.

If the communication device does not have RMA support, then the communication has to be performed over the traditional message passing mechanisms, and the target process has to know the receive datatype to un-marshal the data. In this case, we use the same scheme as in the case of the previous OSC version. We pack the datatype information and send it along with the data.

The last two scenarios have penalties due to the local characteristics of the datatypes. Our future development will address this issue using the datatype caching [9] and signatures [10]. Instead of sending the datatype along with the data, only its signature is sent. If the remote process has a datatype with the same signature, then the data is un-marshaled using that datatype, otherwise the remote process has to request the datatype. If the datatype is not present at the remote site, then the operation has a considerable penalty, however the most common case is that the user has specified the datatype in the remote process. If the datatype has to be sent to the remote process, techniques to efficiently pack it, like the one presented in [11], will be used.

4   Performance

To benchmark this implementation, we compared the current with the previous OSC implementation [3]. Although the underlying programming models differ, some comparisons were also made between the MPI 1 Pt2Pt and MPI 2 OSC performance results. The program used to benchmark was a third party application, the Pallas' MPI Benchmarks [12]. This is the only available set of benchmarks that support one-sided communications. The runs were performed on a Windows 2000 dual PIII (for shared memory) and on a Windows 2000 PIII (for TCP) connected by a switched 100Mb/s Ethernet. All the results refer to two processes. The time is the average of 1000 operations synchronized with MPI_Win_fence. The benchmark has both an aggregated and a non-aggregated version. On the aggregated benchmark, the synchronization happens at the start and finish of the 1000 operations, hence being called twice. On the non-aggregated the synchronization happens twice for every operation. The aggregated results were chosen. This would be the normal operation of a real world OSC application.

Figure 1 shows the results gathered for shared memory and figure 2 shows the results for TCP.

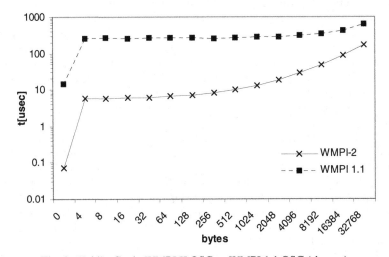

Fig. 1. Unidir_Get in WMPI II OSC vs WMPI 1.1 OSC (shmem)

As can be observed, the performance of the WMPI II OSC implementation is faster than that of WMPI 1.1 on shared memory. This is mainly due to the use of direct RMAs on the shared memory device. The performance for TCP is also better on the new implementation, except for some oscillations that are still occurring at this stage. So far tests have shown that the problem lies in the utilization of the winsock library. We are investigating this issue to discover the cause and solve the problem.

n figure 3 we show the results for the comparison between the MPI 1 and MPI 2 Pallas' benchmarks, using shared memory. We compare the unidir_get to the pingpong and the bidir_get to the pingping. The results show that the performance is very similar, with the one-sided communications taking the lead up to some extent. This shows that this implementation of OSC brings no overhead to the communication operations and can, with a properly implemented synchronization scheme, bring many enhancements to distributed applications.

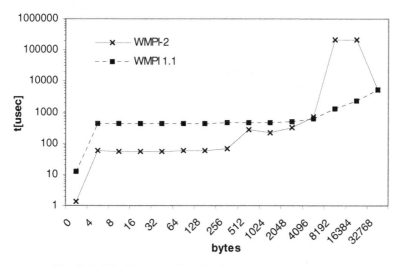

Fig. 2. Unidir_Get in WMPI II OSC vs WMPI 1.1 OSC (tcp)

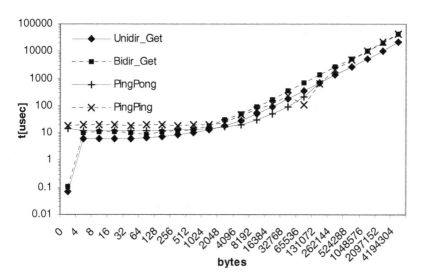

Fig. 3. Comparison between Unidir_Get/Bidir_Get with PingPong/PingPing on shared memory

5   Conclusion

This implementation of OSC is in the testing and tuning phase, but still, some very interesting performance results were shown. The decision to layer this implementation directly on top of WMPI's WML allowed to effectively explore all the new features of WMPI and of the underlying Operating System. Due to the object-oriented paradigm followed by this implementation, all this was implemented easily and any future enhancements to the WML or Operating Systems will be straightforward to explore.

Due to the inexistent loss in performance from double sided to single sided operations, users can try different programming models, using OSC, without fearing loss of performance.

References

1. Snir, M., Otto, S., Huss-Lederman, S., Walker, D., Dongarra, J.: MPI – The Complete Reference, Volume 1, The MPI Core. MIT Press, second edition (1998)
2. Gropp, W., Huss-Lederman, S., Lumsdaine, A., Lusk, E., Nitzberg, B., Saphir, W., Snir, M.: MPI – The Complete Reference, Volume 2, The MPI Extensions. MIT Press (1998)
3. Mourão, F., Silva, J.G: Implementing MPI's One-Sided Communications for WMPI. Proc. of 6th European PVM/MPI Users' Group Meeting, pp 231-238, Springer-Verlag, Barcelona, Spain (September 1999)
4. University of Notre Dame: LAM-6.3 release notes. (1999) http://www.mpi.nd.edu/lam/
5. Pedroso, H., Silva, J.G.: An Architecture for Using Multiple Communication Devices in a MPI Library. Proc. of 8th High-Performance Computing and Networking Europe, pp. 688-697, Springer-Verlag, Amsterdam, The Netherlands (May 2000)
6. Marinho, J., Silva, J.G.: WMPI – Message Passing Interface for Win32 Clusters. Proc. of 5th European PVM/MPI User's Group Meeting, pp. 113-120 (September 1998)
7. Gropp, W., Lusk, E., Doss, N., Skejellum, A.: A High-Performance, Portable Implementation of the MPI Message Passing Interface Standard. Parallel Computing Vol. 22, No. 6 (Sptember 1996)
8. Pedroso, H., Silva, J. G.: MPI-2 Process Creation & Management Implementation for NT Clusters. Proc. of 7th European PVM/MPI Users' Group Meeting, pp. 184-191, Springer-Verlag, Balatonfüred, Hungary (September 2000)
9. Booth, S., Mourão, E.: Single sided MPI implementations for SUN MPI. In Supercomputing 2000 (2000) http://www.sc2000.org/techpapr/papers/pap.pap182.pdf
10. Gropp, W.: Runtime Checking of Datatype Signatures in MPI. Proc. of 7th European PVM/MPI Users' Group Meeting, pp 160-167, Springer-Verlag, Balatonfüred, Hungary (September 2000)
11. Träf, J.,Hempel, R., Ritzdorf, H., Zimmermann, F.: Flattening on the Fly: Efficient Handling of MPI Derived Datatypes. Proc. of 6th European PVM/MPI Users' Group Meeting, pp 109-116, Springer-Verlag, Barcelona, Spain (September 1999)
12. Pallas GmbH: Pallas MPI Benchmarks – PMB. (April 2001) http://www.pallas.de/pages/pmb.htm
13. Critical Software S.A. – WMPI. http://www.criticalsoftware.com/wmpi

Assessment of PVM Suitability to Testbed Client-Agent-Server Applications

Mariusz R. Matuszek

Faculty of Electronics Telecommunications and Informatics,
Dept. of Computer Systems Architecture,
Technical University of Gdańsk,
G.Narutowicza 11/12 st., 80-952 Gdańsk, Poland
Mariusz.Matuszek@pg.gda.pl

Abstract. Client-agent-server is a relatively new processing paradigm, in which mobile code objects 'agents' are utilised. In this paper we look at PVM features and try to assess its suitability to testbed client-agent-server applications, thus trying to determine if our existing PVM experience can be leveraged to this new computing field. It is hoped that this paper will be of most interest to readers already proficient in PVM, who want to be introduced to concepts and issues present in mobile-agent environments.

1   Introduction

While most of us are long familiar with client-server computing model [10], where the client computer(s) connect to server(s) to request various actions being performed and resulting data returned, the idea of mobile agents may still be new to some. In this new model mobile code objects called agents are used, carrying with them both the controlling algorithms and data they might have gathered along the way. Perhaps the biggest distinction from client-server model is this introduction of mobile code – algorithms which used to reside on server and client computers are now able to move freely between the supporting nodes and execute at them at will. The possibilities of this model seem exciting and many dedicated environments were created to support it [6], however, as with all new concepts they have their own learning curve and require some effort to master. In this paper we look at PVM features to determine, whether it is possible to utilise one's existing experience to enter the world of agents and to what extent.

2   The Agent World

Imagine you have just finished writing your newest, breakthrough search algorithm, full of bizzare intelligence, able to index the contents of the whole Internet and give you the Answer[1] to The Ultimate Question of Life, The Universe and

---

[1] which, as all readers of [16] will agree, is 42.

Y. Cotronis and J. Dongarra (Eds.): Euro PVM/MPI 2001, LNCS 2131, pp. 69–74, 2001.

Everything [16]. Trembling with expectation you are about to give the 'start search' command, when sudden realisation dawns on you: downloading and processing all this data is going to take a while. But using the search engines present on remote servers is out of question, they seem oh-so unintelligent to you. If only you could run your search program on all the remote servers, ideally in multiple copies running in parallel. . .

It is time to enter the agent world. In the following parts of this article we will take a look at typical features offered by agent environments and their system-level issues, compare them to what PVM has to offer and try to conceptually implement a simple agent in both environments.

2.1   Agents Walk

An inherent property of mobile agents is their ability to move autonomously between different nodes of their environment. This functionality is provided by agent environment and is one of its basic properties. Agents may also be moved from node to node against their will, for example the owner of the agent may request the agent being sent back (retracted) to the node it originated from. Another feature exhibited by agents and agent environments is their ability to operate in loosely-coupled networks, i.e. networks which may be down part of the time.

To facilitate location transparency, name services of some kind are provided by most agent environments. These services allow agents to refer to remote and local resources by means of some logical name rather then by their absolute paths and URL's. Additionally, some agent environments can inter operate with the CORBA [9] model of accessing remote objects.

2.2   Agents Talk

Most distributed tasks except for the simplest ones require some amount of communication and coordination between parties involved and agents are no exception to this rule. The ultimate goal is to provide agents with mechanisms allowing communication with absolute location-transparency. Currently several mechanisms exist in agent environments to facilitate such communication to a greater or lesser degree. Proposed solutions [1] range from simple communication – either direct or through proxies, shared data spaces (blackboards), meeting points, up to associative tuple spaces. Depending on whether a given communication method requires the agent to be aware of the location of its peer or not it is classified as space-coupled or space-uncoupled. The same analogy applies to requirement for agents to time-synchronise their communication. A given method is then said to be either time-coupled or time-uncoupled.

In any commercial application of agents technology security of communication must be taken into account and indeed, mechanisms allowing encryption and signing [8] of communicated data or code are being provided by some agent environments.

2.3   Agents Compute

This is an obvious statement, however, it must be realised that modern networked environments are highly heterogeneous, and special measures had to be taken to allow agents to execute their code on any supporting node and to move their execution state safely. In the early days of agents a scripting approach was proposed and it was implemented in several languages, like Perl, Python and Tcl. An example of such agent environment is Agent Tcl [12]. Another idea tried by designers of agent environments was to use interpreted forms of general-purpose programming languages. Both approaches suffered efficiency problems and were difficult to apply to large projects. Some were type-unsafe as well and it was very difficult to capture the execution state of such agents. Once Java became popular it was adopted as a de-facto standard programming language for agents and several Java-based agent environments appeared[2]. Examples of such environments are Aglets [13], Voyager [15], Concordia [14].

3   The PVM Way

Since most readers of this article are well familiar with working of and with the PVM environment we will only briefly summarise its features corresponding to our description of agent environments in section 2.

3.1   Code Migration

The only function in PVM which allows to start a task on a remote node is pvm_spawn(). However, the underlying assumption is that the actual executable for a given architecture is already present on the remote node, either physically or by means of a shared file space. There are no provisions in standard PVM to actually send a copy of a task to a remote node for running it, and the proposed extensions [3] of moving running processes between nodes are mostly aimed at dynamic load balancing. A more flexible approach could be implemented on top of the Harness project [4] environment, however, the discusion of this environment is beyond the scope of this article.

3.2   Interprocess Communication

PVM sports a nice, clean and type-safe interface for passing information between pairs and groups of processes. By having two processes join a common group it is possible for them to exchange messages without knowledge of other task's ID. This mechanism may be used to emulate spatially-uncoupled communication between agents. On the other hand, before PVM 3.4 there was no support for

---

[2] It should be noted that, due to properties of Java Virtual Machine, all of these environments share a common deficiency which is their inability to save agent execution state on a thread level. However, this is rarely a problem since most of the time agents migrate of their own will and they get notified of impending relocations.

leaving messages for later pickup by other, perhaps yet nonexistent, tasks. Starting with the release of PVM 3.4 message boxes were introduced [7]. They are implemented as internal associative tuple space inside the virtual machine. This mechanism allows for very nice, time- and spatially-uncoupled communication between processes.

3.3   Platform-Independent Solutions

As it was noted in section 3.1 standard PVM does not allow for code migration between nodes, but let us assume for a moment, that this limitation is waived, what then? An answer might be to write a portable multiplatform code and this currently means writing in Java. Actually, there are two well known Java-based PVM solutions. The jPVM [11] is a native-method interface to PVM for programs written in Java, and the JPVM [5] is a complete PVM-like class library implementation in Java. Still, the problem of lack of support for loosely-coupled networks remains unsolved.

4   A Simple Agent Job Description

To introduce the concept of agents in PVM we describe possible implementation of a very simplistic[3] agent in both a hypothetical agent environment and PVM. Our agent will have a very menial task of visiting each node present on its itinerary and ticking an imaginary mark in an imaginary box next to node's name once it gets there. After visiting all nodes on the list it will return home. Intelligence of this agent will be nil in that it will follow a preset path and should the next node be unavailable it will just wait for it to become available.

While our example may seem oversimplified it should be realised, that a similar agent, visiting a specified set of nodes and gathering load/status information from them, could be actually useful as a monitoring/diagnostic agent in a cluster of workstations.

4.1   Native Implementation

The following pseudo code illustrates a possible implementation in native agent environment:

```
while ((node = get_next_node(itinerary)) != NULL) {
    migrate_to_node(node); // this blocks
    tick_box(node);
}
migrate_to_node(HOME);
```

It is assumed, that migrate_to_node() function does not return until a successful migration has been made and that it waits indefinitely for the node to become available.

---
[3] perhaps Vogon-like would be a better description.

4.2   PVM Implementation

As we mentioned in section 3.1 there are no provisions for code migration in standard PVM. Therefore it is not possible to repeat the scenario seen in native implementation (section 4.1). Instead, we will start our PVM agent task on each node to be visited and have it wait for its turn to do the job. However, we will circulate the itinerary as a token synchronising agents' actions.

The first pseudo code snippet starts agent-tasks and prepares the itinerary, it is also responsible for passing the itinerary to the first agent-process and getting it back after it travelled through all nodes:

```
for (i = 0; i < NUMHOSTS; ++i) {
    hostname = hostnames[i];
    pvm_spawn(''agent'', NULL, PvmTaskHost, hostname, 1, &tid);
    add_to_itinerary(hostname, tid);
}
send_itinerary_to_first_agent();
wait_for_itinerary();
```

and the second fragment illustrates the inner logic (or rather lack thereof) of agent-tasks:

```
while (TRUE) {
    wait_for_itinerary();
    tick_box(my_node);
    send_itinerary_to_next_agent();
}
```

The behaviour of the two implementations is shown in Figure 1.

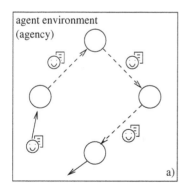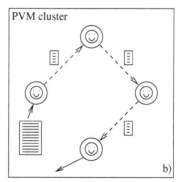

Fig. 1. Behaviour of native (a) and PVM (b) implementations.

It is readily visible, that the agent approach presented on Fig. 1(a) is based on binary object flow while the solution proposed for PVM (Fig. 1(b)) is based on procedural data flow.

5   Conclusion

The main goal of this paper was to introduce existing PVM users to the issues present in agent environments. As it can be seen from this simple comparison of features of PVM and agent environments it is not directly possible to implement any complex type of mobile agent in PVM because it is not flexible enough and lacks features required for agent movement between the nodes, although some work has been done to implement process migration in PVM. Also, the ability to form a consistent environment over loosely-coupled networks is missing in PVM, which is a big obstacle in real world applications but is of much less importance in laboratory conditions.

This conclusion is hardly surprising, as PVM was designed with a different set of features and uses in mind. However, there are many similarities and it is possible to implement some simple agent-like programs in the PVM environment based on data flow. Therefore, our final conclusion is that knowledge gained by using PVM helps in understanding the issues present in agent environments and can be directly applied to speed up the learning process, as well as PVM itself can be used to experiment with agent concepts on a limited scale.

References

1. Cabri G., Leonardi L., Zambonelli F.: *Mobile-Agent Coordination Models for Internet Applications*. IEEE Computer, February 2000, pp. 82–89.
2. Weiss G. (editor): *Multiagent Systems: A Modern Approach to Distributed Artificial Intelligence*. The MIT Press, Massachusetts Institute of Technology, 2000.
3. Czarnul P., Krawczyk H.: *Dynamic Assignment with Process Migration in Distributed Environments*. Recent Advances in Parallel Virtual Machine and Message Passing Interface, Lecture Notes in Computer Science no. 1697, pp. 509–516, Springer Verlag, 1999.
4. Migliardi M., Sunderam V.: *PVM Emulation in the Harness Metacomputing System: A Plug-in Based Approach*. Recent Advances in Parallel Virtual Machine and Message Passing Interface, Lecture Notes in Computer Science no. 1697, pp. 117–124, Springer Verlag, 1999.
5. Ferrari A.J.: *JPVM: Network parallel computing in Java*. In Proc. of the ACM Workshop on Java for High-Performance Network Computing, March 1998.
6. Karnik N.M., Tripathi A.R.: *Design Issues in Mobile Agent Programming Systems*. IEEE Concurrency, July–Sept. 1998, pp. 52–61.
7. Geist G.A.: *Advanced Capabilities in PVM 3.4*. Recent Advances in Parallel Virtual Machine and Message Passing Interface, Lecture Notes in Computer Science no. 1332, pp. 107–115, Springer Verlag, 1997.
8. Schneier B.: *Applied Cryptography*. John Wiley, 2nd edition, 1996.
9. Ben-Natan R.: *CORBA: a Guide to Common Object Request Brooker Architecture*. McGraw-Hill, 1995.
10. Umar A.: *Distributed Computing: A Practical Synthesis*. Prentice-Hall, Inc., 1993.
11. Thurman D.: *jPVM*. http://www.isye.gatech.edu/chmsr/jPVM/
12. Agent Tcl: http://www.cs.dartmouth.edu/~agent/
13. IBM Aglets: http://www.trl.ibm.co.jp/aglets/
14. Mitsubishi Concordia: http://www.meitca.com/HSL/Projects/Concordia/
15. Object Space Voyager: http://www.objectspace.com/products/prodVoyager.asp
16. Adams D.: *The Hitch Hiker's Guide to the Galaxy*. Heinemann:London.

TH-MPI: OS Kernel Integrated Fault Tolerant MPI

Yu Chen, Qian Fang, Zhihui Du, and Sanli Li

Department of Computer Science, Tsinghua University, Bei Jing,
P.R. China
chenyu@tirc.cs.tsinghua.edu.cn

Abstract. Consisting of large numbers of computing nodes, parallel cluster systems have high risks of individual node failure. To overcome the high overhead drawbacks of current fault tolerant MPI systems, this paper presents TH-MPI for parallel cluster systems. Being integrated into Linux kernel, TH-MPI is implemented in a more effective, transparent and extensive way. With supports of dynamic kernel module and diskless checkpointing technologies, our experiment shows that checkpointing in TH-MPI is effectively optimized.

1   Introduction

The clusters of PCs have become popular platforms for computationally intensive distributed applications. This situation is due to the high performance-price ratio, rich applications, open source operation system, such as Linux, and the availability of standard message passing systems, such as Message Passing Interface (MPI). More and more parallel applications are based on MPI currently, since MPI standard has proven effective and sufficient for most of high-performance applications. However, the MPI standard concerns little of fault tolerance. But unfortunately, the failure rate of workstation clusters is much greater than that of other distributed computer systems. That is to say, Long-lived distributed computations with large number of computing nodes have especially high risks of failure.

To meet with the availability demands of parallel applications, fault tolerance have been integrated into some MPI systems, such as CoCheck-MPI[1], startfish[2], etc. However, the checkpointing overheads in these systems are much heavier than those of normal MPI programs without checkpointing, and some of them even need to change the application or MPI library's source code.

To overcome the drawbacks of these systems, we present a fault tolerant MPI system, TH-MPI, which has been integrated into Linux operating system kernel. With the support of fault detecting and fault tolerance in OS level, TH-MPI is implemented in a more effective, transparent and extensive way. We adopted Linux as the OS platform, because of not only its popularity and high performance, but also its open source essentiality. TH-MPI is based on Kool MPI [3], a fundamental implementation of the Message Passing Interface (MPI) standard. With dynamic kernel module and diskless checkpointing technologies, checkpointing overheads are effectively reduced in TH-MPI. Furthermore, TH-MPI supports programs with shared dynamic library, whereas other fault tolerant systems couldn't support them. Being tested in our prototype environment, the initial performance results of TH-MPI are stirring.

Y. Cotronis and J. Dongarra (Eds.): Euro PVM/MPI 2001, LNCS 2131, pp. 75–82, 2001.

The paper is organized as follows: architecture of TH-MPI is figured out in section 2; implementation and practical results are separately presented in section 3 and 4; comparisons with other related works are introduced in section 5. And finally in section 6, we make some conclusions and address our future work.

2   Architecture

2.1 Goals and Schemes

Fault tolerant MPI systems, such as CoCheck MPI may result in lots of overhead and even needs to change parallel application or MPI library source code. The main reason is that the failover processes in these systems usually involve accessing external shared storage, such as shared disks. When a node fails, its checkpointed data must be reloaded from the shared disk into memory before continuing the application. Therefore, the performance of applications is always limited by slow disk access. Another reason is that most fault tolerant MPI systems do checkpointing and recovering in user space, so their implementations are not so effective and transparent as in kernel space.

According to the drawbacks, the main goal of TH-MPI is to reduce the checkpointing overhead and make no change on the application source code, including the MPI library's source codes. Without changing MPI libraries' main source codes, we could transparently add fault tolerant abilities to MPICH and LAM, the mainstreams of MPI implementations, and thus will facilitate development of new MPI library in the future.

Another important feature of TH-MPI is separating communication checkpointing from process checkpointing. This separation makes checkpointing for different communication subsystems easier, and is very important for MPI based applications.

Conclusively, TH-MPI adopts the following schemes to reduce overheads:
1. Checkpointing and recovery process space in OS kernel. With the support of OS kernel, the information of user and kernel space in any process could be acquired in a more effectively way, and the unnecessarily memory copy will be avoided. Simultaneously with the dynamic kernel module technology, we make as little as possible changes on Linux kernel to support fast development of Linux OS kernel.
2. Diskless checkpointing. Diskless checkpointing will store the checkpointed image of parallel process to other nodes through network. Using diskless checkpointing technology will reduce the overheads of directly writing and reading hard disk.
3. TCP communication checkpointing and recovery in OS Kernel space. We checkpoint the TCP communication subsystem in kernel, so the MPI library based on TCP/IP needn't take care of how to checkpoint high-level communication subsystems. In the future, other communications subsystems will be supported.
4. Coordinated checkpointing. It requires processes to orchestrate their checkpoints in order to form a consistent global state. Coordinated checkpointing simplifies recovery and is not susceptible to the domino effect. Also, coordinated checkpointing requires each process to maintain only one permanent checkpoint on storage, reducing storage overhead and eliminating the need for garbage collection.

2.2   Structure of TH-MPI

TH-MPI system includes four parts: Process Fault Tolerant Daemon (PFTD), Communication Fault Tolerant Daemon (CFTD), Monitor Daemon (MD) and MPI library subsystem. Figure1 shows TH-MPI architecture in a computing node.

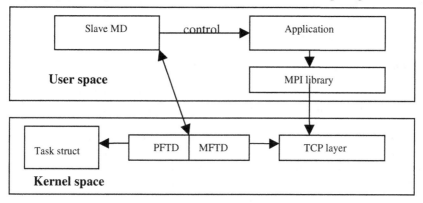

Fig. 1. TH-MPI Architecture in computing node

MD concerns with starting and restarting processes. It has two types: Master MD and Slave MD. Master MD is a main control and monitor process in a main control node. We use Master MD to start and control parallel processes. Slave MD is located in computing node, and its duties are receiving commands from Master MD, running parallel processes in local computing node, sending heartbeat messages to Master MD, and contacting with PFTD and CFTD in local computing node.

PFTD's main jobs are checkpointing and recovering process space of parallel processes. The information being processed, from PFTD's point, includes the memory image of the process, the relevant process/kernel data structures, the state of the CPU registers, and signal handler information, etc.

CFTD takes charge of coordinated checkpointing TCP connections for parallel processes. Because we use coordinated checkpointing method, we have to flush all the TCP sending communication channels and store TCP receiving communication channels in kernel space of process. When process is recovered, CFTD reestablish TCP connection and restore TCP receiving communication channels in kernel space.

3   Implementation

3.1   MPI Communication Library

The TH-MPI communication library subsystem is based on Kool MPI. It provides MPI functions and directly uses the socket system calls with TCP protocol to realize the message passing interface standard. So, there is no middle layer between MPI functions and socket system calls. To implement TH-MPI, we made little changes on Kool MPI library to let it support fault tolerance, and some daemons, such as MDs, PFTDs and MFTDs are introduced to manage and checkpoint parallel processes.

3.2   Daemon Internal

3.2.1 Monitor Daemon

The daemons in TH-MPI are classified into user daemon and kernel daemon. MD is a user daemon. The first job of MD is to exchange information with PFTD and CFTD, and its second job is to control and monitor the MPI-based parallel applications. One slave MD is located per one computing node. A master MD is located in a controlling node, which also may be a computing one. The main processes of MD are shown below:

• *Starting and checkpointing process*: First, we assume that the executive files of parallel applications are already in computing nodes. When parallel application runs the first time, the master MD reads from a configuring file that includes position information of the executive file images in every computing node, and then sends the configuring information of this parallel application to every slave MD in the computing nodes. When slave MD receives the information, it forks and executes the parallel application in local node, and then informs PFTD and CFTD to prepare checkpointing the parallel application.

• *Restoring process*: MD also has the duty to restart the checkpointed parallel application when it failed. When some nodes failed, the master MD creates a new configure file and sends recover message to the new backup and other old running computing nodes. When receiving the recovering information, slave MD gets checkpointed executive process image from PFTD and creates a new TCP communication with the help of CFTD. Therefore, the parallel application can keep on running.

3.2.2 Communication Fault Tolerant Daemon

CFTD and PFTD make up of a kernel module, which is also called as a kernel daemon. It is dynamically loaded into kernel as a Linux kernel module. We made little changes on Linux kernel with dynamic kernel module technology, so CFTD and PFTD could be easily updated to accord with the fast development of Linux kernel.

The main job of CFTD is to manage the parallel application's low level communication subsystem (such as TCP). The main information that CFTD concerns is shown in Figure 2. The main processes of CFTD are shown below:

• *Checkpointing process*: When it's time to checkpoint parallel process, the master MD broadcasts a request message to all slave MDs, asking them to checkpoint local parallel process, and then the slave MDs send messages to local CFTD through /proc directory in Linux system. When CFTD receives these messages, it first reads the all TCP sock structs of paralle process in kernel space, then flushes all the TCP sending communication channels in this parallel process, and sends an acknowledgment message back to the local slave MD, and slave MDs send back acknowledgments to master MD in turn. After the master MD receives acknowledgments from all slave ones, it broadcasts a commit message that completes the two-phase checkpointing protocol. After receiving the commit message, each slave MD asks PFTD and CFTD atomically makes the diskless checkpointing. Finally, CFTD stores the parallel process's receiving queues and buffers of TCP sockets in Linux kernel, and consturct a checkpointed communication image and sends it to another computing node.

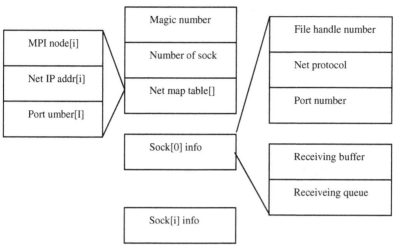

Fig. 2. The checkpointed communication image

• *Restoring process*: When recovering the running parallel processes, MD sends network information of the new computing nodes to CFTD. The network information contains the new corresponding relationship of logical MPI identity number and IP address with real process identity number of MPI based processes. Therefore, CFTD first gets the corresponding checkpointed comm-unication image of parallel process from other nodes, and then creates new TCP connection between new computing nodes. CFTD finally changes checkpointed sock data in corresponding sock structure with the same file handler number. The main changes are TCP head information of TCP packages in receiving queue and buffer, including port number, sequence number, acknowledgement number, etc. After finishing these changes, the restored parallel process could communicate with remote parallel process in other nodes again.

3.2.3 Process Fault Tolerant Daemon

Excluding net context, the main job of PFTD is to checkpoint the parallel application's process space context. The main information that PFTD concerns is shown in Figure 3. The main processes of PFTD are shown below:

• *Checkpointing process*: When PFTD gets checkpointing information from the slave MDs, it checkpoints the parallel process in Linux kernel. First, it stops parallel process's execution and makes the parallel process sleeping; second, it reads the parallel process's task structure from Linux OS kernel, and stores register values, signal info and process's virtual memory content into a checkpointed process image; finally, through network it sends this image and checkpointed communication image of this process to the backup node's memory.

• *Restoring process*: When recovering the execution of parallel process, MD forks a victim process, then calls a special system call dynamically provided by PFTD to restart process. When PFTD received this system call, it first gets process image from the backup nodes. Second, PFTD deletes the victim process's virtual memory image, puts the checkpointed process image into victim process's task structure, and restores checkpointed process's registers value, signal info, etc. Third, it waits CFTD to recreate net connection and restore net information. Finally, it returns to the checkpointed parallel process.

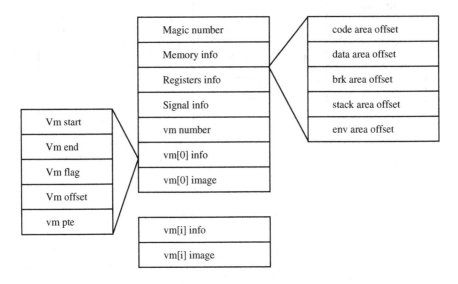

Fig. 3. The checkpointed process image

4   Performance

We use two MPI based examples to test checkpointing performance in TH-MPI. The experiment environment consists of two 700MHZ PIII PCs with 128MB memory connected by 100Mb/s Ethernet, which have local disk and run on Linux kernel 2.2.14. One example is a ping-pong code, and another is a two dimension matrix multiply code. To simulate real parallel applications, both examples adopt large size buffer variable leading to large memory sizes of executive image.

Disk checkpointing methods are classified into ones with synchronization and ones without synchronization. The former insures checkpointed image is written to disk; whereas, the latter does not. The performance of disk and diskless checkpoint methods are separately tested in our experiment, and the comparing results of checkpointing performance are shown in table 1. The problem sizes in two-dimension matrix multiply MPI code are changed to test the different checkpointing performances, and the performance results are shown in table 2.

From the tables below, we could find performance of diskless checkpointing is better than that of disk checkpointing. Compared with the real cluster execution environment, our experiment system is merely a prototype. Therefore, in the real cluster systems with larger disk cache memory and higher speed network, such as Giga net or Myrinet, the diskless checkpointing performance will be much more perfect than that of diskless.

Table 1. Checkpointing performances of Ping Pong and Matrix Multiply program

MPI Program name	Checkpoint image size (byte)	Disk check-point without sync (10^{-6}sec)	Disk check-point with sync(10^{-6}sec)	Diskless checkpoint (10^{-6}sec)
Ping Pong	63421858	23216859	27323567	10982648
Matrix Multiply	75450208	26457786	32065448	13120610

Table 2. Checkpointing performances of Matrix Multiply program with differenct problem scales

Problem scale	Checkpoint image size(byte)	Disk check-point without sync(10^{-6}sec)	Disk check-point with sync (10^{-6}sec)	Diskless checkpoint (10^{-6}sec)
512x512	2158912	30058	846114	183018
1024x1024	5304640	69877	1691590	746566
2048x2048	17887552	259343	7266360	2952648
4096x4096	75450208	26457786	32065448	13120610

In Table 2, the performance of disk checkpointing without synchronization is the best in some situations. The main reason lies in that the checkpointed image size is not very large, so when PFTD writes image to disk, the large part of image is firstly located in Linux disk cache memory. If the checkpoint image is much larger than the size of disk cache memory, the performance of disk checkpointing with synchronization will drop quickly. Whereas, the problem size makes less influences on the diskless checkpointing, which could maintain good performance in most situations.

In the realization of TH-MPI, we design the communication of checkpointed image data in Linux kernel. Apparently, it is no need to copy data to and from user space if communication is realized in OS kernel, and reduces two times memory copy overheads, which cannot be avoided if communication checkpointed data in user space.

From the tests and analysis, we could draw the conclusion that if the cluster system is connected by a high performance network such as Myrinet, and each computing node has a big memory, then the OS kernel supported diskless checkpointing technology is very attractive and acceptable.

5   Related Work

Condor [4] is a distributed system running on cluster of workstations. It supports checkpoint/restart in order to provide fault tolerance and process migration. But, unlike TH-MPI, Condor requires the programs to be linked with a special library, and it couldn't support checkpointing dynamic linked libraries and a group of parallel processes, such as MPI-based applications.

Libckpt [5] is a transparent checkpointing library on uniprocessors running UNIX. It implements most optimizations that have been proposed to improve the performance of checkpointing, including incremental checkpointing, forked checkpointing and copy-on-write checkpointing. However, Libckpt is realized in user space, and it couldn't support diskless and dynamic linked libraries checkpointing.

A project close to TH-MPI is EPCKPT [6]. It realizes checkpointing ability in Linux kernel. But it has to change some Linux kernel file, and could not be dynamic loaded into kernel as a kernel module. As far as we know, the newest version of EPCKPT only based on Linux kernel 2.2.1, doesn't support diskless checkpointing and doesn't consider checkpointing MPI based programs.

6   Conclusions and Future Work

TH-MPI is a new experiment to provide application programmers with different methods of dealing failures within MPI applications. Different from other fault tolerant MPI systems, TH-MPI is implanted in a more effective, transparent and extensive way. In the future, it is hoped that by experimenting with TH-MPI, new applications methodologies and algorithms will be developed to realize both high performance and high availability required for the next generation parallel cluster machines. Furthermore, TH-MPI could also be used to experiment with other common low level communication subsystem, such as VIA, which is our next aim, and to support the mainstream MPI implementation, such as MPICH and LAM.

References

1. G. Stellner, "CoCheck: Checkpointing and Process Migration for MPI", In Proceedings of the Int'l Parallel Processing Symposium, pp 526-531, 1996.
2. A. Agbaria and R. Friedman, "Starfish: Fault-Tolerant Dynamic MPI Programs on Clusters of Workstations", In the 8th IEEE Int'l Symposium on High Performance Distributed Computing, 1999.
3. M. Kim and S. Kim, "(Kool MPI): Toward an optimized MPI implementation for the Linux clusters", Technical Report, Sejong University, Korea, 2000
4. M. Litzkow, M. Livny, and M. Mutka, "Condor: A hunter of idle workstations", In Proc. of the 8th Int'l Conference on Distributed Computing Systems (ICDCS'88), 1988.
5. J. S. Plank, M. Bech, G. Kingsley, and K. Li, "Libckpt: transparent Checkpointing Under UNIX", In Usenix inter 1995 Technical Conference, pp 220-232, 1995.
6. E. Pinheiro, "Truly-Transparent Checkpointing of Parallel Applications", Technical Report, Rutgers University, 1999

CPPvm – C++ and PVM

Steffen Görzig

DaimlerChrysler Research and Technology 3,
Software Architecture (FT3/SA),
P.O. Box 23 60, 89013 Ulm, Germany
steffen.goerzig@daimlerchrysler.com

Abstract. CPPvm is a C++ class library for message passing. It provides an easy–to–use C++ interface to the parallel virtual machine software PVM. CPPvm closes the gap between the design of object-oriented parallel programs in C++ and the underlying message passing possibilities of PVM. Although PVM can be used directly in C++ programs due to its C–functions, it does not support C++ specific features. CPPvm enlarges PVM with such features as classes, inheritance, overloaded operators, exception handling and streams. CPPvm also hides some details of PVM from the user and thus makes it easier to write parallel programs.

This paper describes the concepts of CPPvm. An example will explain how to transfer C++ objects between processes. CPPvm is available for many architectures, from Windows to several UNIX derivatives.

1 Introduction

Software libraries for cluster computing like the Parallel Virtual Machine (PVM, [1]) or the Message Passing Interface (MPI, [2]) mainly support procedural programming languages such as Fortran or C. Since object-oriented programming has become more and more popular in the last decade, several projects have started to develop class libraries based on existing message passing software. Examples for PVM are Para++ [3], PVM++ [4], and EasyPvm [5]. Examples for MPI are Para++, OOMPI [6], and the MPI-2 C++ bindings for MPI [7].

This paper describes CPPvm (C Plus Plus Pvm, [8]). CPPvm is a C++ message passing class library built on top of PVM. CPPvm enlarges PVM with C++ features such as classes, inheritance, overloaded operators, exception handling and streams. The main functionality of CPPvm is to pass the contents of C++ objects between several processes running in parallel. The processes can run on a network of computers with heterogenous architectures and different operating systems (e.g. Windows and the most UNIX derivatives). The parallel virtual machine connects these hosts to build just one machine. This allows a transparent message passing between all CPPvm processes running on these hosts.

Y. Cotronis and J. Dongarra (Eds.): Euro PVM/MPI 2001, LNCS 2131, pp. 83–90, 2001.

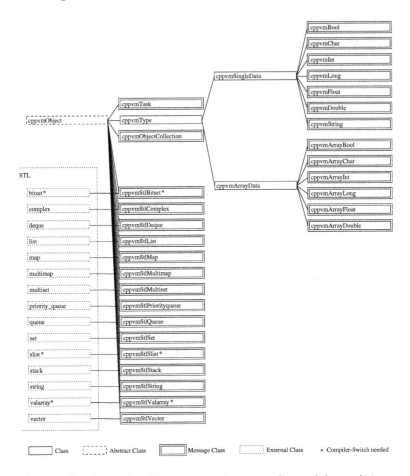

Fig. 1. Message class hierarchy. The message classes can be used for explicit message passing as well as for distributed and mailbox objects.

2   Concepts

The fundamental concept of CPPVM is identical to that of PVM: a heterogeneous collection of hosts hooked together by a network can be used as a single large parallel virtual machine. Processes running on these hosts can become part of the virtual machine system. The processes can also spawn other processes on every host in the system and exchange data among each other.

This scenario directly leeds to a set of classes: Processes must first be linked to the parallel virtual machine. This is done by CPPVM classes for process handling which allow processes to connect to PVM and spawn child processes. "Send and receive" stream classes are furthermore needed for explicit message passing between processes. The stream classes are the transport channels for message objects. It is also possible to use message objects as distributed or mailbox objects. Figure 1 shows the hierarchy of available message classes.

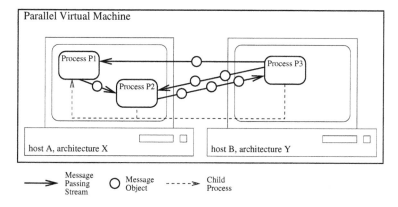

Fig. 2. Explicit Message Passing.

2.1 Explicit Message Passing

Explicit message passing is used to transfer data between objects of different processes. Processes within CPPVM usually have a parent–child relationship. For example in figure 2 process P1 has spawned process P2 on host A and process P3 on host B.

These relationships are used to define CPPVM message passing streams. In the example process P1 has opened a send stream to its child process P2. The message passing streams are used to send or receive C++ objects. In contrast to PVM objects can be sent with or without waiting until the other process has received the object (blocking or nonblocking). The modes of the receiving stream are blocking, nonblocking and "timeout receive". The latter tries to receive an object within a specified time range, otherwise the process continues without having received the object.

2.2 Distributed Objects

Another possibility to exchange data between processes are "distributed objects" (see figure 3). A distributed object is a C++ object which has instances in several processes.

A global data base contains one instance for every distributed object in the system. Processes can have a local instance of this object. A "read method" of the object is used to update the local object, a "write method" updates the corresponding object in the data base. When using distributed objects there is no need for send/receive streams, since the instances of a distributed object are matched by a unique identifier.

2.3 Message Mailbox Objects

Message mailbox objects are a superset of distributed objects. Mailbox objects allow to generate more than one instance of one object in the global data base

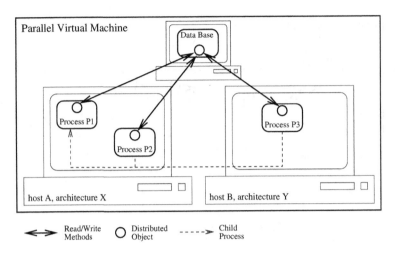

Fig. 3. Distributed Objects.

(see figure 4). Whilst for distributed objects the connection between the local object and the instance in the data base is one-to-one, for message mailbox objects, the local object can randomly access all instances of an object in the data base (1:n connection).

2.4   User Defined Classes

Standard CPPVM message classes (see figure 1) should be sufficient for most kinds of parallel applications. However an interesting feature of CPPVM is the possibility to write user-defined message classes. These classes can be used for explicit message passing as well as for distributed and message mailbox objects. It is also possible to enhance existing classes with this ability, e.g. transform an existing program into a parallel program.

To create a CPPVM message class it must be derived from the class cppvmObject. Thereafter few modifications have to be made on this class:

– Modify/create the constructor and virtual destructor.
– Implement the virtual method cppvmTransfer: This method defines all message passing class variables.
– Add the macro CPPvmMethodsDeclaration(msgtag) to the class declaration.

The constructor of the class must call the constructor of the class cppvmObject (directly or indirectly). The virtual destructor is needed e.g. for deleting a collection of objects.

The method cppvmTransfer specifies all data used for message passing. Only the data defined in this method is transfered. This method is called for explicit message passing as well as for distributed objects and message mailbox objects. The message passing data defined in the method cppvmTransfer can be:

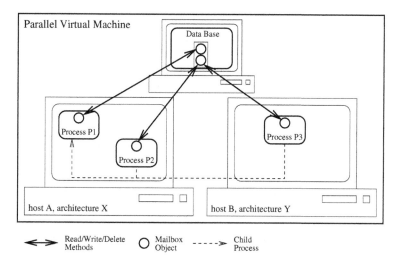

Fig. 4. Message Mailbox Objects.

- Standard C/C++ types (single values and arrays).
- Other CPPvm message objects derived from cppvmObject.
- The data of the base class.

2.5   Advanced Topics

CPPvm contains many more concepts for parallel programming than those described above. These concepts cannot be explained in detail in this paper, the following list might however help to give an impression of the possibilities opened by CPPvm:

Semaphores CPPvm includes an implementation of the semaphore concept proposed by Dijkstra [9]. Semaphores are used to synchronize parallel processes. They can also be used for critical sections. A critical section is a set of instructions which shares resources with other parallel processes. The result of a critical section can change unpredictable when these processes are running at the same time. A critical sections can be controlled by semaphores allowing only one process to enter the section at one time.

Multi-Spawn In CPPvm more than one child process can be spawned at one time. Messages can be broadcasted to all subprocesses. This feature supports architectures of parallel programs dividing a complex problem into less complex, identical subproblems.

Catchout The output of a child process can be redirected to cout/cerr of the master process or into a file.

Forward When using explicit message passing incoming messages can be directly forwarded to other processes.

Groups Processes can be combined to form groups. Messages can be broadcasted to all members of a group. A process can be a member of several groups simultaneously.

Context Processes can be spawned into a special context. Messages sent within one context cannot be received in another context. Therefore a context can help to avoid misleading messages. This is useful e.g. for the design of large parallel programs using old code or parallel libraries stemming from other developers.

Exceptions CPPVM uses C++ exceptions to indicate internal errors. Every error which occurs in CPPVM throws an exception object which can be caught by the user and handled individually.

Notifications Notification classes give informations about modifications of the virtual machine: adding new hosts, a host is deleted or crashed, a process exits or is killed.

Templates CPPVM also supports message passing for C++ template classes.

3   Example for Explicit Message Passing

Imagine a greeting ceremony among sportsmen called "give-me-five". The program consists of the two processes coach and player (see figure 5).

The process coach is started on the host beckenbauer and spawns the process player on the host klinsmann. player sends a message to coach. coach prints the message and PVM is halted. The source code is then the following:

coach.cpp

```
#include "cppvm.h"

int
main()
{
  int value;

  // spawn child 'player' on host
  // 'klinsmann'
  cppvmSpawnConnection child("player", "",
    PvmTaskHost, "klinsmann");

  // receive descriptor
  // (blocking receive
  // from child process)
  cppvmReceiveStream recStrm(child,
    CPPvmRBchild);

  cout << "coach: give me five!" << endl;

  // receive value from player
  recStrm >> value;

  cout << "player: " << value << endl;

  // halt the virtual machine
  child.halt();

  return 0;
}
```

player.cpp

```
#include "cppvm.h"

int
main()
{
  int value=5;

  // connect to pvm
  cppvmConnection pvmConn;

  // send descriptor
  // (nonblocking send
  //  to parent process)
  cppvmSendStream sendStrm(
    pvmConn, CPPvmSNBparent);

  // send value
  sendStrm << value;

  return 0;
}
```

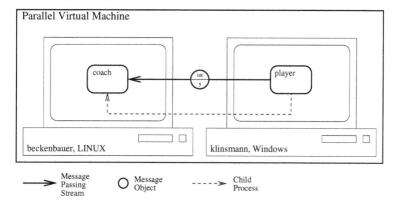

Fig. 5. Give me five!

For explicit message passing following types can be used:

- all CPPvm message classes (see figure 1)
- the standard C++ types bool, char, double, float, int and long as well as constants[1]
- the standard template library (STL) classes bitset, complex, deque, list, map, multimap, multiset, priority_queue, queue, set, slist, stack, string, valarray, and vector

4   Conclusion

Object-orientation is state-of-practice in the field of programming. CPPvm was designed to support object-oriented programming in C++ for cluster computing.

CPPvm is based on the Parallel Virtual Machine (PVM) and is published under the GNU Library General Public License (LGPL) [8]. As shown, CPPvm enlarges PVM with C++ features as classes, inheritance, overloaded operators, exception handling and streams. CPPvm also hides some details of PVM (e.g. starting PVM daemons or adding hosts) from the user and thus makes it easier to write parallel programs. CPPvm is available for many architectures, from Windows to several UNIX derivatives. CPPvm allows to:

- combine a heterogenous collection of computers
- spawn and kill processes dynamically
- detect failed processes and hosts
- send/receive C++ objects
- use distributed C++ objects
- use C++ objects together with a message mailbox
- write user-defined message C++ classes

---

[1] Constants can of course only be sent and not received because they cannot change their values.

- use C++ templates
- use standard template library (STL) classes
- use semaphores
- use CPPVM together with existing PVM software

CPPVM has a very detailed documentation in several formats (Postscript, PDF and HTML). Many examples – from the very simple "hello world" program to more complex application/server programs – help the user to make his way from his very first parallel steps to the innermost secrets of MIMD programming.

References

1. A. Geist, A. Beguelin, J. Dongarra, W. Jiang, R. Manchek, and V. S. Sunderam: *PVM: Parallel Virtual Machine A Users' Guide and Tutorial for Network Parallel Computing*, MIT Press, 1994, http://www.epm.ornl.gov/pvm/pvm_home.html/
2. *MPI: Message Passing Interface*, http://www.erc.msstate.edu/labs/hpcl/projects/mpi/
3. O. Coulaud and E. Dillon: "PARA++ : C++ Bindings for Message Passing Libraries", in *The EuroPvm '95 Users Meeting*, Lyon, France, http://www.loria.fr/projets/para++/
4. *PVM++*, http://goethe.ira.uka.de/~wilhelmi/pvm++/
5. *EasyPvm*, http://www.brunel.ac.uk/~mepghfb/pvm_c++_wrapper.htm
6. J. M. Squyres, B. C. McCandless, and A. Lumsdaine: "Object Oriented MPI: A Class Library for the Message Passing Interface", in *The 1996 Parallel Object-Oriented Methods and Application Conference (POOMA '96)*, Santa Fe, New Mexico
7. *MPI-2: Extensions to the Message-Passing Interface*, July 1997. Message Passing Interface Forum, http://www.mpi.nd.edu/research/mpi2c++
8. S. Görzig: *CPPvm: C++ Interface to PVM (Parallel Virtual Machine)*, 1999. http://www.informatik.uni-stuttgart.de/ipvr/bv/cppvm
9. E. W. Dijkstra: "The Structure of the THE Multiprogramming System", in *Commun. of the ACM 11*, pp. 341–346, May 1968

Persistent and Non-persistent Data Objects on Top of PVM and MPI

G. Manis

University of Ioannina
Department of Computer Science
P.O. Box 1186
GR-45110 Ioannina, Greece
manis@cslab.ece.ntua.gr

Abstract. PVM and MPI have been widely accepted as standards in message passing programming. Although message passing seems to be the most attractive paradigm for distributed and parallel application development, many extensions for both PVM and MPI have been suggested towards other programming philosophies. The proposed system implements persistent and non-persistent data objects shared by any participating process. Shared objects are organized based on concepts derived from the fields of object-oriented programming. Inheritance is implemented through an object hierarchy in which each object inherits information from the ancestor objects, located higher in the hierarchy tree. Mechanism for object migration and replicated objects are supported.

1   Introduction

Object oriented design is described by its use of objects, classes and inheritance. Objects are software components that include data and methods related to this data. A class is a prototype that defines the variables and the methods of all objects instantiated by this class. Inheritance is the key in object oriented design. Classes can inherit variables and methods from other classes located higher in a class hierarchy. Systems that do not support inheritance are characterized as "object based" rather than "object oriented".

Although classes seem to be an integral part of object oriented design, the truth is different. In [1] an alternative approach is presented in which objects do not derive from classes but are instantiated as clones of other objects. Inheritance is implemented forming an object, rather than a class, hierarchy. An object may inherit data and methods from another object, the latter forwarding the invocations it cannot serve, to the former. The parent object may forward the invocation to its parent object, if this is necessary, until an object is finally able to respond or there is no other parent object in the hierarchy tree. In the latter case, the invocation fails.

An example is illustrated in figure 1, where an object hierarchy for a card game is presented. There are three objects "Player" and one object "Dealer". One of the three players will be the dealer. This player (only) inherits from the dealer the method "deal" and the "cards" to deal. An object oriented programming language based on this idea is "SELF"[2].

Y. Cotronis and J. Dongarra (Eds.): Euro PVM/MPI 2001, LNCS 2131, pp. 91–97, 2001.

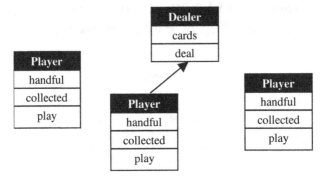

Fig. 1. Object hierarchy in a card game. Objects are created without classes and inheritance is based on the forward of invocations.

In the PVM and MPI extension presented in this paper, the shared objects are data entities that can be created and accessed through methods supported by the system (create, read, write). For each invocation a new message is created and sent to the appropriate server. Shared objects are organized based on concepts derived from the field of object oriented programming. Inheritance is implemented through an object hierarchy. All other object oriented programming concepts are preserved or adapted to the philosophy of the proposed system. Objects are managed by servers. Mechanisms for both object migration and replication are also supported.

The objects of the proposed system can be characterized as persistent and non-persistent. Non-persistent objects live as long as the application runs. These objects reside in memory. Persistent objects remain available even after the application has been terminated. Persistent objects are located in system disks. In the proposed system there is no distinction in the way that persistent and non-persistent objects are created and accessed. Both kind of objects are created and accessed using the same system calls. What differentiates persistent from non-persistent objects is the server where the object resides. Servers for persistent objects store objects in files, while servers for non-persistent objects keep objects in local memory. Since objects in this system can migrate from one server to another, it is possible a non-persistent object to become persistent and vice versa. In addition, using the mechanism of replication, non-persistent object can be used as cached data of a persistent object that resides in a system disk.

A short presentation of this work in earlier stages is described in [3], where a distributed shared memory system followed similar ideas is outlined. In the rest of this paper, the basic concepts of the proposed model will be discussed in detail. Implementation issues will be presented in the third part, while some conclusions will be discussed in the last section.

2   Basic Concepts

In many DSM systems (e.g. Orca [4]) objects represent shared variables that can be accessed by any participating process. In Chorus [5], objects are passive entities, while in Emerald [6] objects can be passive or active. An Amoeba [7] object is a more general concept. Objects in Amoeba can be data, processes or operating system resources. In the Apertos operating system [8] an object can define (a part of) the behavior of another object. All the above are different conceptions of the concept "object".

The object is a data type that can be accessed only by well-defined operations. In this system, objects are data structures that are accessed strictly through a set of supported operations (read, write, copy etc.). Objects in this system are not instantiated by classes. They follow the concepts of "delegation" [1]. In this philosophy, inheritance is expressed through a hierarchy of objects. Each invocation that cannot be performed by an object is forwarded to the parent object, according to the ideas of classless object-oriented programming, as presented in the introduction. It must be noticed that the mechanisms for the implementation of this kind of hierarchy are very simple.

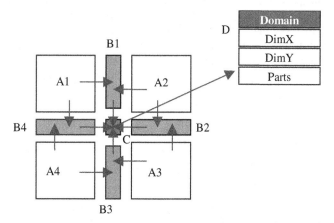

Fig. 2. The decomposition of the problem of boundary values into an object hierarchy. Object A1 inherits from object B1 and B4. Object B1 inherits from object C and object C inherits from object D. Object B4 is the marginal values between object A1 and object A4. Object D is an array with general information about the domain.

An example, such as the solution of a classic problem (e.g. solution of differential equations) based on the technique of boundary values, may be used to illustrate the above concepts. The domain of this problem is separated into a number of parts, equal to the number of available processors, and each part is assigned to one processor. Each processor makes a certain number of iterations. In all iterations, every processor needs to communicate with the neighboring processors in order to exchange boundary values. Supposing there are four processors available, the domain of the problem can be decomposed as shown in fig. 2, according the design philosophy of the proposed system. The big white squares A1, A2, A3 and A4 are the non-shared parts of the array, each one of which is assigned to one processor. The light dark rectangular shapes B1, B2, B3 and B4 are the marginal values accessed by two processors. The object A1 that belongs to processor 1 inherits from objects B1 and B4. In a similar way, objects A2, A3, A4 inherit from B1 and B2, B2 and B3, B3 and B4 respectively. All B1, B2, B3 and B4 inherit from C (this is not necessary – used to enrich the example) and C inherits from D, an object that contains some general information about the domain. A read request for "DimX", sent to object A1, will be forwarded to object B1, from there to C and then to D, which will reply to the request.

The idea is simple, but let's have a look at how this design keeps up with the main concepts of object-oriented programming: abstraction, encapsulation, polymorphism, inheritance and reusability. We will look at each one of them in detail below.

According to the concept of abstraction, the object appears for the rest of the world as a simple and coherent entity. In the previous example, the domain of the problem has been divided into several pieces. The fact that the information is decomposed and located in different places does not affect the user. The user makes all the invocations to objects A1, A2, A3 and A4 and does not know anything about the mechanisms involved in recovering the information.

In the case of encapsulation, things are more complicated. Shared data are accessed through a well-defined interface, the available operations. This is compatible with the concept of encapsulation, especially because this is done through messages: for each operation a message is created and sent to the proper object. However, the lack of user-defined methods and the restrain of access to the supported operations only, narrows the ability of the programmer to hide the implementation details. Since we limit the design to data only, the problem is concentrated on the fact that the programmer cannot change the name of the shared variables or the fields of the shared structures.

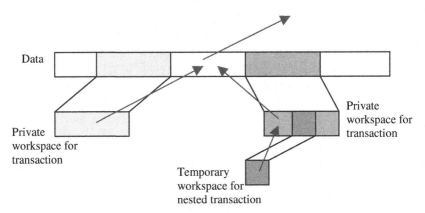

Fig. 3. Data overriding in a transaction system.

Three of the main concepts of polymorphism are data overriding, method overriding and method overloading. This is not a great loss, since the main focus of this proposed method is shared data access through the use of the supported operations. What is really interesting is what happens with overriding of data. An example is shown in figure 3. Suppose there are two concurrent transactions acting on the same data. Each transaction creates a private workspace in which all writes are performed until the transaction commits or aborts. The private workspace replies to requests related to data located in the private space, and forwards the rest of the requests to the real workspace. Each transaction "sees" its virtual workspace, without affecting or being affected by other concurrent transactions.

The role of inheritance and how it is supported by this system have been discussed above. Objects inherit information from other objects located higher in the hierarchy tree. Information on the parents is located inside each object and not in a class, since there are no classes in this model. This characteristic allows the building of different hierarchy trees with the same objects, according to each specific application. In addition, it is possible to change the hierarchy tree, even during run-time. In the example of figure 1, it is possible to change the dealer of the game simply by moving the arrow from the second to another player.

Apart from the added flexibility to the hierarchical structure, it is possible to build different hierarchy trees using the same objects, increasing the possibility of using the same object in more that one application (reusability). Reusability increases for another reason: the deconstruction of complex objects into simpler entities gives more flexibility to the programmer, who can select among a larger number of objects, which can be used separately or in conjunction as an object hierarchy.

3   Implementation Issues

In this paragraph will be presented some implementation details for the basic mechanisms supported by the system: creation, location, migration of the shared data, as well as creation and access of read replicas. All functions have been implemented on both PVM and MPI using only message passing and process creation facilities.

A read or write request that cannot be performed by the invoked object is sent to the parent object. This process continues until one object is able to reply or there are no other objects higher in the hierarchy tree, in which case the invocation fails.

New objects are created as clones of other objects. Cloning has been suggested for the creation of new objects in classless object-oriented systems. Although this is not a crucial issue for the proposed system, cloning is supported. A new object is created as a copy of an existing object. The new object is an identical copy of the original object and by default has the same ancestors, since this information is stored inside the object.

Many systems (especially the distributed shared memory systems) support read replicas. Read replicas are read-only copies of the original objects, and are located at the processors from which invocations are performed frequently. When local copies are kept in many processors, the overall system performance is usually improved. In this system, all read requests are addressed to local copies, if a local copy for the object exists in the processor, while all write requests are addressed directly to the original object. The original object keeps track of all read-only copies of the object. Upon receiving a write request, the object updates itself and brings the read-only copies up to date. Updating the original first object and then the replicas allows a high degree of memory consistency (sequential consistency).

Access to all objects (original or read replicas) is possible through a location server. All objects are registered in this server, which provides the necessary information for their access. Requests for the location of an object should include the name of the object and the information whether the object is to be accessed by a read or a write operation. When a write operation is performed, the location server returns the location of the original object. If the request is related to a read operation, the location server returns the address of the local copy of the object, if a local object exists in the specific processor, otherwise, the location server returns the location of the original object.

The location server performs the main operations for object migration as well. Objects are located by their names. When an object migrates, it copies itself to the new location and informs the location server. The location server re-directs all the read and write requests to the new location. The location server addresses from now on, all new read and write requests to the new location of the object.

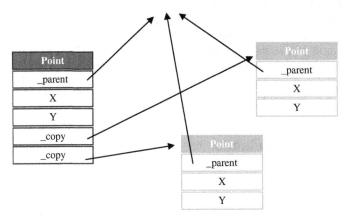

Fig. 4. The structure of an object. The left object is the original. The last two fields of the original object are pointers to the read-only replicas. The rest of the fields are the same. All three has the same parent.

A sample object and its read-only replicas are shown in fig 4. This object represents a point. The first field points to the parent object. The next two fields are the data (x and y coordinates of the point). The last two are pointers are to the read-only replicas of the object. These fields are neither copied to the replica object, nor cloned when a new object is created. A write operation to the field "X" of the object will update its own "X" and then, following the links of the last two fields, it will update the "X" fields of the two read-only replicas.

Finally, there is the issue that has been implied earlier but not discussed yet: multiple inheritance. In the example of figure 2, objects A inherit from more than one of objects B, actually from two. Multiple inheritance is very useful and is not contrary to the proposed concept of simplicity ruling this whole system. The implementation of multiple inheritance is very easy due to the recursive nature of the problem. When an invocation that has been forwarded to the first parent of the object fails, then the request can be forwarded to the second parent, and so on, until no more parents are left or one of the ancestors of the object is able to reply.

4   Conclusions

In this paper an extension to PVM and MPI has been proposed. It is based on persistent and non-persistent data objects. The main idea of the system is to design hierarchies of objects, so that the objects inherit information from other objects located higher in the hierarchy tree. It follows the classless approach: no classes exist, new objects are created through a mechanism called cloning, while the information for the parent objects is stored inside the objects. Object hierarchy can change even during run-time. Simplicity in the design and implementation is a significant characteristic of the system.

References

[1] H. Lieberman, "Using Prototypical Objects to Implement Shared Behavior in Object Oriented Systems," *Proc. Conf. Object Oriented Programming Stsems, Languages and Applications; reprinted in Signal Notices, 21(11)*, pp 214-223, 1986

[2] D. Ungar, R. Smith, "Self: The Power of Simplicity," *Lisp and Symbolic Computations,* 1991

[3] G. Manis, "Data-Object Oriented Design for Distributed Shared Memory," *High Performance Computing and Networking,* 2001

[4] H. E. Bal, M. F. Kaashoek, A. S. Tanenbaum, "Experience wit Distributed Programming in Orca," *Proc. IEEE Int. Conf. On Computer Languages,* pp. 79-89, 1990

[5] M. Rozier, U. Abrossimov, F. Armand, I. Boule, M. Gien, M. Guillemont, F. Herrmann, C. Kaiser, S Langlois, P. Leonard, W. Neuhauser, "Overview of the Chorus Distributed Operating System," *Tech. Report CS/TR-90-25-1,* Chorus Systemes

[6] E. Juil, H. Levy, N. Hutchinson, A. Black, "Fine-Grained Mobility in the Emerald System," *ACM Trans. on Computer Systems,* vol 6, no 1, pp. 109-133, 1988

[7] A. S. Tanenbaum, R. van Renesse, H. van Staveren, G. Sharp, S. Mullender, J. Jansen, G. van Rossum, "Experiences with the Amoeba Distributed Operating System," *Communications ACM,* vol 33, no 12, pp. 46-63, 1990

[8] Y. Yokote, "The Apertos Reflective Operating System: The Concepts and Its Implementation," *Proc. Int. Conf. on Object-Oriented Programming, Systems, Languages, and Applications,*1992

System Area Network Extensions to the Parallel Virtual Machine

Markus Fischer

Department of Computer Science and Engineering, University of Mannheim,
Germany
fischer@ti.uni-mannheim.de

Abstract. This paper describes the design and implementation of an
interface to allow a simple integration of different (high speed) network
interconnects for a message passing environment. In particular, a common SAN layer has been developed and tested with an improved PVM
version. The latter has been extended to provide a pluggable interface
for low level SAN layer. In this context GM and PM for Myrinet and
SISCI for Scalable Coherent Interface (SCI) have been implemented.
With this pluggable interface an approach such as the channel device
for MPICH has been made to easily integrate existing and new networking devices. Multi Protocols which handle several interconnects are
supported as well. This allows for harnessing heterogeneous clusters with
different high speed interconnects using the fastest available communication device when possible. The given results show PVM's capability of
achieving low latency, high bandwidth using appropriate devices.

1   Introduction

Since a few years, the platform for parallel computing is changing. A major shift
from massively parallel systems to clusters of workstations / PC's can be seen.
The latter are built up from commercial off the shelf (COTS) components, which
partly offer high performance known only from supercomputers, however at a
fraction of the cost (e.g.: CPU performance).

One of the last special products for high performance clustering is the network interface card (NIC). Currently, the PCI bus is the standard for IO being
supported by any OS and thus a very portable solution for networking.

However, communication intensive applications are hampered by the standard networking type (Ethernet) with its overburdened protocol (TCP/IP). In
particular, the high latency for small messages as well as the low bandwidth for
bulk messages slow down the application.

Within a cluster, the distance between nodes is rather small and error rates
are extremely low. Thus, a light weight protocol for message transfer is more
practical.

Since several years, new networking devices exist to build up a system area
network (SAN), which delivers performance in the range of Gigabits/s. As of
today, popular high speed interconnects are Myrinet, GigaNet, Scalable Coherent
Interface, Servernet, ... , all plugging into the PCI bus.

Y. Cotronis and J. Dongarra (Eds.): Euro PVM/MPI 2001, LNCS 2131, pp. 98–105, 2001.

1.1   Current Implementations

With the variety of different networking devices and multiple low level API's for the same device (GM, PM and BIP for Myrinet for example), it would be efficient to provide a clean API for a higher message passing environment to allow for an easy integration of existing or new interconnects.

The portable MPI implementation MPICH [2], which provides a Chameleon device, is such an approach. Another message passing environment is PVM which provides more flexibility than current MPI implementations. Some of its advanced features are dynamic creation of processes and communication partners, or fault tolerance. But also a collection of very heterogeneous machines can be harnessed to form one single entity, transparent to the application through the message passing library. The MPICH implementation of MPI provides an abstract device for communication layers and several channel devices have been implemented.

As a contrast, PVM [6] does not provide an abstract device, but has a mixed code that differentiates between nodes from MPP's to workstations using Ethernet. For workstations, control is given by additional PVM daemons (PVMd), which are running on each host of the virtual machine. Another part of the PVM system is the PVM library (*libpvm*) to which tasks have to be linked. Messages are sent using the *tid* as destination parameter. Within PVM, two routing policies are given. Using the default routing, messages are transferred via the UDP connected daemons which route the message to the final destination.

To improve communication performance a direct connection between two tasks can be established, leaving the PVMd's outside the transfer.

This mechanism has been used to extend the PVM communication primitives to provide an interface for other network devices not using TCP/IP protocols.

Other research used ATM as a network [10], still relying on TCP/IP and not gaining much more performance than using Fast Ethernet. Also, multi devices which handle several network connections were not supported, but are addressed within this PVM extension. Finally, other ports to SCI are available [4], [3].

The rest of the paper is organized as follows. In the next section generic requirements for message passing stylish communication are observed. Section 3 describes the implementation of the interface as well as different plug-in's for multiple interconnection types and also discusses optimizations and different communication patterns within a device. Section 4 provides performance results of the developed devices. Finally we conclude our efforts.

2   Basic Requirements

For a message passing environment like PVM, several functions are required in order to establish a point to point connection between two tasks. One important feature for a dynamic environment is to allow the establishment of connections during runtime. Obviously, this feature is also resource friendly since it does not set up unnecessary connections which are not used during execution. In order to allow this, a request for setting up a direct connection must be transported

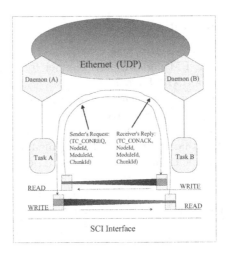

Fig. 1. Establishing a direct SAN connection between two nodes (with SCI as an example)

from one node to another. The initial request structure typically includes basic SAN specific information.

The delivery of requests is performed using an additional standard network such as Fast Ethernet. The standard PVM provides controlling daemons which will deliver data between tasks until a direct connection between two nodes using a faster method is made.

To make this scheme universal to several interconnects, this transfer is split up in a request and a ack/grant phase between two nodes (see figure 1 for details).

This mechanism can be split up in the following phases a

1. Request - a node invokes a new pt2pt connection, passes the request to the daemon and waits for an acknowledgement after which the connection is finalized.
2. Delivery - a controlling daemon holds the request and the data is passed via conventional communication mechanisms to the receiver
3. Ack and Grant - a node receives the request when it enters a library function, performs steps to setup and finalize the connect and returns a ACK if resources are available

After this exchange of SAN information further communication takes place only using the fastest interconnect available. Typical content of a setup message can be a port number, node id, or memory addresses.

When a direct communication is established, low level calls to the API of the interconnect can then be used to exchange data. For this, a sender holds a struct for each communication partner in which the type of connection is stored. However, for a receiving node, message arrival has to be detected (and received) from different networking devices.

3   Implementations on Different Interconnect Devices

In the following the extensions to PVM are explained which allow the plug in of different interconnects. To allow a common interface for several interconnects, a plug in has to register its functions in a header file which is included into the lpvm.c file which implements the setup of new connections as well as send/recv functions and detects message arrival. As described above, the following functions are prototyped *san_request(), san_ack(), san_send(), san_recv() and san_poll()*.

When enabling direct connections, a PlugIn references and calls the registered functions (e.g: the request and ack, the send/recv and select functions) from the header file. For them it is required to be autonomous functions not requiring additional control (for example the *request* function provides **all** required information for the receiver, or a *send* function does only return when a (possibly later) message delivery can be guaranteed.) Thus, the extended PVM does not differentiate between various devices but calls the registered functions in the same manner.

3.1   SISCI Plugin for SCI

SCI (Scalable Coherent Interface) is the international standard IEEE #1596-1992 for computer-bus-like services on a ring-based network [9]. It supports distributed shared memory (DSM) and message passing for loosely and tightly coupled systems. Optionally cache coherent transactions are supported to implement CC-NUMA (Cache Coherent Non-Uniform Memory Access) architectures. However, using the PCI bus as commodity interface to build a SAN, major key features of SCI, such as the cache coherency, no longer exist since it is not possible to snoop on the host memory bus. Thus, for each message transaction, data has to be explicitly memcopy'd into memory mapped regions.

Another disadvantage is the performance difference for the so called *put* or *get* scenario. Writing to the network is an order of magnitude faster than reading from the network (*writing to* or *reading from* remote memory with 73MB/s to 12MB/s respectively). This must kept in mind by developing and implementing an efficient message passing plug in for a SCI network.

Details on PlugIn Functions. When establishing a direct connection, a basic information is the SCI-node ID, the SCI memory location and the chunk id to map exported memory. This setup is performed by the initiator **and** the recipient, both exporting to (during setup) and mapping from (when finalizing the direct connection) the communication partner (the remote task). Thus, the initiator receives necessary information within the ACK/GRANT reply.

– For the send function, a ring buffer has to be implemented (SCI exports memory but does not offer function to manipulate it). The remote side updates a read pointer indicating the last position of the ring buffer read. This way, a sending process does not overwrite unread data and flow control is guaranteed. An earlier version of the plug in used SCI's remote interrupters, however the performance (it took about $60\mu s$) was too slow to achieve any performance. When a sent has been made, the number of messages sent is updated on the remote side [3].

- The recv function also implements a ring buffer and updates the read pointer of the last position on the remote side.
- select/poll functionality is implemented by querying the number of messages received for each communication partner. A local structure stores the numbers read so far and if for any communication partner this number is higher, the function returns with a pointer to the new message. To achieve efficiency, the interrupter method could not be used since context switches hampered the performance. Also the notification mechanisms had to keep in mind the enormous performance differences for the put and get scenario. Thus, for example, the *number of messages* sent is stored locally at receiver side, while the *read* pointer is stored in physical memory of the sender.

3.2   GM Plugin for Myrinet

Myrinet [8] is a high speed interconnection technology made by Myricom. It uses source path routing and is capable to transfer variable message lengths at 2 GBit/s. The (open source) GM driver including a Myrinet Control Program comes from Myricom and implements DMA only. GM provides a connection less protocol in which sender and receiver are identified by so called host id's and gm_ports. A process gets access to the Myrinet by opening a GM port. However in order to send or receive data, a process must provide a pool of bin's (of different) sizes which can be accessed by Myrinet's DMA engines. First, data is then transfered from host memory to network memory (SRAM on Myrinet card), then injected into the network. Obviously incoming data should be transfered to host memory as soon as possible clearing the network and the limited SRAM memory on the Myrinet card. Thus the PlugIn has to provide enough bin's so that the DMA engines always find a slot to put the data into.

Details on PlugIn Functions. When establishing a direct connection, GM port information is exchanged, but an establishment of a connection is not necessary.

- For implementing the send function, the gm_send_* mechanisms which are already provided by GM can be used. For a message transfer, data is first copied into an appropriate bin. To achieve better performance, this data is pipelined (see section 3.4 for details). One key feature is the flow control coming with GM. In particular, a function can be given to the gm_send_* function which is called after the send has completed. This is of importance since with the efficient usability of DMA's, overlapping of computation and communication can be implemented easily. A bin can then be re-used when the function has been called, signaling the end of the message transfer.

- Implenting the recv function for the PlugIn is also straight forward by using the gm_receive function which returns the first entry from a FIFO queue. Into this queue all incoming messages are inserted and when finding an entry in the queue message data already has been DMA'd from network memory to host memory into one of the provided bins. Further GM functions provide the source of the message to be identified for higher level message passing systems.
- A select/poll mechanism is provided as well through GM. The gm_receive function returns a NO EVENT tag, if no message has arrived or directly provides the message. Thus, the function is non blocking by default.

3.3   PM2 Plugin for Myrinet

PM2 is another low level API for Myrinet. It is developed by the RWCP Japan [7] and is open source. PM2 runs as API but is specifically designed to run in the SCore environment which also provides other message passing environments such as MPICH-PM, but also a set of tools to manage a cluster. A program using PM2 must first open the PM device and then open a context which has to be bound to a channel. Buffers are provided within PM2 functions. A process call pmGetSelf to identify itself in the PM context. Other nodes may use this information to send messages. A key feature of PM2 is the availability for different platforms! Not only Myrinet is supported but Fast Ethernet and Gigabit Ethernet as well. This way, also for conventional interconnects, the overburdened TCP/IP protocol is left out of the way and performance is greatly improved ([5]). For example, this features allows to run MPICH applications on hundreds of nodes, since the protocol is connectionless.

Details on PlugIn Functions. When establishing a direct connection, the PM port number information (0, .. , n -1) is exchanged but like GM, a setup of a connection is not necessary.

- For implementing the send function, the pmSend mechanism can be used. It uses only an internal context value as a parameter, which has been initialized by PM when providing a send buffer which was tagged with the destination node information.
- The Recv Function can be easily implemented with the pmReceive function.
- Select/Poll - Like GM, the behavior within PM2 is the same, the pmReceive function is non blocking.

3.4   Optimizations Left within PlugIn

Since the PlugIn provides only a simple interface for compatibility reasons, optimizations have to be coded into the device.

Optimized Memcpy Functions. Recent research focussed on zero copy message transfers which avoids unnecessary copies of data. This often requires a lot of changes to higher level code, but does not result in dramatic gain of performance (2-5 per cent [1]). When developing different PlugIn's for different network interconnects, it became clear, that for each selected host architecture optimized memcpy functions can increase the performance compared to C lib memcpy function calls. A major gain can be seen when using SCI. Here a bandwidth limiting factor is the performance with which data is copied to exported memory. Using the standard memcpy call the performance peaked at only 22MB/s. With optimized memcpy routines (for example using the FPU's 64bit operands) the peak value achieved on a PII 450 system was 73 MB/s, compared to 280MB/s when copying data from shared memory to local memory. Thus the maximal performance for higher level message passing environments such as PVM or MPI is limited by the put performance of 73MB/s. Further optimizations such as MMX, SSE, SIMD or VIS can be considered as well.

Data Pipelining. When using DMA capable hardware such as the Myrinet network card, a major performance increase can be made by pipelining / interleaving message data. In this case, separating larger messages into multiple chunks, which can be chained by the DMA, increase the performance, since a memcpy of a shorter message takes less time and a first DMA transaction can already start while other memcpy's follow. This simple technique for example increased the performance from 32MB/s to 79 MB/s.

Registering Memory. To avoid data copies into an extra buffer some operating systems like LINUX provide the functionality of registering memory. In this case data is pinned down and can be accessed by DMA to be injected into the network directly. This enables an optimization for some operating systems to switch between different communication patterns depending on the message size. Registering memory however involves a costly kernel trap. In particular, on a PIII 600 with Linux 2.2.14, we measured a break even of only 11KBytes, so that a memcpy of data performed less than registering memory. Thus, for larger messages better performance was achieved by switching the message protocols.

4   Performance

In the following the three currently available PlugIns are compared. With raw performance we define the maximal performance possible at SAN API level. We took the nntime application coming with PVM, which measures latency and bandwidth.

4.1   PVM with SISCI PlugIn vs Raw SISCI on SCI

The tests were made on a PII 450 System with Dolphins 32 Bit/33Mhz D313 cards. The half round trip latency for sending a null message is 60usec. The bandwith reaches 55 MB/s.

4.2   PVM with GM PlugIn vs Raw GM on Myrinet

These performance measurements were made using a PIII 600 system equipped with LANai7 plugged into a 32bit/33Mhz PCI slot running LINUX. The half round trip latency for sending a null message is 55usec. The bandwith peaks at 70MB/s

4.3   PVM with PM PlugIn vs Raw PM

For the PM plug in we had access to a PIII 700 system equipped with LANai9 plugged into a 64bit/33Mhz PCI slot running LINUX. The half round trip latency for sending a null message is 40usec. The bandwith peaked at 110MB/s.

5   Conclusion and Future Work

We have designed and implemented a PlugIn interface to allow the usage of different popular interconnects and their various low level API's. An important side effect of this research was to step into the different interconnection types and their API's in every detail. The stability of GM, which attaches to the system as a kernel module, makes this SAN API our favorite protocol. The modul makes the integration of Myrinet very easy, unlike PM which requires a rudimentary kernel patch. In addition, the GM plug in also works for Windows NT/2000.

Further development will be to integrate the VIA interface into this common SAN layer.

References

1.   T. Warschko et al. On the Design and Semantics of User-Space Communication Subsystems. In *PDPTA 1999*, Las Vegas, 1999.
2.   Message Passing Interface, MIT Press, 1994
3.   M. Fischer and J. Simon. Embedding SCI into PVM. In *EuroPVM97*, Krakow, Poland, 1997.
4.   I. Zoraja, H. Hellwagner and V. Sunderam SCIPVM: Parallel distributed computing on SCI workstation clusters In Concurrency: Practice and Experience, Vol. 11, 1999
5.   H. Tezuka et al. Pin-down Cache: A Virtual Memory Management Technique for Zero-copy Communication. In IPPS/SPDP'98, IEEE, 1998.
6.   J. Dongarra et al. *PVM: Parallel Virtual Machine. A User's Guide and Tutorial for Networked Parallel Computing*. MIT Press, Boston, 1994.
7.   http://pdswww.rwcp.or.jp/
8.   N.J. Boden et al. Myrinet: A Gigabit-per-second Local Area Network. IEEE Micro, 15(1):29–36, February 1995
9.   *IEEE Std for Scalable Coherent Interface (SCI)*. Inst. of Electrical and Electronical Eng., Inc., New York, NY 10017, IEEE std 1596-1992, 1993.
10. H. Zhou and A. Geist Faster Message Passing in PVM. Technical Report, Oak Ridge National Laboratory, Oak Ridge, TN 37831, 1995.

Adding Dynamic Coscheduling Support to PVM

A. Gaito[1], M. Rak[2], and U. Villano[1]

[1]Università del Sannio, Facoltà di Ingegneria, C.so Garibaldi 107, 82100 Benevento, Italy
anogaito@libero.it, villano@unisannio.it
[2]DII, Seconda Università di Napoli, via Roma 29, 81031 Aversa (CE), Italy
maxrak@iol.it

Abstract. This paper deals with the profitability, the design and the implementation of coordinated process scheduling under PVM. Firstly the principal coscheduling techniques proposed in the literature are reviewed, paying particular attention to those that can be effective in network of workstations. Then the problems linked to the design of a coscheduling scheme for PVM are discussed, and a prototypal implementation is presented. The obtained results show that coscheduling support can reduce significantly the response time of PVM programs, without affecting in a meaningful way the performance of sequential workload.

1 Introduction

In a network of workstations (NOW), the execution of parallel programs is obtained by exploiting the independent computing facilities available in each node. The processes making up the parallel program compete for computing resources (CPU time, memory, I/O devices) with the sequential load of each machine and (possibly) with processes belonging to other parallel programs. The illusion of a single parallel machine obtained through the use of suitable run-time environments (e.g., PVM, MPI) in fact relies upon the exploitation of independent and possibly different operating systems. Therefore, the parallel application processes are scheduled without any notion of parallel program global state, in a completely uncoordinated way.

While it is commonly agreed upon that shared memory parallel programs can obtain significant performance advantages by coordinating local OS schedulers, the profitably of such a technique for message-passing environments is still a research issue. As a matter of fact, following up the seminal work by Ousterhout [1], numerous research contributions have appeared in the literature [2, 3], pointing out that shared-memory programs are severely affected by the global scheduling policy and proposing solutions for coordinated scheduling (*coscheduling* or *gang scheduling*), which exploit centralized or fully distributed algorithms.

It is relatively clear that in a shared-memory program, just to mention a trivial example, the response time of a process waiting for acquiring a lock will be affected by any scheduling decision regarding the process holding the lock. But it is not equally clear whether the same is true for the coarse-grained and rather loosely-coupled processes that make up a parallel program executed in a NOW. To the Authors' knowledge, only two recent papers try to address this issue. The first one is

Y. Cotronis and J. Dongarra (Eds.): Euro PVM/MPI 2001, LNCS 2131, pp. 106–113, 2001.

due to Wong *et al.* [4], who show that local scheduling decisions can nevertheless impact MPI program performance, and propose a method to coordinate local schedulers using implicit information (i.e., information obtained through the communication events normally generated during program execution). The second contribution [5], instead, has the same finalities as this paper, namely to study the effect of coscheduling on PVM parallel programs. Asking whether local scheduler coordination, successful for MPI programs, can be useful also for PVM applications makes certainly sense. Although PVM and MPI may look similar, PVM is more oriented than MPI to coarse-grained parallel applications running on heterogeneous networks of workstations. Solsona *et al.* have implemented both explicit and implicit coscheduling under PVM, obtaining interesting results. Even if the implementation details are not dealt with in their paper, it is reasonable to wonder if other coscheduling schemes, maybe more canonical than the ones they employed, could also be successful in obtaining high performance from PVM applications.

In practice, proving that global parallel program performance is sensitive to local scheduler choice is far from trivial. In the literature, this task has been traditionally approached from two different directions, resorting to analytical or simulation models of distributed system operation, or by comparing the behavior of a parallel program executed in isolation (where all schedulers are trivially coordinated, because they have no other processes to schedule) to the behavior of the same in the presence of computing load. Both of them can lead to incorrect conclusions, or, at least, to conclusions that cannot be easily extended to a wide range of hardware/software environments. Therefore, we decided to follow a more pragmatic approach, namely to implement a coordination scheme for PVM, and to compare the obtained performance results to those measured in normal, uncoordinated system operation. In other words, the usefulness of coscheduling is proven by showing that it allows to obtain better performance results than local, uncoordinated process scheduling.

This paper is organized as follows. In the next Section, we review the principal coscheduling techniques proposed until now, paying particular attention to those that can be effectively adopted in a NOW. Then the problems linked to the design of a coscheduling scheme for PVM are discussed, and a prototypal implementation of our solution is presented. After showing a set of tests exploring the performance benefits and the scheduling fairness of our proposal, the conclusions are drawn.

2   Coscheduling Techniques

Parallel scheduling is logically composed of two independent phases, the allocation of processes to processors (*space-sharing*) and the scheduling of processes assigned to processors over time (*time-sharing*). The latter is by far the most important issue in NOWs, due to the distinctive characteristics of their typical workload (the use of large sets of idle or lightly loaded workstations to execute computation-intensive parallel applications). The traditional and customary solution to time-sharing of processes is to ignore completely the problem, letting processes be scheduled in a completely uncoordinated way by the native operating systems of the workstations in the NOW. In 1982, Ousterhout firstly introduced the concept of *coscheduling* (also known as *gang scheduling*): the simultaneous scheduling of processes making up a parallel application across all involved processors, which gives the illusion of a distributed dedicated machine [1]. Using coscheduling, the destination process of a message is guaranteed to be scheduled, and so the time spent for synchronization is exactly the same as in a dedicated environment. Processes can spin (instead of blocking and let other process obtain the CPU) while waiting for messages or synchronization events.

The use of local OS schedulers (and hence no coordination and no exploitation of global knowledge) and coscheduling (and hence full coordination based on global system knowledge) are at the extreme ends of the spectrum of possible solutions to the time-sharing scheduling problem. In between there are techniques, such as *implicit* and *dynamic* coscheduling, which exploit information available locally to deduce what is scheduled on other nodes, and do not require any explicit exchange of additional information.

With *local scheduling*, each workstation independently schedules its processes. This is a simple technique, but the performance of fine-grained parallel applications (and not only, as shown in the rest of this paper) is significantly lower than the one that can be attained using coscheduling techniques [6, 7]. This is due to the lack of coordination of the local schedulers across the workstation cluster.

Explicit coscheduling [1] relies on a close coordination of the communicating processes making up the parallel program. A static global list of the order in which the processes are to be scheduled is constructed, and a global, simultaneous context switch is performed on all involved processors. Implementing this technique is no easy task, due to scalability and reliability issues, not to mention the difficulties linked to the precomputation of the schedule of communicating processes [8]. In addition to this, explicit coscheduling is particularly unfair, in that it penalizes heavily the performance of interactive and I/O-bound sequential processes [2, 9].

Using *dynamic* or *implicit coscheduling*, the independent schedulers of each workstation are coordinated by means of local events occurring naturally during the parallel application execution [10, 11]. The rationale of *implicit coscheduling* [3, 4, 10] is that a process waiting for a reply message should receive it "soon" if the sender is currently scheduled on the remote node. In all the above-mentioned papers, the primary mechanism adopted to achieve coordination is *two-phase spin-blocking*. With two-phase spin-blocking, a process waiting for a message spins for some predetermined time before relinquishing the CPU. If the response is received before the time expires, it continues executing, otherwise it blocks and another process is scheduled. On the other hand, *dynamic coscheduling* (DCS) [12, 13] deduces from the arrival of a message that the sender is scheduled on a remote node and schedules the receiver accordingly. Processes wait for messages by busy-waiting; all message arrivals are explicitly treated as demand for coscheduling, and the destination process is scheduled through explicit control of scheduler priorities.

In [14], Nagar *et al.* compare the performance of nine implicit/dynamic scheduling algorithms, obtained by handling message arrival information in all possible ways (no reschedule, immediate reschedule, periodic boost of the waiting process) in conjunction with all known coordination mechanisms (busy-wait, spin-block, spin-yield). Almost always, the best results are obtained by the schemes based on *Periodic Boost* (PB). Using PB, a thread within the kernel periodically examines the processes that are communication endpoints and possibly boosts their priorities.

In our knowledge, the only paper dealing with coscheduling of PVM parallel programs is reference [5]. The authors have implemented both explicit and implicit coscheduling in a cluster of PC running Linux. Even if the implementation details are not dealt with and the experimental results are relatively limited, the proposed schemes seem worth of further investigation.

3   The Design of a Dynamic Coscheduling Scheme for PVM

The implementation of a coscheduling scheme under PVM is a challenging task. This is due to the particular characteristics of the virtual machine implementation, which come from the traditional socket-TCP/IP messaging interface. When the virtual machine is started, a backbone of system processes (PVM daemons, *pvmd*s) is set up. The set of daemons (actually, one for machine) manages the virtual machine, launches the execution of cooperating tasks, assigns them unique identifiers (*tid*s), manages groups, and introduces some degree of fault-tolerance, recognizing faulty machines and excluding them dynamically from the PVM.

Besides performing management functions, the network of *pvmd*s receives messages from local tasks, routing them through other *pvmd*s to the destination node. At the destination node, the local *pvmd* forwards the message to the receiving task. In fact, indirect routing through *pvmd*s is the default PVM communication behavior; direct routing is available as an option, to be used when the number of communicating tasks is not high and communication performance is more than an issue. In practice, most existing PVM codes use *pvmd*-routed communication.

The three main issues in the design and implementation of a coscheduling service are recognizing the processes that need to be coscheduled, providing mechanisms for achieving the coordination, and assessing the performance impact of the scheme adopted [13]. Let us consider the first two points, deferring performance considerations to the next Section. Little, if not nothing, of the studies summarized in the previous section applies to the case of PVM programs running in a NOW. Due to the peculiar characteristics of PVM tasks (coarse-grain and loosely coupled), to the high communication latencies, and, above all, to the interposition of the network of *pvmd*s between communicating processes, the issue here is not to find a way to coschedule communicating tasks, thing which would be of little use. In theory, when *pvmd*-routed communications are used, both the communicating tasks and all the daemons between them should be contemporarily scheduled to guarantee fast message delivery, and this is clearly impossible in uniprocessors. The only reasonable coordination technique is to speed up as much as possible the transfer and service of messages. Often, the service of a message involves the transmission of further messages (forwarding, replies, ...). Almost inevitably, some of these message transfers will be on critical program execution path, and therefore speeding up message handling will have direct consequences on program response time.

As regards the mechanisms used to achieve coordination, the message transmission through the *pvmd* network makes it not reasonable to spin within a PVM task, waiting for an incoming message, as in the majority of implicit coscheduling schemes. If a task is spinning and does not relinquish the CPU, the *pvmd* cannot be scheduled (at least in uniprocessors), making it possible to receive the message the task is waiting for. Even if direct routing is used, the momentary unavailability of the daemon would stop any virtual machine housekeeping (group management, periodic communication with other PVM daemons, ...). For the above-mentioned reasons, the only existing implementation of implicit coscheduling under PVM [5] does not even try to keep the receiving task spinning during the whole message waiting phase, but just during the reading of the multiple fragments making up the received message.

In light of all the above, we decided to adopt a message-based dynamic coordination scheme similar to the one used by DCS or PB, using explicit control of the priority of receiver tasks. Whenever a message for a local task is available, the task priority is boosted for a fixed interval of time (*HPS*, which stands for *High Priority time Slice*). If the receiver task is already waiting for the received message, it is scheduled very soon and runs in preference to other processes until the *HPS* expires. The rationale is to let it reach a possible send communication primitive before being descheduled. In fact, very often a received message involves a fast reply, or has simply to be forwarded. If instead the destination task is not waiting, because it has not yet executed a *receive* primitive, the priority boost received for a time *HPS* gives it a chance to reach soon the communication end-point. It should be noted that, rigorously speaking, this scheme does not necessarily lead to coscheduling, i.e., to the contemporary execution of communicating tasks on different processors. Surely does not in the case of indirect communication through daemons. However, it is a (loose) coordination scheme, which should hopefully help the completely uncoordinated local O.S. schedulers to find a better arrangement of activities on the basis of locally-available knowledge on the status of the currently executed PVM program.

An issue of paramount importance is the fairness of the resulting scheduling scheme. It is clear that any parallel application speed-up is not obtained for free, in that local interactive response is traded off for parallel program performance. The problem is balancing at best the needs of parallel and sequential workloads. In fact, the problem is more complicated, as the scheduler fairness should be considered with respect to both sequential load and competing parallel applications. As regards the effect on sequential workload, the proposed scheme makes it possible to reach a reasonable compromise simply by varying the parameter *HPS*. A short *HPS* limits the boost given to parallel programs and reduces the detrimental effect of coscheduling over sequential applications. On the other hand, large values of *HPS* tend to favor parallel jobs. As far as the effect of coscheduling among parallel applications running at the same time is concerned, the proposed technique can also be reasonably fair, provided that all the workloads have similar communication frequencies. In fact, it clearly tends to favor processes communicating often. At least in theory, a process steadily receiving messages at a frequency higher than $1/HPS$ could be entirely executed at high priority. This behavior is analogous to the policy of traditional Unix schedulers, which favor I/O-bound and interactive jobs.

4   Experimental Results

We think that an implementation of a coscheduling scheme for PVM should satisfy the following requirements:

- to be completely transparent for the programmer, so as not to entail any modification to existing code;
- to be relatively light-weight, in that the behavior of the "standard" PVM is not significantly altered;
- to be easily portable, at least as the original PVM is. This requirement has a number of important implementation consequences. The most fundamental of them all is the necessity to modify PVM so as to include local scheduler priority control, instead of implementing a modified scheduler for the host operating system.

In the case of indirect (*pvmd*-routed) communication, a solution with the above characteristics can be obtained fairly easily by modifying the *pvmd* code. When a message for a local task is received, the modified PVM daemon must boost the receiver's priority, starting a timer that allows to reset the priority when the interval *HPS* expires. Just to give the reader an idea of the daemon modifications involved, we have obtained this behavior by about 500 lines of code, on a total of more than 5000 lines making up the original *pvmd* source. Implementing the same priority management in the case of direct communication is instead relatively complex, unless the coordination scheme is modified in such a way that priority is boosted only for waiting processes, and not for any process with an outstanding received message. Since the most immediate objective of this paper is just to explore the profitability of coscheduling techniques for PVM and to assess their performance impact, for the time being the priority boost has been implemented only for indirect communications.

The current implementation has been written and tested for the Linux architecture, but it is easily portable to any target provided with a POSIX-compliant scheduler. In our modified PVM version, daemons are executed within the *round-robin* real-time scheduling class. Every time that a message for a local task is received, the daemon moves the task from the standard time-sharing class to the *round-robin* real-time scheduling class, at a priority level just under that of the daemon itself. After a time *HPS* from priority increase, the task is moved back to the standard time-sharing class.

Quantifying the impact of the scheduling scheme on system performance is not trivial, being linked to many factors, such as number and type of sequential and parallel workload, host operating system(s), hardware, communication hardware and software. In fact, most of the research work in this field is based on performance evaluations by simulation on synthetic workload [2, 12], or by running only one parallel program in the presence of competing sequential workload [13]. Here we will present the results obtained in some of our first tests, performed on a small-sized cluster of three Pentium-class PCs running Linux and connected by Fast Ethernet. In all the experiments leading to the results presented below, the presence of sequential load has been simulated by executing a sequence of five programs, three runs of a ray tracing program (POV-Ray) producing three images at different resolutions, interleaved with a sort code and a program performing a search in a long vector. The parallel code used as benchmark is an old test code available in many PVM code repositories, a Cholesky factorization program working on a $N*N$ data matrix [15].

Table 1 shows the Cholesky code response time (in the presence of sequential load) as N varies, with the standard PVM (no coscheduling), and with the coscheduling support activated, *HPS* equal to 5 and 10 ms, respectively. The speed-up obtained by coscheduling (percentage reduction of response time as compared to standard PVM) is represented graphically in Fig. 1. It varies from about 6 to 50% depending on the value of the problem dimension N and hence on the relative weight of computation and communication. Table 2 shows the performance penalty imposed to sequential workload by the unfairness of the coscheduling scheme. It contains the response times for all the five programs making up the test sequential load, as measured in isolation (no PVM load), and concurrently with the Cholesky code (with no coscheduling, *HPS* = 5 and 10 ms, respectively). In all our experiments, the performance loss of the sequential workload (as compared to the response times measured in the presence of parallel load running under standard PVM) is contained between 4 and 12 %.

Table 1. Parallel load response times (sec.)

N	no coscheduling	HPS = 5ms	HPS = 10ms
100	0.863	0.507	0.435
200	2.367	1.522	1.466
300	3.740	2.449	2.373
400	4.585	3.572	3.497
500	7.669	6.185	6.015
600	10.570	8.657	8.404
700	12.686	10.993	10.675
800	17.563	15.863	15.572
900	22.890	21.310	20.901
1000	29.107	27.403	27.157

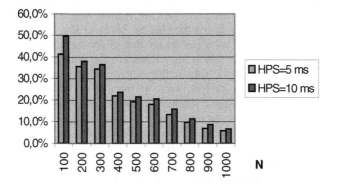

Fig. 1. Parallel load speed-up due to coscheduling

Table 2. Sequential load response times (sec.)

	no parallel load	parallel load, no coscheduling	HPS = 5ms	HPS = 10ms
POV-Ray 320*200	21.751	35.469	39.268	39.946
POV-Ray 640*480	102.696	170.108	188.653	188.724
POV-Ray 800*600	160.364	264.246	294.095	294.212
Sort	21.425	36.888	39.207	39.323
Search	0.279	0.419	0.437	0.448

5   Conclusions

In this paper we have presented a dynamic coscheduling scheme specially designed for PVM. It relies on explicit boost of task priorities under the control of *message-received* events. We have implemented it, obtaining a prototype that does not alter the programming interface and the typical behavior of PVM, and is easily portable across different architectures. The very first set of experiments performed shows that it can reduce significantly parallel response times. The unfairness and the consequent loss in responsiveness of sequential jobs is typically reasonable, and can in any case be controlled by varying the coscheduling parameter *HPS*. Also in the light of other recent research contributions, it is possible to conclude that communication costs are not the unique impediments to high performance in NOW, and that schemes capable of scheduling efficiently the workload of sequential and communicating parallel processes executed in each node deserve further attention.

References

1. Ousterhout, J. K.: Scheduling Techniques for Concurrent Systems. Proc. 3rd Int. Conf. on Distributed Computing Systems (1982) 22-30
2. Vahdat, A. M., Liu, L. T., Anderson, T. E., Patterson, D. A.: The Interaction of Parallel and Sequential Workloads on a Network of Workstations. Proc. of 1995 ACM Sigmetrics/Performance Joint Int. Conf. on Measurement and Modeling of Computer Systems (1995) 267-278
3. Culler, D. E., Mainwaring, A. M.: Scheduling with Implicit Information in Distributed Systems. Proc. of 1998 ACM Sigmetrics Conf. on the Measurement and Modelling of Computer Systems (1998) 233-243
4. Wong, F. C., Dusseau, A. C., Building MPI for Multi-Programming Systems using Implicit Information. In: Dongarra, J., Luque, E., Margalef, T. (Eds.): Recent Advances in Parallel Virtual Machine and Message Passing Interface, LNCS, Vol. 1697 215-222
5. Solsona, F., Giné, F., Hernández, P., Luque, E.: Implementing Explicit and Implicit Coscheduling in a PVM Environment. In: Bode, A., Ludwig, T., Karl, T., Wismüller, R. (eds.): Euro-Par 2000 Parallel Processing, LNCS, Vol. 1900. Springer-Verlag (2000) 1165-1170
6. Feitelson, D. G., Rudolph, L.: Gang Scheduling Performance Benefits for Fine-Grained Synchronization. Journal of Parallel and Distributed Computing **16** (1992) 306-318
7. Gupta, A., Tucker, A., Urushibara, S.: The Impact of Operating System Scheduling Policies and Synchronization Methods on the Performance of Parallel Applications. Proc. of 1991 ACM Sigmetrics Conf. (1991) 120-132.
8. Feitelson, D. G., Rudolph, L.: Coscheduling Based on Run-Time Identification of Activity Working Sets. International Journal of Parallel Programming **23** (1995) 136-160
9. Efe, K., Schaar, M. A.: Performance of Co-Scheduling on a Network of Workstations. Proc. of the 13th Int.Conf. on Distributed Computing Systems (1993) 525-531
10. Dusseau, A. C., Arpaci, R. H., Culler, D. E.: Effective Distributed Scheduling of Parallel Workloads. Proc. of 1996 ACM Sigmetrics Int. Conf. on Measurement and Modeling of Computer Systems (1996) 25-36
11. Sobalvarro, P. G.: Demand-based Coscheduling of Parallel Jobs on Multiprogrammed Multiprocessors. PhD Thesis, MIT, Cambridge, MA (January 1997)
12. Sobalvarro, P. G., Weihl, W. E.: Demand-based Coscheduling of Parallel Jobs on Multiprogrammed Multiprocessors. Proc. of IPPS '95 Workshop on Job Scheduling Strategies for Parallel Processing (1995) 63-75
13. Sobalvarro, P. G., Pakin, S., Weihl, W. E., Chien, A. A.: Dynamic Coscheduling on Workstation Clusters. SRC Technical Note 1997-017, Digital, Palo Alto, CA (1997)
14. Nagar, S., Banerjee, A., Sivasubramaniam, A., Das, C. R.: A closer look at coscheduling approaches for a network of workstations. Proc. of 11th ACM symposium on Parallel Algorithms and Architectures (1999) 96-105
15. Aversa, R., Mazzeo, A., Mazzocca, N., Villano, U.: Analytical Modeling of Parallel Applications in Heterogeneous Computing Environments: a Study of Cholesky Factorization. In: Malyshkin, V. (ed.): Parallel Computing Technologies, LNCS, Vol. 1662, Springer-Verlag (1999) 1-12

A Model to Integrate Message Passing and Shared Memory Programming

J.A. González, C. León, C. Rodríguez, and F. Sande

Centro Superior de Informática
Departamento de E.I.O y Computación
Universidad de La Laguna, Tenerife, Spain
`casiano@ull.es`

Abstract. During the last decade, and with the aim of improving performance through the exploitation of parallelism, researchers have introduced, more than once, *forall* loops with different tastes, syntaxes, semantics and implementations. The High Performance Fortran (HPF) and OpenMP versions are, likely, among the most popular. This paper presents yet another *forall* loop construct. The One Thread Multiple Processor Model presented here aims for both homogeneous shared and distributed memory computers. It does not only integrates and extends sequential programming but also includes and expands the message passing programming model. The compilation schemes allow and exploit any nested levels of parallelism, taking advantage of situations where there are several small nested loops. Furthermore, the model has an associated complexity model that allows the prediction of the performance of a program.

1   Introduction

Since the early stages of parallel computing, one of the most common solutions to introduce parallelism has been to extend a sequential language with some sort of parallel version of the *for* construct, commonly denoted as *forall* construct. Although similar in syntax, these *forall* loops differ in their semantics and implementations. The High Performance Fortran (HPF) and OpenMP versions are, likely, among the most popular [8], [10]. This paper presents yet another *forall* loop extension for the C language. These extensions are aimed both for homogeneous distributed memory and shared memory architectures.

There are several differences with most current versions of HPF and OpenMP. One is that the parallel programming model introduced has associated a complexity model that allows the analysis and prediction of the performance. The other is that it allows and exploits any nested levels of parallelism, taking advantage of situations where there are several small nested loops: although each loop does not produce enough work to parallelize, their union suffices. Perhaps the most paradigmatic example of algorithms using several levels of nested parallelism are recursive divide and conquer algorithms. Although the last language specification of both HPF and OpenMP allow some form of nested parallelism, most current implementations lack these features.

Y. Cotronis and J. Dongarra (Eds.): Euro PVM/MPI 2001, LNCS 2131, pp. 114–125, 2001.
© Springer-Verlag Berlin Heidelberg 2001

There is a third interesting feature of the model: it does not only integrates and extends the sequential programming model but also includes and expands the message passing programming model.

The next section introduces a reduced and idealistic version of the model, considering the simplified case where the number of available processor is larger than the number of processors required by the algorithm. Additionally, the section introduces the associated complexity model. Section three deals with the complementary scenario. Load balancing issues, mapping and scheduling policies are addressed here. A set of examples, covering parallel versions of the Quicksort, FFT, matrix products and Convex Hull are presented in section five. These are accompanied of their corresponding computational results. Finally, the last section presents the conclusions. Although the gains for distributed memory machines are far from what can be obtained using raw MPI, there are features that makes worth to consider the feasibility of this model. Among those, let us remark its equivalence with the OpenMP approach for shared memory machines and the fact that it does not only extend the sequential but also the Message Passing Programming Model

2   Syntax and Semantic

As OpenMP, the programming model being introduced, extends the classic sequential imperative paradigm with two new constructs: parallel sections and parallel loops. In contrast with OpenMP, the model aims for both distributed and shared memory architectures. The implementation on the last can be considered the "easy part" of the task.

> **forall**(i= e_1; i<= e_2)
>   **result** $(r_{1i}\, s_{1i})$, $(r_{2i}\, s_{2i})$ ... $(r_{mi}\, s_{mi})$
> $compound\_statement_i$

Fig. 1. *forall* simplified syntax

A simplified version of the current syntax of parallel loops appears in figure 1. The programmer states that the different iterations i of the loop can be performed independently in parallel. The results of the execution of the i-th iteration are stored in the variables pointed by r_{1i}, r_{2i}, etc. Their sizes are respectively s_{1i}, s_{2i}, etc. To establish the semantic, let us imagine a machine composed of a number of infinite processors, each one with its own private memory and a network interface connecting them. We will call it the One-Thread Multiple-Processor Machine, abbreviated *OTMP*. The processors are organized in groups. At any time, the memory state of all the processors in the same group is identical. An *OTMP* computation assumes that they also have the same input data and the same program in memory. The initial group is composed of all the processors in the machine. Each processor is a *RAM* machine [1], the only difference among them is an internal register, containing an integer, the *NAME* (or *NUMBER*) of the processor. This register is not available to the programmer. When reaching the former *forall* loop, each processor decides in terms of its *NAME* the corresponding initialization of i and its subsequent value of the register *NAME*. Each independent thread *compound_statement*$_i$ is executed by a subgroup. The simple equations performed by any processor *NAME* ruling the division process are:

$$\text{Number of iterations to do: } M = e_2 - e_1 + 1 \tag{1}$$

$$\text{Iteration to do: } i = e_1 + NAME \% M \tag{2}$$

$$\text{for } (j = 1; j < M; j++) \quad neighbour[j] = \Phi + (NAME + j) \% M \tag{3}$$

$$\text{where } \Phi \text{ is given by: } \Phi = M \times (NAME \, / \, M)$$

$$\text{New value of register: } NAME = NAME \, / \, M \tag{4}$$

These equations decide the mapping of processors to tasks, and how the exchange of results will take place. After the *forall* is finished, the processors recover their former value of *NAME*.

Each time a *forall* loop is executed, the memory of the group, up to that point contains exactly the same values. At such point the memory is divided in two parts: the one that is going to be modified and the one that is not changed inside the loop. Variables in the last set are available inside the loop for reading. The others are partitioned among the new groups. The clause *result* next to the *forall* has the purpose to inform the compiler of the new "ownership" of the part of the memory that is going to be modified. It announces that the group performing thread i "owns" (and presumably is going to modify) the memory areas delimited by $[r_{ij}, s_{ij}]$. r_{ij} is a pointer to the memory area containing the j-th result of the i-th thread. The size of this result is s_{ij}.

To guarantee that after returning to the previous group, the processors in the father group have a consistent view of the memory and that they will behave as the same thread, it is necessary to provide the exchange among neighbors of the variables that were modified inside the *forall* loop. Let us denote the execution of the body of the i-th thread (*compound_statement*$_i$) by T_i. The semantic imposes two restrictions:

1. Given two different independent threads T_i and T_k and two different result items r_{ij} and r_{kt}, it holds:

$$[r_{ij}, s_{ij}] \bigcap [r_{kt}, s_{kt}] = \emptyset \; \forall i, j, k, t$$

2. For any thread T_i and any result j, all the memory space defined by $[r_{ij}, s_{ij}]$ has to be allocated previously to the execution of the thread body. This makes impossible the use of non-contiguous dynamic memory structures like lists or trees in the threads results.

The programmer has to be specially conscious of the first restriction: it is mandatory that the address of any memory cell written during the execution of T_i has to belong to one of the intervals in the list of results for the thread.

```
 1    forall(i=1; i<=3)  result (ri+i, si[i]); {
 2      ...
 3      forall(j=0; j<=i)  result (rj+j, sj[j]); {
 4        int a, b;
 5        ...
 6        if (i % 2 == 1)
 7          if (j % 2 == 0) send(j+1, a, sizeof(int));
 8          else receive(j-1, b, sizeof(int));
 9        ...
10      }
11      ...
12  }
```

Fig. 2. Two nested foralls

As an example, consider the code in figure 2. Initially, the infinite processors are in the same group, represented by the root of the tree in figure 3. All the processors are executing the same thread and have identical values stored in their local memory. Applying equations 1, 2, 3 and 4, the parallel loop in line 1 of figure 2 divides the group in three. After the execution of the loop, and to keep the coherence of the memory, each processor exchanges with its two neighbors the corresponding results. Thus, processor 0 in the group executes iteration i=1 and sends si[1] bytes starting at the address pointed by ri+1 to processors 1 and 2. Furthermore, it receives from processor 1 si[2] bytes and stores those bytes in its local address starting at ri+2. Analogously, receives from processor 2 si[3] bytes starting at the address pointed by ri+3. The same exchange is repeated among the other corresponding triplets $((3,4,5),(6,7,8),\ldots)$. You can easily visualize the operation if you realize that any new nested *forall* creates/structures the current subgroup according as a *M-ary* hypercube, where M is the number of iterations in the parallel loop and the neighborhood relation is given by formula 3. Thus, this first *forall* produces "the face" of a ternary hypercubic dimension, where every corner has two neighbors.

The second nested *forall* at line 3 requests for different number of threads in the different groups. The first group (i = 1) executes a loop of size 2, and so the group is divided following a binary hypercube. The loop for the second group (i = 2) is of size 3, and the processors in this subgroup are accordingly divided in a ternary dimension. The last group (i = 3) executes a *forall* of size 4, and consequently the group is partitioned in 4 subgroups. In this 4-ary dimension, each processor is connected with 3 neighbors in the other subgroups. Therefore, at the end of the nested compound statement, processor 17 will send data pointed by rj+1 to processors 14, 20 and 23 and will receive from them data pointed by rj+0, rj+2 and rj+3. The same will occur with any of the quartets involved.

The complexity $\mathcal{T}(\mathcal{P})$ of any *OTMP* program $\mathcal{P}$ can be computed in what refer to sequential ordinary constructs (*while, for, if, ...*) as in the RAM machine [1]. The cost of the *forall* loop is given by the recursive formula:

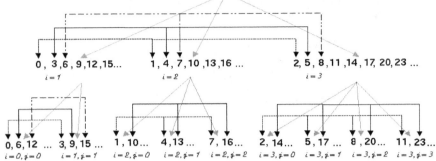

Fig. 3. The mapping associated with the two nested *forall* in figure 2

$$\mathcal{T}(forall) = A_0 + A_1 \times M + \max_{i=e_1}^{e_2}(\mathcal{T}(T_i)) + \sum_{i=e_1}^{e_2}(g \times N_i + L) \qquad (5)$$

where T_i is the code of *compound_statement$_i$*, M is the size of the loop and N_i is the size of the message transferred in iteration i,

$$M = (e_2 - e_1 + 1) \quad N_i = \sum_{k=1}^{m} s_{ki}$$

A_0 is the constant time invested computing formulas 1, 2 and 4. A_1 is the time spent finding the neighbors (formula 3). Constant g is the inverse of the bandwidth, and L is the startup latency.

Assuming the rather common case in which the volume of communication of each iteration is roughly the same

$$N \approx N_i \approx N_j \ \forall i \neq j = e_1, e_2$$

From formula 5 and from the fact that in current machines the scheduling and mapping time $A_0 + A_1 \times M$ is dominated by the communication time $\sum_{i=e_1}^{e_2}(g \times N_i + L)$, it follows the result that establishes when an independent *for* loop is worth to convert in a *forall*:

Theorem 1.

$$\mathcal{T}(forall) \leq \mathcal{T}(for) \Leftrightarrow \left(M \times (g \times N + L) + \max_{i=e_1}^{e_2} \mathcal{T}(T_i) \ll \sum_{i=e_1}^{e_2} \mathcal{T}(T_i) \right) \qquad (6)$$

A remarkable fact is that the *OTMP* machine not only generalizes the sequential RAM but also the Message Passing Programming Model. Lines 6-8 in figure 2 illustrate the idea. Each new *forall* "creates" a communicator (in the MPI sense). The execution of lines 7 and 8 in the fourth group (i=3) implies that thread j=0 sends a to thread j=1. This operation carries that, at the same time that processor 2 sends its replica of "a" to processor 5, processor 14 sends its copy to processor 17 and so on. To summarize: any send or receive is executed by the infinite couples involved. Still, the two aforementioned constraints have to be true. Any variable modified inside the loop and non local to the loop has to be allocated before the loop and has to appear inside the return clause.

The *OTMP* model admits several coherent extensions, including reduction clauses (with syntax and semantic similar to the ones in HPF and OpenMP) and an equivalent to the OpenMP *section* construct. We will not treat them here, for sake of briefness, and since the emphasis of this paragraph is in the (simplified) model rather than in the implementation details.

3   Mapping and Scheduling

Unfortunately, an infinite machine, such as that described in the previous section, is only an idealization. Real machines have a restricted number of processors. Each processor has an internal register *NUMPROCESSORS* where stores the number of available processors in its current group.

Three different situations have to be considered in the execution of a *forall* construct with $M = e_2 - e_1 + 1$ threads:

- *NUMPROCESSORS* is equal to 1;
- M is larger or equal than *NUMPROCESSORS*;
- *NUMPROCESSORS* is larger than M;

The first case is trivial. There is only one processor that executes all the threads sequentially, and there is not opportunity to exploit any intrinsic parallelism. In this situation, the generated code could be improved by eliminating any attempt to exploit parallelism.

The second case has been extensively studied as flat parallelism. The main problem that arises is the load balancing problem. To deal with it, several scheduling policies have been proposed [10]. Many assignment policies are also possible: *block*, *cyclic-block*, *guided* or *dynamic*.

The third case was studied in the previous section. But the fact that the number of available processors is larger than the number of threads introduces several additional problems: the first is load balancing. The second is that, not anymore, the groups are divided in subgroups of the same size.

If a measure of the work w_i per thread T_i is available, the processors distribution policy viewed in the previous section can be modified to guarantee an optimal mapping [3]. The syntax of the *forall* is revisited to include this feature (figure 4).

If there are not weights specification, the same work load is assumed for every task. The semantic is similar to that proposed in [2]. Therefore, the mapping is computed according to a policy similar to that sketched in [3]. There is, however, the additional problem of establishing the neighborhood relation. This time the simple $one - to - (M-1)$ hypercubic relation of the former section does not hold.

$$\textbf{forall}(i = e_1; i <= e_2; w_i)$$
$$\textbf{result}\ (r_{1i}\, s_{1i}),\ (r_{2i}\, s_{2i}),\ \ldots$$
$$compound\_statement_i$$

Fig. 4. *forall* with weights

Instead, the hypercubic shape is distorted to a polytope holding the property that each processor in every group (corner of the polytope) has one and only one incoming neighbor in any of the other groups.

The current software system implementing the $OTMP$ model consists of a C compiler and a run time library, built on top of MPI [9]. Input operations are broadcasted to all the processors in order to keep the coherence of memory.

4   Examples

Four examples have been chosen to illustrate the use of the $OTMP$ model: Matrix Multiplication, Fast Fourier Transform, Quicksort and the Convex Hull. Results are presented for six different platforms: Cray T3E, IBM SP2, Hitachi SR2201, Sun Enterprise HPC 6500, SGI Origin 2000, Compaq Alpha Server and a Beowulf Linux cluster.

4.1   Matrix Multiplication

The problem to solve is to compute *tasks* matrix multiplications ($C^i = A^i \times B^i$ $i = 0, \ldots tasks - 1$). Matrix A^i and B^i have respectively dimensions $m \times q_i$ and $q_i \times m$. Therefore, the product $A^i \times B^i$ takes a number of operations $w[i]$ proportional to $m^2 \times q_i$. Figure 5 shows the algorithm. Variables A, B and C are arrays of pointers to the matrices. The loop in line 1 deals with the different matrices, the loops in lines 5 and 7 traverse the rows and columns and finally, the innermost loop in line 8 produces the dot product of the current row and column. Although all the *for* loops are candidates to be converted to *forall* loops, we will focus on two cases: the parallelization of only the loop in line 5 (labelled FLAT in the results in figure 6) and the one shown in figure 5 where additionally, the loop at line 1 is also converted to a *forall* (label NESTED).

This example illustrates one of the common situation where you can take advantage of nested parallelism: when neither the *inner* loop (lines 5-10) nor the external loop (line 1) have enough work to have a satisfactory speedup, but the combination of both does. We will denote by $SP_R(\mathcal{A})$ the speedup of an algorithm $\mathcal{A}$ with R processors and by $\mathcal{T}_P(\mathcal{A})$ the time spent executing algorithm $\mathcal{A}$ on P processors. Let us also simplify to the case when all the inner tasks take the same time, i.e. $q_i = m$. Under this assumptions, the previous statement can be rewritten more precisely:

Lemma 1. *Let be tasks* $< P$, $SP_P(inner) < SP_{P/tasks}(inner)$ *and tasks* $\times$ $(g \times m^2 + L) \leq \mathcal{T}_{P/tasks}(inner)$ *then* $SP_P(FLAT) < SP_P(NESTED)$

proof Let us first remark that the hypothesis $tasks < P$ says that there is not enough work for a satisfactory exploitation of the outer loop. Analogously, the condition $SP_P(inner) < SP_{P/tasks}(inner)$ implies that P processors are too many for the work in $inner$. From the definitions and the code in figure 5,

$$\mathcal{T}_P(FLAT) = tasks \times \mathcal{T}_P(inner)$$
$$\mathcal{T}_1 = tasks \times \mathcal{T}_1(inner)$$
$$SP_P(FLAT) = \frac{tasks \times \mathcal{T}_1(inner)}{(tasks - 1) \times \mathcal{T}_P(inner)} = SP_P(inner)$$

From formula 5, neglecting the time spent in scheduling and mapping, we have:

$$SP_P(NESTED) = \frac{tasks \times \mathcal{T}_1(inner)}{\mathcal{T}_{P/tasks}(inner) + tasks \times (g \times m^2 + L)}$$
$$SP_P(NESTED) \geq \frac{tasks \times \mathcal{T}_1(inner)}{2 \times \mathcal{T}_{P/tasks}(inner)} = \frac{tasks}{2} \times SP_{P/tasks}(inner)$$

Since we have $tasks > 2$ and from the hypothesis $SP_P(inner) < SP_{P/tasks}(inner)$,

$$SP_P(NESTED) > SP_P(FLAT)$$

```
1   forall(i = 0; i < tasks; w[i])
2   result(C[i], m * m) {
3     q = ... ;
4     Ci = C+i; Ai = A+i; Bi = B+i;
5     forall(h = 0; h < m)
6     result(Ci[h], m) {
7       for(j = 0; j < m; j++)
8         for(r = &Ci[h][j], *r = 0.0, k = 0; k < q; k++)
9           *r += Ai[h][k] * Bi[k][j]
10  }
11 }
```

Fig. 5. Exploiting 2 levels of parallelism

When the size of the problems to solve is large, the speedup reached is linear as it shows the right side of figure 6. In the next examples we will concentrate in the more interesting case where the nested loops are small and, according to the complexity analysis, there are few opportunities for parallelization.

4.2 Fast Fourier Transform

Linear transforms, especially Fourier and Laplace transforms, are widely used in solving problems in science and engineering. The Discrete Fourier Transform (DFT) is defined as

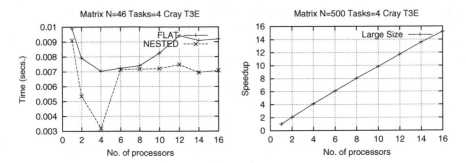

Fig. 6. On the left: Nested versus Flat parallelism. On the right: Speedup reached for large size problems

$$A(j) = \frac{1}{N} \sum_{k=0}^{N-1} a(k)e^{-2\pi ikj/N} \tag{7}$$

The *OTMP* code in figure 7 implements the Fast Fourier Transform (FFT) algorithm developed by Tukey and Cooley [5]. Parameter $W_k = e^{-2\pi ik/N}$ contains the powers of the N-th root of the unity. The computational results for such implementation on a Cray T3E appear in the left hand side of figure 8. The speedup curve behaves according to what a complexity analysis following formula 5 predicts: a logarithmic increase in the speedup.

4.3 QuickSort

The well-known quicksort algorithm [7] is a divide-and-conquer sorting method. As such, it is amenable to a nested parallel implementation. This and the next example are specially interesting, since the size of the generated sub-vectors are used as weights for the *forall*. Remember that, depending on the goodness of the pivot chosen, the new subproblems may have rather different weights.

The right hand side of figure 8 presents the speed-ups on a digital Alpha Server, an Origin 2000, an IBM SP2, a CRAY T3E and a CRAY T3D. The size of the problem was 1M integers.

4.4 QuickHull

For a subset S of $\mathbf{R}^n$, the convex hull $ch(S)$ is defined as the smallest convex set in $\mathbf{R}^n$ containing S. The QuickHull algorithm [6] is a fast way to compute the convex hull of a set of points on the plane. The algorithm shares a few similarities with its namesake, QuickSort: QuickHull is also recursive and each recursive step partitions data into several groups.

Figure 9 presents the speedups for different platforms: A digital Alpha Server, a Hitachi SR2201, an Origin 2000, an IBM SP2, a CRAY T3E and a CRAY T3D. The size of the problem was 2M points.

```
1  void llcFFT(Complex *A, Complex *a, Complex *W, unsigned N,
                unsigned stride, Complex *D) {
2    Complex *B, *C, Aux, *pW;
3    unsigned i, n;
4
5    if(N == 1) {
6      A[0].re = a[0].re;
7      A[0].im = a[0].im;
8    }
9    else {
10     n = (N >> 1);
11      forall(i = 0; i <= 1)
12      result(D+i*n , n) {
13        llcFFT(D+i*n, a+i*stride, W, n, stride<<1, A+i*n);
14      }
15     B = D;                      /* Combination phase */
16     C = D + n;
17     for(i = 0, pW = W; i < n; i++, pW += stride) {
18       Aux.re = pW->re * C[i].re - pW->im * C[i].im;
19       Aux.im = pW->re * C[i].im + pW->im * C[i].re;
20       A[i].re = B[i].re + Aux.re;
21       A[i].im = B[i].im + Aux.im;
22       A[i+n].re = B[i].re - Aux.re;
23       A[i+n].im = B[i].im - Aux.im;
22     }
25   }
26 }
```

Fig. 7. The *OTMP* implementation of the FFT

5 Conclusions

The current compiler is a prototype, lacking many available optimizations. Even incorporating these improvements, the gains for distributed memory machines will never be equivalent to what can be obtained using raw MPI. However, the combination of the following properties,

- the easiness of programming,
- the improving obtained in performance,
- the guarantee of portability to any platform,
- the fact that it does not introduce any overhead for the sequential case,
- the existence of an accurate performance prediction model,
- its equivalence with the OpenMP approach for shared memory machines and therefore, its portability to these kind of architectures and
- the fact that it does not only extend the sequential but the MPI programming model

makes worth the research and development of tools oriented to this model.

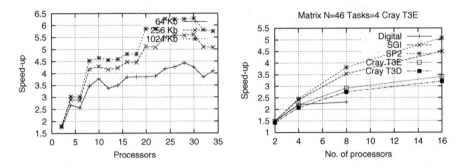

Fig. 8. Left: FFT. Cray T3E. Different sizes. Right: QuickSort. Size of the vector: 1M

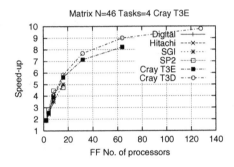

Fig. 9. QuickHull. Number of points: 2M

Acknowledgments

We wish to thank to the Edinburgh Parallel Computing Centre (EPCC), to the Centro de Investigaciones Energéticas, Medioambientales y Tecnológicas (CIEMAT), Centre Europeu de Parallelisme de Barcelona (CEPBA) and Centre de Supercomputació de Catalunya (CESCA). During this research, Dr. de Sande was on leave at th EPCC with a grant from the European Community (contract No HPRI-CT-1999-00026). This research also benefits from the support of Secretaría de Estado Universidades e Investigación, S EUI, project MaLLBa, TIC1999-0754-C03.

References

1. Aho, A. V. Hopcroft J. E. and Ullman J. D.: The Design and Analysis of Computer Algorithms, Addison-Wesley, Reading, Massachusetts, (1974).
2. Ayguade E., Martorell X., Labarta J., Gonzalez M. and Navarro N. Exploiting Multiple Levels of Parallelism in OpenMP: A Case Study Proc. of the 1999 International Conference on Parallel Processing, Aizu (Japan), September 1999.

3. Blikberg R., Sørevik T.. Nested parallelism: Allocation of processors to tasks and OpenMP implementation. Proceedings of The Second European Workshop on OpenMP (EWOMP 2000). Edinburgh, Scotland, UK. 2000
4. Brigham, E. Oren: The Fast Fourier Transform and Its Applications, Prentice-Hall, Inc. (1988)
5. Cooley, J. W. and Tukey, J. W.: An algorithm for the machine calculation of complex Fourier series, Mathematics of Computation, **19**, 90, (1965) 297–301.
6. Eddy, W.: A New Convex Hull Algorithm for Planar Sets, ACM Transactions on Mathematical Software 3(4), (1977) 398–403.
7. Hoare, C. A. R.: Quicksort, Computer Journal, 5(1), (1962) 10–15.
8. High Performance Fortran Forum: High Performance Fortran Language Specification. Version 2.0 http://dacnet.rice.edu/Depts/CRPC/HPFF/versions/hpf2/hpf-v20/index.html (1997)
9. MPI Forum: MPI-2: Extensions to the Message-Passing Interface, http://www.mpi-forum.org/docs/mpi-20.ps.Z (1997).
10. OpenMP Architecture Review Board: OpenMP Specifications: FORTRAN 2.0. http://www.openmp.org/specs/mp-documents/fspec20.ps (2000).

An Architecture
for a Multi-threaded Harness Kernel*

Wael R. Elwasif, David E. Bernholdt, James A. Kohl, and G.A. Geist

Computer Science and Mathematics Division, Oak Ridge National Lab
Oak Ridge, TN 37831, USA
{elwasifwr,bernholdtde,gst,kohlja}@ornl.gov

Abstract. Harness is a reconfigurable, heterogeneous distributed meta-computing framework for the dynamic configuration of distributed virtual machines, through the use of parallel "plug-in" software components. A parallel plug-in is a software module that exists as a synchronized collection of traditional plug-ins distributed across a parallel set of resources. As a follow-on to PVM, the Harness kernel provides a base set of services that plug-ins can use to dynamically define the behavior of the encompassing virtual machine. In this paper, we describe the design and implementation details of an efficient, multi-threaded Harness core framework, written in C. We discuss the rationale and details of the base kernel components – for communication, message handling, distributed control, groups, data tables, and plug-in maintenance and function execution – and how they can be used in the construction of highly dynamic distributed virtual machines.

Keywords: Harness, Parallel Plug-ins, PVM, Virtual Machines, Multi-threaded

1 Introduction

Next-generation high-performance scientific simulations will likely depart from present day parallel programming paradigms. Rather than relying on monolithic platforms that export a large, static operating environment, applications will need the flexibility to dynamically adapt to changing functional and computational needs. The application itself will dictate the ongoing configuration of the parallel environment, and will customize the run-time subsystems to suit its varying levels of interaction, roaming connectivity, and distinct execution phases.

Harness [2, 4, 9] is a heterogeneous distributed computing environment, developed as a follow-on to PVM [8]. Harness is a dynamically configurable distributed operating environment for parallel computing. It is designed with a

* This material is based upon work supported by the Department of Energy under grant DE-FC0299ER25396, and the Mathematics, Information and Computational Sciences Office, Office of Advanced Scientific Computing Research, Office of Science, U. S. Department of Energy, under contract No. DE-AC05-00OR22725 with UT-Battelle, LLC.

Y. Cotronis and J. Dongarra (Eds.): Euro PVM/MPI 2001, LNCS 2131, pp. 126–134, 2001.

lightweight kernel as a substrate for "plugging in" necessary system modules based on runtime application directives. Plug-ins can be coordinated in parallel across many networked machines, and can be dynamically swapped out during runtime to customize the capabilities of the system. Plug-in components can control all aspects of a virtual parallel environment, including use of various network interfaces, resource or task management protocols, software libraries, and programming models (including PVM, MPI and others).

Harness is a cooperative effort among Oak Ridge National Laboratory, the University of Tennessee, and Emory University, and involves a variety of system prototypes and experiments that explore the nature and utility of parallel plug-ins and dynamically configured operating environments. This paper describes the architecture and implementation of an efficient, multi-threaded Harness kernel. Previous implementations of Harness have relied heavily on Java features and RMI-based communication approaches. Our Harness kernel is written in C to be more efficient, and accommodates fully distributed control modules. This kernel exports the core interface for interacting with the Harness system, and provides a platform for the development of applications and plug-ins for use with Harness. The core functions provided by the kernel can be utilized to control the loading and unloading of plug-ins, the execution of arbitrary functions exported by plug-ins, and manipulation of system state information via a simple database interface.

While the merit of a "pluggable" design has been recognized in web browsers and desktop productivity suites, it has yet to manifest itself in collaborative heterogeneous computing environments for distributed scientific computing. PVM supports some limited dynamic pluggability, with "hoster", "tasker", and "resource manager" tools that can be instantiated at runtime to take over handling of certain system functions [10]. Traditional distributed computing platforms usually provide a static model in which the set of services available to applications cannot be easily altered. Globus [6] provides an integrated set of tools and libraries for accessing computational Grid resources. Legion [11] provides an object-based interface for utilizing arbitrary collections of computers across the global internet. NetSolve [3] consists of a front-end library and back-end server for farming out scientific computations on resources available over the network. The Harness model is distinct from these approaches in that it centers on the concept of the dynamic virtual machine as pioneered by PVM. User-level plug-ins can be added to Harness without modification to the base system, and Harness supports interactive loading and unloading of pluggable features during execution. The Harness (and PVM) virtual machine model provides flexible encapsulation and organization of resources and application tasks, allowing groups of computers and tasks to be manipulated as single unified entities.

Harness offers a perfect testbed for high-performance component-based systems such as the Common Component Architecture (CCA) [1, 7]. Both CCA and Harness build applications and system functionality using pluggable modules. Harness incorporates the dynamic capability to load and unload these components, where the CCA does not require this flexibility. For Harness to be a

CCA-compliant framework, it need only load an appropriate "CCA Services" plug-in that implements the base CCA interfaces.

This paper's main contribution is the introduction of the design of a high performance, multithreaded, modular Harness kernel that allows for maximum flexibility and extensibility through add-on plugins. The kernel is designed to provide a minimal platform for the construction of distributed virtual machines, while accommodating the use of plugins with different communication and cpu utilization profiles.

2   The Architecture of the Harness Daemon

The Harness daemon is a multi-threaded application that provides a runtime environment under which threadsafe plug-in functions run. The Harness daemon described here is being implemented in C-language at ORNL. C was chosen for this prototype to provide a high performance implementation with a small application footprint for more efficient use of resources. This design choice requires special handling that is implicitly taken care of in object-based languages, such as plug-in instantiation and function interface description. A Java-based prototype of Harness [14] has been under development at Emory University for the past year. Our use of C minimizes the runtime resource requirements, while still guaranteeing a high degree of portability in the production system, though with some increased complexity. Our resulting system architecture and interface are compatible with all source language implementations of Harness, given appropriate syntactic variations. Work such as the SIDL (Scientific Interface Description Language) [15, 13], as a cooperative project to the CCA, will likely provide the necessary language-specific processing to handle such variations.

The Harness daemon, in and of itself, is not intended to be a high-performance messaging system, nor do we expect many applications to make direct calls to the Harness APIs. The basic Harness APIs are intended primarily for use by plug-ins, which in turn provide the bulk of the actual user-level functionality. The services at the core of Harness are intended to serve two basic purposes: to provide the means for Harness daemons to exchange command and control information, and to act as a substrate for the loading/unloading and invocation of plug-in functionality. The design is intentionally minimalist to remain flexible and avoid unnecessary constraints on what plug-ins or user applications can do.

Figure 1 depicts a Harness daemon with several plug-ins, connected to a user application and to several other Harness daemons. The core of the Harness daemon, represented as the heavy box, contains three functional units with distinct roles: the local HCore functionality, the Controller (HCtl), and the Message Handler (HMsg). The HCore API functions are not directly exposed to the Harness user (e.g. the plug-in writer). All calls are routed through the Controller (HCtl) unit, which validates commands and coordinates their execution with the overall Harness virtual machine. The Message Handler (HMsg) unit exposes a second, self-contained API for processing and routing all incoming and outgo-

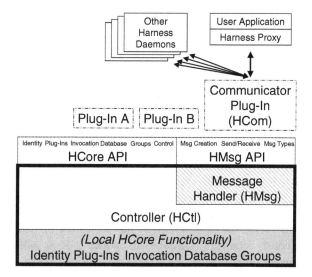

Fig. 1. A Schematic Representation of a Harness Daemon.

ing messages. The HCore and HMsg APIs are exported for use by plug-ins and applications, represented in Figure 1 as dashed boxes.

The interactions between the HCtl and HMsg units are simple and well-defined, so modular implementations of each can be exchanged without impacting the other. These are not plug-ins, however, because *some* implementation of each unit must always be present, and they cannot be dynamically replaced. Yet this modularity provides an important flexibility, particularly with respect to the Controller.

2.1   The Controller

The Controller mediates the execution of all user-level calls, satisfying security requirements and insuring that all local actions are appropriately coordinated with the rest of the Harness virtual machine. Various choices of control algorithms can be imagined in different circumstances. Strictly local control, ignoring the existence of other Harness daemons, can be useful in a context where Harness is integrating several plug-in functionalities on a single host (e.g. performance monitoring, debugging, resource management). A master/slave algorithm could be used to mimic the behavior and performance characteristics of PVM. Perhaps the most interesting is a distributed, peer-based controller using an algorithm such as [5] to provide fault tolerance and recovery capabilities. Such algorithms involve multiple daemons working together to arbitrate commands and store the state information for the virtual machine. This eliminates the single point of failure in a master/slave control algorithm.

Within each of these broad classes of control algorithms, there are many possible variations in the details of how particular commands behave within the collective environment. The Controller *can* provide its own interface to allow tuning or modification of its behavior, likely via configuration parameters passed through the environment or a configuration file. This mechanism will be implementation-specific and is not considered part of the main Harness APIs.

2.2   The Local HCore Functionality

The HCore API functions can be divided into six distinct encapsulated modules: *Identity, Plug-Ins, Invocation, Database, Groups* and *Control*.[1] Each Harness daemon and application task has an opaque Harness ID, or HID, which is locally generated and globally unique. This HID is returned by h_myhid() and is used in many other functions. Harness provides functions to load and unload plug-ins and allow them to register exported methods. Function pointers to these plug-in methods can be acquired locally for direct invocation, or remote invocations can constructed and executed in blocking or non-blocking fashion. All registered plug-in functions are expected to accept an H_arg (essentially C's argv) as input. Functions in libraries or other software modules that export non-H_arg arguments can easily be converted to Harness plug-in functions by enclosing them in simple argument-parsing function wrappers.

A simple database interface is available both to the Harness core and to plug-ins, and is the fundamental subsystem that the local Harness daemon and the virtual machine use to maintain their state. Database tables can be public or private. An h_notify function makes it possible for plug-ins to request notification on certain events which manifest themselves as changes to database tables.

Nearly all HCore functions take a "group" argument, specifying a subset of the current Harness virtual machine on which the functions are to be executed. Three groups are always available: *local, global* and *control*. The *local* and *global* groups indicate that the operation should take place on the local daemon or on *all* daemons participating in the virtual machine, respectively. The *control* group designates use of the selected decision-making daemon(s) in whatever control algorithm is running. Knowledge of this group is needed, for example, to allow "leaf" daemons (i.e. those daemons not directly participating in the decision-making control algorithm) to recover from faults or partitioning of the virtual machine. Users can also create other groups for their convenience through functions provided in the API.

Currently the only control function in the API (aside from any exported HCtl interface) is h_exit(), which terminates the local Harness daemon.

2.3   The Message Handler and Communication Modules

The core message processing of the Harness daemon is handled by two distinct but closely cooperating units: communication plug-ins and the message handler. Communication (HCom) units are plug-ins that are responsible for sending and

---

[1] The full API for these functions is published on the Harness web site [12].

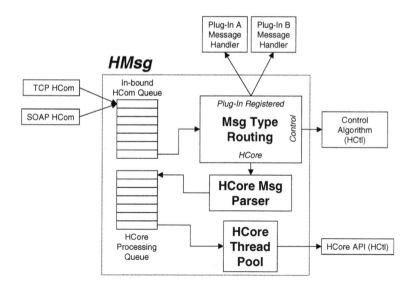

Fig. 2. Internal Message Handler Organization (HMsg).

receiving messages. HComs are not mandatory – it is possible to use Harness with strictly local functionality. However, most interesting uses of Harness involve inter-machine communication and therefore HComs. HComs treat message traffic as opaque, and are only responsible for putting messages out on the wire and receiving them from other nodes. Simultaneous use of multiple HCom plug-ins is possible. We expect that each HCom will handle all traffic of a given protocol, so that one might process TCP traffic, and another, for example, transmissions using the SOAP protocol.

The Message Handler (HMsg) unit is part of the daemon itself, and provides a uniform interface between the communication plug-ins and the daemon HCore processing. HMsg's interface to other kernel components is defined through communication queues and an HCore processing queue. As shown in Figure 2, HMsg queries an inbound queue for messages that are received through one or more HCom's. Three main message categories define the behavior of the kernel. Control messages deal with the coordinated maintenance of global state and are routed to relevant entry points in HCtl (e.g. notification of a new kernel joining the virtual machine). Messages that invoke a local HCore function (e.g. loading a plugin or calling a local plugin function) are parsed and put on the HCore processing queue for execution. Access to local HCore functionality is **only** done through an HCtl layer that guarantees consistent updates to the global state of the virtual machine. Any HCore command that passes through HCtl can trigger a global communication among the control daemons to coordinate the global state change. The third message type that is handled by HMsg is plugin-defined messages. These messages do not go through the parsing phase; instead they are delivered as opaque objects to message handler functions which are defined when the message type is registered with the kernel.

HMsg provides an API for the construction and parsing of HCore command messages, and assumes symmetric and compatible HMsg implementations on both ends of any HCom messaging connection. Plug-ins can also provide their own functions to create and parse messages (for example those with their own message types). HMsg contains a user-level function to actually send the resulting message buffers. HMsg's role here is a small but crucial one: it determines which of the several possible HComs should be used to transfer the message, and then places the message on the outbound queue for that HCom.

The HCom/HMsg system is designed to be a flexible, low-overhead mechanism for the processing of messages. However we expect that high performance communication plug-ins will be developed and utilized for actual production scale data transfer in application tasks. The Harness infrastructure will be applied to set up customized *out-of-band* data channels. The internal Harness communication paths should be used for command and control functions but need not be used for actual data transmission.

3 Multi-threaded Architecture

The Harness kernel is designed to be a multi-threaded application. Plug-in and HCore functions are executed in separate threads that are allocated from a worker pool managed by the kernel. In addition to these worker threads, plug-ins and other kernel components can also have multiple threads of execution. For example, the communication plug-in HCom can have one thread to handle system-level communication calls, and one or more threads to manage the interface to other components of the Harness kernel (e.g. the output queue fed by HMsg).

Using a multi-threaded Harness kernel protects the daemon from blocking unnecessarily, such as when a plug-in function waits for I/O. It is therefore not necessary to restrict plug-in functions to be non-blocking and return control immediately to the Harness kernel. The basic design philosophy of Harness is to afford plug-ins maximum operational flexibility. Plug-ins should be allowed to exploit blocking and/or non-blocking communication modalities, independent of the native kernel implementation. In addition, providing separate threads of execution within the Harness kernel enables "long-running" plug-in functions, such as those typically found in parallel applications which loop indefinitely in I/O-bound states. Separating these invocations leaves the Harness kernel available to process other requests. While much of this same functionality can be accomplished by instantiating each plug-in in its own process, this design incurs extra overhead for inter-plug-in communication and synchronization. Such issues are likely for cooperating parallel plug-in systems, and overheads could be significant for closely-coupled plug-ins.

As a result of the multi-threaded nature of the Harness kernel, it is necessary to require that all plug-in functions are threadsafe. To alleviate the potential complexity in converting non-threadsafe libraries into plug-ins, Harness provides a utility suite that includes basic threading synchronization operations and "mutexes" that assist plug-in functions in protecting critical sections and handling mutual exclusions. A trivial solution for a given library is to acquire a single mutex lock before executing *any* of its plug-in functions, however more complex (and efficient) solutions are possible using a set of distinct mutexes for dependent function groups. For plug-ins that cannot easily be made threadsafe through the use of simple mutex wrapper functions, a threadsafe *plug-in proxy* can coordinate plug-in function invocations with an external process that implements the given functions. The protocol by which such proxies would communicate is not dictated by Harness. As a side benefit, the Harness utility interface also serves as an abstraction layer that hides the implementation-specific details of the underlying thread implementation on a given platform (e.g. pthreads in Unix-based systems and Windows threads in Win32 systems).

4   Conclusion

We have described the multi-threaded architecture of the Harness kernel. Harness is being developed as a follow-on to PVM to explore research issues in reconfigurable heterogeneous distributed computing. Harness daemons are intended to provide basic command and control communication among themselves, and a substrate for dynamically plugging in user or system modules. These plug-ins are expected to be the main source of user-level functionality.

The Harness kernel presents two APIs to the user (plug-in developer). The HCore API provides the basic functionality of the daemon itself: managing plug-ins, invoking functions, and maintaining the database (which allows both the daemon and plug-ins to store state information). The HMsg API has a concise interface focused on the processing and sending of messages to other Harness tasks. The HMsg unit of the kernel, together with specialized HCom communication plug-ins, is responsible for all communication among Harness daemons, and provides a means for point-to-point communication among plug-ins as well. We have described the structure of the HMsg unit in some detail, as well as the multi-threaded nature of the Harness kernel.

References

1. Rob Armstrong, Dennis Gannon, Al Geist, Katarzyna Keahey, Scott Kohn, Lois McInnes, Steve Parker, , and Brent Smolinski. Toward a common component architecture for high-performance scientific computing. In *Proceedings of the The Eighth IEEE International Symposium on High Performance Distributed Computing*, 1998.
2. M. Beck, J. Dongarra, G. Fagg, A. Geist, P. Gray, J. Kohl, M. Migliardi, K. Moore, T. Moore, P. Papadopoulos, S. Scott, and V. Sunderam. HARNESS: a next generation distributed virtual machine. *Special Issue on Metacomputing, Future Generation Computer Systems*, 15(5/6), 1999.

3. Henri Casanova and Jack Dongarra. NetSolve: A network-enabled server for solving computational science problems. *The International Journal of Supercomputer Applications and High Performance Computing*, 11(3):212–223, Fall 1997.

4. J. Dongarra, G. Fagg, A. Geist, and J. A. Kohl. HARNESS: Heterogeneous adaptable reconfigurable NEtworked systems. In IEEE, editor, *Proceedings: the Seventh IEEE International Symposium on High Performance Distributed Computing, July 28–31, 1998, Chicago, Illinois*, pages 358–359. IEEE Computer Society Press, 1998.

5. Christian Engelmann. *Distributed Peer-to-Peer Control for Harness*. M.Sc. thesis, University of Reading, 2001.

6. I. Foster and C. Kesselman. Globus: A metacomputing infrastructure toolkit. *The International Journal of Supercomputer Applications and High Performance Computing*, 11(2):115–128, Summer 1997.

7. Dennis Gannon, Randall Bramley, Thomas Stuckey, Juan Villacis, Jayashree Balasubramanjian, Esra Akman, Fabian Breg, Shridhar Diwan, and Madhusudhan Govindaraju. Developing component architectures for distributed scientific problem solving. *IEEE Computational Science & Engineering*, 5(2):50–63, April/June 1998.

8. Al Geist, Adam Beguelin, Jack Dongarra, Weicheng Jiang, Robert Manchek, and Vaidy Sunderam. *PVM: Parallel Virtual Machine: A Users' Guide and Tutorial for Networked Parallel Computing*. Scientific and engineering computation. MIT Press, Cambridge, MA, USA, 1994.

9. G. A. Geist. Harness: The next generation beyond PVM. *Lecture Notes in Computer Science*, 1497:74–82, 1998.

10. G. A. Geist, J. A. Kohl, P. M. Papadopoulos, and S. L. Scott. Beyond PVM 3.4: What we've learned, what's next, and why. *Special Issue on Metacomputing, Future Generation Computer Systems*, 15(5/6):571–582, 1999.

11. Andrew S. Grimshaw, William A. Wulf, and the Legion team. The legion vision of a worldwide virtual computer. *Communications of the ACM*, 40(1):39–45, January 1997.

12. ORNL Harness Home Page. http://www.csm.ornl.gov/harness/

13. S. Kohn, G. Kumfert, J. Painter, and C. Ribbens. Divorcing language dependencies from a scientific software library. In *Proceedings, 10th SIAM Conference on Parallel Processing*, March 1999.

14. M. Migliardi and V. Sunderam. Plug-ins, layered services and behavioral objects application programming styles in the harness metacomputing system. *Future Generation Computer Systems*, 17:795–811, 2001.

15. B. Smolinski, S. Kohn, N. Elliott, N. Dykman, and G. Kumfert. Language interoperability for high-performance parallel scientific components. In *Proceedings, Int'l Sym. on Computing in Object-Oriented Parallel Environments (ISCOPE '99*, 1999.

Parallel IO Support for Meta-computing Applications: MPI_Connect IO Applied to PACX-MPI

Graham E. Fagg[1], Edgar Gabriel[2], Michael Resch[2], and Jack J. Dongarra[1]

[1] Department of Computer Science, University of Tennessee, Suite 413,
1122 Volunteer Blvd, Knoxville, TN-37996-3450, USA
fagg@cs.utk.edu
[2] Performance Computing Center Stuttgart,
Allmandring 30, D-70569 Stuttgart, Germany
gabriel@hlrs.de

Abstract. Parallel IO (PIO) support for larger scale computing is becoming more important as application developers better understand its importance in reducing overall execution time by avoiding IO overheads. This situation has been made more critical as processor speed and overall system size has increased at a far greater rate than sequential IO performance. Systems such as MPI_Connect and PACX-MPI allow multiple MPPs to be interconnected, complicating IO issues further. MPI_Connect implemented Parallel IO support for distributed applications in the MPI_Conn_IO package by transferring complete sections of files to remote machines, supporting the case that all the applications and the file storage were completely distributed. This system had a number of performance drawbacks compared to the more common usage of metacomputing where some files and applications have an affinity to a home site and thus less data transfer is required. Here we present the new PACX-MPI PIO system based initially on MPI_Connect IO, and attempt to demonstrate multiple methods of handling MPI PIO that cover a greater number of possible usage scenarios. Given are some preliminary performance results as well as a comparison to other PIO grid systems such as the Asian Pacific GridFarm, and Globus gridFTP, GASS and RIO.

1. Introduction

Although MPI [13] is currently the de-facto standard system used to build high performance applications for both clusters and dedicated MPP systems, it is not without some issues. Initially MPI was designed to allow for very high efficiency and thus performance on a number of early 1990s MPPs, that at the time had limited OS runtime support. This led to the current MPI-1 design of a static process model. While this model was possible to implement for MPP vendors, easy to program for, and more importantly something that could be agreed upon by a standards committee. During the late 90s a number of systems similar to MPI_Connect [1] and PACX-MPI [3] were developed that allowed multiple MPPs running MPI applications to be inter-

Y. Cotronis and J. Dongarra (Eds.): Euro PVM/MPI 2001, LNCS 2131, pp. 135–147, 2001.

connected while allowing for internal MPP communications to take advantage of vendor tuned software rather than using a TCP socket based portable implementation only.

While MPI_Connect and PACX-MPI solved some of the problems of executing application across multiple platforms they did not solve the problem of file distribution and collection or handling multiple specialized file subsystems. Initially this was not a problem as many parallel applications were developed from sequential versions and they continued to utilize sequential IO operations. In many cases only the root processor would perform IO. This state of affairs was compounded by the initial version of MPI not having any Parallel IO API calls, thus forcing users of Parallel IO to use non portable proprietary interfaces. The MPI-2 standard did however include some Parallel IO operations, which were quickly implemented by a number of vendors. The work presented here is built on the MPI-2 Parallel IO interface and the appication supported are expected to use the MPI-2 Parallel IO interface to maintain portability.

1.1   MPI-2 Parallel IO

Parallel I/O can be described as multiple access of multiple processes to the same, shared file. The goal of using parallel File I/O within MPI applications is to improve the performance of reading or writing data, since I/O is often a bottleneck, especially for applications, which are frequently check-pointing intermediate results. With the MPI-2 standard, an interface is designed which enables the portable writing of code, which makes use of parallel I/O.

MPI–I/O provides routines for manipulating files and for accessing data. The routines for file manipulation include opening and closing files, as well as deleting and resizing files. Most file manipulation routines have to provide as an argument an MPI communicator, thus these operations are collective.

For reading or writing data, the user has several options. MPI-I/O gives the application developer the possibility to work either with explicit offsets, with individual file pointers or with shared file pointers. The operations are implemented as blocking and non-blocking routines. Additionally, most routines are available in both collective and non-collective versions. Taking it all together, the user may choose between more than 28 routines, which fits best to his applications IO access pattern.

1.2   MPI_Connect and MPI_Conn_IO

MPI_Connect is a software package that allows heterogeneous parallel applications running on different Massively Parallel Processors (MPPs) to interoperate and share data, thus creating Meta-applications. The software package is designed to support applications running under vendor MPI implementations and allow interconnection between these implementations by using the same MPI send and receive calls as developers already use within their own applications. This precludes users from learning a new method for interoperating, as the syntax for sending data between applications is identical to what they already use. The support of vendor MPI implementations means that internal communications occur via the optimized vendor versions and thus incur no performance degradation as opposed to using only TCP/IP, the only other option previously available. Unlike systems such as PACX-MPI, each application maintains its own MPI_COMM_WORLD communicator, and this allows each application to connect and disconnect at will with other MPI applications as shown in figure 1.

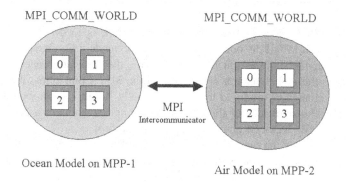

Fig. 1. MPI_Connect overview

The MPI_Conn_IO API was developed to support multi-site multi-application execution using shared files. The three primary requirements were to provide:

(A) Local parallel file system access rates

(B) Global single name space naming for all files

(C) Single points of storage

To provide (A) the system had to utilize the MPI-2 PIO API for file access. To provide (B) which is important for GRID applications where location independence is required to handle uncertainties caused by runtime scheduling, a dedicated naming/resolution service was created. (C) single point of storage is where the input and output files are stored in well known locations as known by the naming service so that users do not have to distribute their files at every location where an execution is possible. This required the use of dedicated file serving daemons. An overview of the MPI_Conn_IO system is given in figure 2.

To allow applications the ability to handle this distributed access to Parallel IO the users had to add two additional library calls, one to gain access to the global/shared file and the other to release it. The call to access the file was one of two variations:

(A)   MPI_Conn_getfile(globalname, localname, outsize, comm)

(B)   MPI_Conn_getfile_view(globalname, localname, my_app, num_apps, dtype, outsize, comm)

Both calls would access the naming/resolution service to locate the file server with the required file. Variant (A) would then copy the complete file onto all the systems in the MPI_Connect system. The individual applications themselves would then have to handle the portioning of the file so that they each accessed the correct sections of data. In variant (B) the call is a combination of opening a file and setting the access pattern in the terms of access derived data types and relative numbers of nodes in each part of the MetaApplication. This allows the file daemons to partition the files at the source so that file distribution and transmission costs are reduced. For example, if two applications of sizes 96 and 32 nodes opened the same file with the first setting *my_app*=0, *num_apps*=2 and *dtype* to MPI_DOUBLE, and the second opened the file with *my_app*=1, *num_apps*=2 and *dtype* = MPI_DOUBLE then the file would get stripped between the two sites. The first site would get the first 96 doubles with the second site getting then next 32 doubles and so on. The files on the local Parallel IO subsystems would be named by the user supplied *localfname* parameter.

Once the file has been globally opened it can then be opened via the standard MPI-2 Parallel IO call MPI_File_open (com, *localfname*, mode, info, fhandle).

After the application has finished with the file, the application needs to release its references to the global version of the file via the MPI_Conn_releasefile API call. If the file was created or modified locally its globally stored version would be updated by this call. The semantics of this operation attempt to maintain a coherency in the case of shared file segments across multiple applications.

To further reduce the time it takes to transfer a file, the TCP/IP routing of the files contents can be explicitly controlled by store and forwarding through a number of IBP [15] network caches at various locations throughout the network. This scheme is used to avoid known network bottlenecks at peak times.

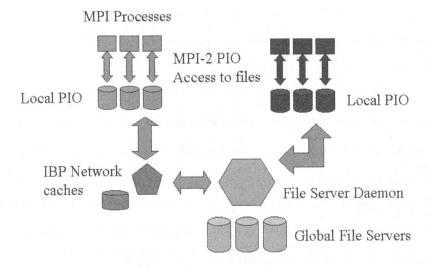

Fig. 2. MPI_Conn_IO overview

1.3   PACX-MPI

PACX-MPI is an implementation of the MPI standard optimized for clustered wide-area systems. The concept of PACX-MPI relies on three major ideas: First, in clustered systems, the library has to deal with two different levels of quality of communication. Therefore, PACX-MPI uses two completely independent layers to handle operations on both communication subsystems. Second, for communication between processes on the same machine, the library makes use of the vendor-MPI, since this allows it to fully exploit the capacity of the underlying communication subsystem in a portable manner. And third, for communication between processes on different machines, PACX-MPI introduces two communication daemons. These daemons allow buffering and packing of multiple wide area communications as well as handling security issues centrally.

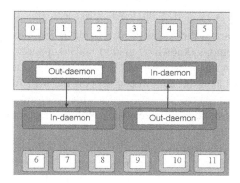

Fig. 3. Concept of PACX-MPI

2. Distributed and Grid Parallel IO issues

The requirements and implementation issues of Distributed and Grid based file systems overlap heavily. Although the topic is very complex and broad it can be simplified to a number of key issues.

Distributed File Systems (DFS) such as AFS [12] primarily focus on performance through mechanisms such as caching, and specialized file locking techniques at the OS layer and are typically aimed at supporting sequential (POSIX style) access patterns across a wide area network rather than an unreliable global one. They assume that restrictions such as single user IDs and a single coherent file name space.

Grid based tools are primarily aimed at simplified transmission of file data, where the actual data IO is done locally from some temporary storage location. Here the issues become those of single point authentication, efficient handling of different backend file services, mapping of multiple user Ids, and handling the naming and resolution of multiple replicated copies of data [6][11]. The replication issue is becoming more important as more systems utilize caching and pre-staging of data to reduce access latencies to remote storage. Systems such as gridFTP and Global Access to Secondary Storage (GASS) [8] exemplify this. GridFTP provides an FTP abstraction that hides multiple access, transmission and authentication methods, and GASS which provides a simple POSIX file IO style API for accessing remote data using URL naming. GASS then utilizes local caching to improve repeated IO access rates. One advantage of GASS over AFS is that the caching is user configurable via an exposed cache management API.

PACX-MPI PIO aims to support functionality in-between these two levels and is very close to systems such as RIO [9], Stampi [4] and the Grid Data Farm [10]. RIO was aimed at supporting MPI Parallel IO access patterns via a server client model. Unlike traditional server client models, the clients are parallel applications and the servers can directly access multiple parallel IO file servers, thus ensuring that performance of the parallel file system was passed onto the client side. The implementation contained forwarder nodes on the client side that translate local IO access requests and pass these via the network to the server side which then accesses the real file servers in parallel. The Grid Data Farm is a Petascale data-intensive computing project initiated in Japan as part of the Asian Pacific Grid. In common with RIO it has specialized file server daemons (known as gfsd's) but it supports a specialized API that uses a RPC mechanism much like that of NFS for actual file access by remote processes. This was needed as it did not directly support the MPI API. An interesting feature of the system is that the file server nodes can also perform computation as well as IO, in which case they attempt to process only on local data, rather than more expensive to access remote data.

MPI_Conn_IO is equivalent to a mix of GASS and RIO. It has additional calls to access remote files, which in effect are pre-staging calls that cache the complete requested data fully onto a local Parallel File System via MPI-IO operations. It has special file handling daemons like the RIO server processes, although it does not support their on demand access operation.

The target implementation of PACX-MPI is hoped to be a system that allows for a mode of operation anywhere between the GASS/MPI_Conn_IO fully cached method and the RIO on demand method. The authors believe that there are many different styles of access pattern to file data within parallel and meta/grid applications and that only by supporting a range of methods can reasonable performance be obtained for a wide range of applications [2].

3. PACX-MPI PIO Aims

The current aims of PACX-MPI-PIO are:
- (A) Support a single MPI-2 Parallel IO file view across any MPI communicator (assuming it is distributed across multiple MPPs)
- (B) Allow access to files via the MPI-2 API using the efficient vendor supplied Parallel IO libraries where possible
- (C) Do not force the application developer to use any additional API calls unless they are in the form of performance hints
- (D) Reduce any file management to a minimum
- (E) Support a number of caching and transmission operations

As PACX allows communicators (A) to be across multiple machines the IO layer has to move file data between sites automatically if they are to support reduced file management (D). A consequence of (D) is that one of the execution sites is considered the 'home' site where the data files will reside but we will not restrict all files to be at the same home site. Allowing PACX-MPI-PIO to work without forcing the use of any additional calls (unlike MPI_Conn_IO) requires either the MPI-2 file PIO calls to be profiled using the MPI profiling interface or the preprocessing of the user application so that the file movement is hidden from the users application. (B) forces the profiled library to call the MPI-2 PIO API directly. (C) and (E) mean that any application level requests to change the system transfer options and caching methods should be in the form of MPI attributes calls, or even compile time flags.

To simplify initial development a number of restrictions are placed on the user applications use of the MPI Parallel IO API. Firstly, the applications open files for reading, writing or appending, but not for mixed read and write operations. Secondly once a file has been opened, it can set its file view but not modify this file view without reopening the file. This restriction maybe removed in a future version if there is a need, but currently the authors have not found any application that need this functionality. Lastly, shared file pointers between multiple MPPs are not supported.

3.1   PACX-MPI PIO File Manipulation and Transmission Modes

Manipulation of files and transmission of file data can be broken into a number of issues.

(A)   Naming. As the target files exist at one of the execution sites, the need for any global naming scheme is negated and the local names can be used instead. The requirement is that users specify which site is the home site to the PACX system. This can be done via an entry in the PACX host file, or over ridden by an MPI attribute call. Any data stored in a cache at another site uses a one time temporary name.

(B)   Caching of data. Is the complete file replicated across all MPPs, or just sections of the file. Is the data stored on the local PIO system or purely in memory.

(C)   Who does the remote access? Are additional daemons required or can the PACX communication daemons do the work.

The initial test implementation of PACX-MPI PIO only supports two modes of remote file access with various options:

1.   Direct access
2.   Proxy access

3.2   PACX-MPI PIO Direct Access

Direct access is where the data on the remote (i.e. non home) site is stored on the PIO system as a data file on disk and the MPI-2 PIO application calls translate directly to MPI-2 PIO library calls. This is just the same as the original MPI_Conn_IO library, except there is not a specialized File server and thus less data needs to be transferred, and there are no API changes or global names needed.

In this mode of operation, when the applications make the collective MPI_File_set_view call, the local (home site) application reads the file data in via parallel MPI IO calls, and then sends the data via the communications daemons to the remote nodes, which then write the data via the MPI IO calls to their local temporary parallel files. The applications can then proceed as normal handling the data with the profiled MPI Parallel IO API calls. The use of MPI IO calls to store the data is important as the authors have found that on some systems such as the Cray T3E, the data is not automatically stripped if just a single processes writes it out and thus performance drops to that of a single disk for IO access.

The data storage and access at the remote sites can be in one of three forms:

1. Complete File Copy (CFC). In this mode all sites have an identical copy of the complete file. This is useful for small files where the cost of moving all or part of the file is negligible compared to the overall execution time. It is also useful for supporting multiple changes to the file view during a single application run. This mode of distribution is show in figure 4.

2. Partial File Copy (PFC) with full extent. This is where only the data needed on each of the distributed systems is copied to them. This reduced the amount of data to be copied. The file is stored with the same extent as the original copy, i.e. it is sparse storage. This mode is useful as it does not require any remote file view translation allowing for multiple file view support. The only disadvantage is that it requires the same amount of storage at each site as the original file which maybe considerable. An example of this is shown in figure 5.

3. Partial File Copy (PFC) with compacted storage. This is the same as above except that the storage is compacted so that no gaps in the remote files exist as shown in figure 6. This has three consequences. The first is the file uses much less space. The second is that it requires all the file view API calls on the remote machines to be translated so that they correctly address the new data layout. This translation is automatic and only performed once. The third consequence is multiple file views are not supportable, and that the temporary file would have to be completely rebuilt each time the file view was changed.

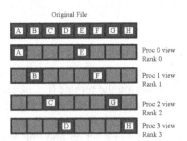

Fig. 4. File view for data stripping on MPPs with complete data files

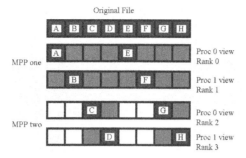

Fig. 5. File view for stripping of data across two MPPs with full file extent

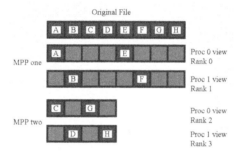

Fig. 6. File view for stripping data across two MPPs with compacted file storage (reduced extent)

3.3   PACX-MPI PIO Proxy Access

Proxy access is when the file data only exists on the home/local site and all other applications pass their IO access requests to the primary application, which then performs the accesses on their behalf. The overall structure is similar to RIO and the Data Grid Farm as shown in figure 7. Due to its nature it is known as On Demand Access (ODA) as IO calls are made dynamically as demanded by remote processes, or cached as the remote system reads and writes data from a memory buffer in the form of a message queue. The number and size of pending IO accesses can be varied dynamically. Another feature is that pre-fetching of read requests can be performed as the file view is known in advance, which leads to more efficient latency hiding.

A number of issues exist which effect overall performance. Firstly, the local application has to perform all the IO operations, which can reduce the level of parallelism within the Parallel IO subsystem. Secondly, the amount of intersystem communication dramatically increases, and even with pre-catching, intersystem bandwidth become a dominating factor.

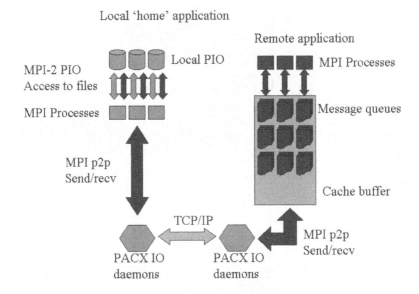

Local 'home' application

Fig. 7. Proxy, On Demand Access (ODA) cached access structure

4. PACX-MPI IO Experiments

Initial experiments were conducted on a Cray T3E-900 with 512 nodes at the University of Stuttgart. Unfortunately the version of MPI-2 PIO installed was based on a version ROMIO that utilized only the POSIX file IO calls and thus the raw PIO performance was poor [5].

We ran several different classes of experiment to characterize the IO subsystem:

(A)   Basic Parallel IO benchmark using only collective blocking calls (MPI_Read/Write_ALL).

(B)   Partition to partition TCP performance (to test interconnection speeds)

(C)   Single machine, single partition tests

(D)   Single machine, multiple partition tests (each partition runs a separately initiated MPI application under PACX-MPI) with a single coherent file view across the whole PACX-MPI job.

Out of the different caching, file transfer and storage methods described in section 3 we present only the partial file copy (PFC) with compacted storage (modified file view) and on-demand access (ODA, system cached) for the distributed multiple MPI application cases as these best represent the current issues in a final implementation. These are compared against the single application in terms of both aggregate IO bandwidth and application execution time.

The data files used were all over 150 Mbytes to avoid multi-level system caching effects, and from previous experience with real CFD applications [2] we varied the granularity of individual node IO accesses from small/fine, i.e. 100 doubles per node per access to very large/course, i.e. 100,000 doubles per node per IO access.

Figure 8 shows the aggregate IO bandwidths for accessing a large (150+ Mbyte) using the MPI-2 parallel IO API using a total of 32 nodes.

The first line is for reading on a single partition with 32 nodes and peaks at around 25 MB/Second. The second line shows the write bandwidth which is much lower with a peak of only 10.5 MB/Second. The third line shows the bandwidth of accessing the file with two separate 16 node partitions using the PFC method. This method includes all file transfers and its performance is dominated by the slow parallel file write IO operations. The forth line shows twice the interconnection bandwidth. This is show as it is the limit for file transfers. (Twice is shown as we are utilizing the compacted file view so only half the file is transferred and we are measuring the aggregate bandwidth being sum of all individual node bandwidths).

The forth line shows the on demand access (ODA/cached) method which performs much better than the PFC method at around 10.5 MB/Second for large accesses. As the IO performance goes up the effects of interconnection will dominate this method. The effects of latency for this method, can be hidden by increasing the pre-fetch buffer size. This technique becomes infeasible for course grain accesses on larger numbers of nodes due to memory constraints.

Figure 9 shows the single application single partition version compared to PFC and ODA/cached methods in terms of time rather than bandwidth for different granularities of access. Here the degradation in performance of using the PFC version is more apparent especially for smaller accesses, although the ODA version appears to perform remarkably well.

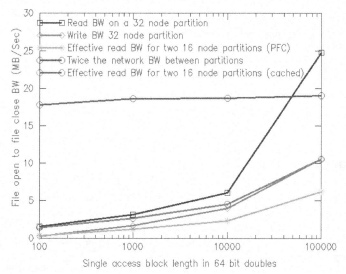

Fig. 8. Aggregate Bandwidth of different method as a function of access granularity

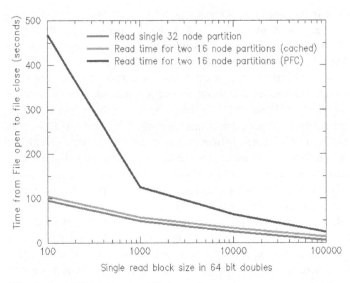

Fig. 9. Overall application IO times for various access methods

5. Conclusions and Future Work

The PACX-MPI PIO initial implementation appears to be a good vehicle to experiment with various caching, data access and storage mechanisms while supporting the portable use of the MPI-2 Parallel IO API. The PIO system is transparent in use and requires no code changes from that of a standalone application even when executing transparently on multiple MPPs. The system also utilizes the concept of a home site for the storage of files, which reduces the work load of the application users as they no-longer have to handle file staging manually when performing multi-site application runs.

Currently, future work is aimed at automatically improving performance, especially for the class of applications that perform regular user directed check-pointing.

References

1. G. E. Fagg, K. S. London, and J. J. Dongarra, "MPI_Connect: managing heterogeneous MPI applications interoperation and process control", in V. Alexandrov and J. Dongarra, editors, *Recent advances in Parallel Virtual Machine and Message Passing Interface,* volume 1497 of *Lecture notes of Computer Science,* pages 93-96. Springer, 1998. 5th European PVM/MPI User's Group Meeting.
2. D. Cronk, G. Fagg, and S. Moore, " Parallel I/O for EQM Applications", Department of Computer Science technical report, University of Tennessee, Knoxville, July 2000.

3. E. Gabriel, M. Resch, T. Beisel and R. Keller "Distributed Computing in a heterogeneous computing environment", in V. Alexandrov and J. Dongarra, editors, *Recent advances in Parallel Virtual Machine and Message Passing Interface,* volume 1497 of *Lecture notes of Computer Science,* pages 180-188. Springer, 1998. 5th European PVM/MPI User's Group Meeting.
4. T. Imamura, Y. Tsujita, H. Koide, and H. Takemiya, "An architecture of Stampi: MPI library on a cluster of parallel computers", in J. Dongarra, P. Kacsuk, and N. Podhorszki, editors, *Recent advances in Parallel Virtual Machine and Message Passing Interface,* volume 1908 of *Lecture notes of Computer Science,* pages 200-207. Springer, 2000. 7th European PVM/MPI User's Group Meeting.
5. Rolf Rabenseifner and Alice E. Koniges, "Effective File-I/O Bandwidth Benchmark", in A. Bode, T. Ludwig, R. Wissmüller, editors, Proceedings of Euro-Par 200, pages 1273-1283, Springer 2000.
6. B. Allock et all, "Secure, Efficient Data Transport and Replica Management for High-Performance Data-Intensive Computing", submitted to IEEE Mass Storage Conference, April 2001.
7. R. Thakur, W. Gropp, and E. Lusk, "On Implementing MPI-IO Portably and with High Performance", in *Proc. Of the Sixth Workshop on I/O in Parallel and Distributed Systems,* May 1999, pages 23-32.
8. J. Bester, I. Foster, C. Kesselmann, J. Tedesco, S. Tuecke, "GASS: A Data Movement and Access Service for Wide Area Computing Systems", in *Sixth Workshop on I/O in Parallel and Distributed Systems, 1999.*
9. I. Foster, D. Kohr, R. Krishnaiyer, J. Mogill, "Remote I/O: Fast Access to Distant Storage", in *Proc. Workshop on I/O in Parallel and Distributed Systems, (IOPADS),* pages 14-25, 1997.
10. O. Tatebe et all., "Grid Data Fram for Petascale Data Instensive Computing", Elechtrotechnical Laboratories, Technical Report, TR-2001-4, http://datafarm.apgrid.org.
11. B. Tierny, W. Johnston, J. Lee, M. Thompson, "A Data Intensive Distributed Computing Architecture for Grid Applications", Future Generation Computer Systems, volume 16 no 5, pages 473-481, Elsevier Science, 2000.
12. J. Morris, M. Satyanarayanan, M. Conner, J. Howard, D. Rosenthal and F. Smith. "Andrew: A distributed personal computing environment", *Communications of the ACM,* 29(3):184-201, 1986.
13. William Gropp, et. Al., MPI – The Complete Reference, Volume 2, The MPI Extensions, The MIT Press, Cambridge, MA 1999
14. William Gropp, Ewing Lusk, and Rajeev Thakur, Using MPI-2, Advanced Features of the Message-Passing Interface, The MIT Press, Cambridge, MA 1999
15. James S. Plank, Micah Beck, Wael R. Elwasif, Terry Moore, Martin Swany, Rich Wolski "The Internet Backplane Protocol: Storage in the Network", NetStore99: The Network Storage Symposium, (Seattle, WA, 1999)

TOPPER: A Tool for Optimizing the Performance of Parallel Applications

Dimitris Konstantinou, Nectarios Koziris, and George Papakonstantinou

National Technical University of Athens
Dept. of Electrical and Computer Engineering
Computer Science Division
Computing Systems Laboratory
Zografou Campus, Zografou 15773, Greece
{dkonst,nkoziris,papakon}@cslab.ece.nua.gr

Abstract. In this paper we present an autonomous and complete tool for optimizing the performance parallel programs on multiprocessor architectures. The concern of TOPPER's users is bound to the construction of two separate graphs, describing the overall application's task partitioning and interprocess communication requirements, as well as the architecture of the available multiprocessor system. TOPPER proceeds with the elaboration of these two graphs and proposes an efficient task mapping, aiming to minimize the application's overall execution time. When the communication between the various tasks is carried out with the use of MPI routines, the tool not only proposes an optimal task allocation but also can execute automatically the parallel application on the target multiprocessing machine.

1   Introduction

In the last years the evolution in the fields of VLSI technology and computer networking expanded the utilization of parallel computing systems. Many computationally intensive applications demand parallel computers, since they provide the users with modularity, scalability and low cost decentralized processing power. Nevertheless, parallel computing has a major drawback, which discourages application developers to use it. The existence of many processing elements is not fully exploited, due to the interprocessor communication overhead. The degradation of optimal speedup when the number of processors increases is caused by the occasionally excessive amount of messages between non-neighboring cells.

Since the parallel application and the target multiprocessing system can be depicted in two separate graphs, this problem can be reduced to the efficient mapping of the task graph onto the processor graph. El-Rewini et H. Ali in [1], [2] proved that the general scheduling problem of an arbitrary task graph with communication delays onto an architecture with fixed size and interconnection topology is NP-complete.

Even though a large body of literature exists in the area of scheduling and mapping the various tasks that constitute a parallel program on the processor network within a reasonable amount of time, these methods are only partially exploited. Some software tools have been proposed, supporting automatic scheduling and mapping, though

Y. Cotronis and J. Dongarra (Eds.): Euro PVM/MPI 2001, LNCS 2131, pp. 148–157, 2001.

most of them are inadequate for practical purposes. Those tools' main function is to provide a simulation environment, in order to help us understand how scheduling and mapping algorithms behave. On the other hand, there are numerous parallelizing tools, but they do not integrate well with scheduling algorithms.

We have developed an integrated software tool, called TOPPER, designed to help the programmers of parallel applications to efficiently execute parallel programs. Firstly, a parallel application is depicted as a directed graph, whose nodes represent the application's tasks and the data exchange is represented by the graph's edges. Additionally, the tool's user forms another similar graph to describe the available multiprocessor system. After these initial steps, TOPPER copes with the task-allocation problem, performing a series of appropriate functions for mapping the given task graph onto the target processor graph. Moreover, TOPPER's assorted methods for task allocation can perform an effective mapping, regardless of the task or the processor topology, guaranteeing portability and substantial speedup for most parallel programs.

2   TOPPER – An Implementation of the Multi-step Approach

In order to find efficient methods to enhance parallel programs' performance, researchers have followed a multi-step approach, addressing separately each step of the general scheduling problem. These successive steps are outlined as follows: *task clustering*, *cluster merging* and *physical mapping*. The emerging demand for the optimization of parallel applications generated the need for developing an efficient environment to incorporate all the above-mentioned operations.

2.1   Task Clustering

The initial step of the multi-step approach is the scheduling of the task graph that represents a parallel application onto a fully connected network of processors. The classical clique architecture is, therefore, used as a target with either limited or unlimited number of processors [3]. In this first step, researchers propose algorithms that minimize the maximum makespan, disregarding the actual processor's topology. The scheduling problem is NP-Complete in the majority of general cases. Papadimitriou et Yannakakis in [4] have proved that the classical scheduling problem of a task graph with arbitrary communication and computation times is NP-Complete and proposed an approximation algorithm which approaches the optimal solution within a factor of two. In addition to this, Sarkar in [5] and Gerasoulis in [6] proposed faster heuristics with efficient performance. All of these algorithms perform the same initial step: the clustering of the tasks into large nodes, generating a graph of clusters. Several tasks are encapsulated in each cluster, they are scheduled in the same processor and have thus zero intracommunication overhead. As far as clusters' intra-communication is concerned, it equals to the aggregate intercommunication cost between all the tasks of the various clusters.

The selected algorithm for task clustering that is used by TOPPER was proposed by Yang et Gerasoulis in [6] and is based on the existence of a Dominant Sequence (DS) in a directed, acyclic, scheduled task graph. This sequence represents the longest path that can be traversed in such a graph. Obviously, parallel execution time (PT) is strongly related with the DS. Dominant Sequence Clustering (DSC) algorithm performs a series of edge eliminations, placing the two endpoint-tasks of the zeroed edge in the same cluster as follows:

- Initially set each task to form a single cluster.
- Compute the initial DS.
- Mark all edges as unexamined.
- While there are still unexamined edges do:
  - Zero an edge in DS if PT does not increase.
  - Mark this edge as examined.
  - Find the new DS.

This algorithm's time complexity is sufficiently low: $O((|V_c|+|E_c|)\log|V_c|)$. Additionally, DSC's performance, as outlined in [6] and also tested in practice in [8], is a guaranty of the tool's total effectiveness in optimizing parallel applications.

Clearly, DSC groups the tasks into clusters, supposing an infinite number of processors are available. The processor graph size and connection network is actually taken into account during the next steps of the multi-step approach.

2.2   Cluster Merging

In this step, the set of clustered tasks is mapped onto a clique of bounded number of processors. Assuming that the set of clusters exceeds the number of available processors, two or more clusters end up being assigned to the same processor, reducing the number of clusters to the exact number of processors. Sarkar in [5] has proposed a scheduling heuristic with $O(|V_c|(|V_c|+|E_c|))$ complexity, where $|V_c|$ stands for the number of nodes and $|E_c|$ for the number of edges of the cluster graph. A lower complexity heuristic, used in Pyrros [7], is the work profiling method, which merges clusters offsetting their diverging arithmetic load. The arithmetic load of each cluster is the aggregate of the load of all tasks belonging to this cluster. Liou and Palis in [11] have demonstrated that, when task clustering is performed prior to scheduling, load balancing is the preferred approach for cluster merging, producing better final schedules than other methods such as minimizing communication traffic.

A load balancing algorithm, proposed by Konstantinou et Panagiotopoulos in [8], is implemented in TOPPER. Let $|V_c|$ be the number of resulting clusters from the task clustering step and $|V_p|$ the number of actual processors available for the execution of the parallel application, with $|V_p| < |V_c|$. This algorithm performs the following steps:

- Compute the arithmetic load of each cluster.
- Sort the $|V_c|$ clusters in decreasing order of their loads.
- Assign the first $|V_p|$ clusters to the $|V_p|$ free processors.
- While there are still clusters not assigned to any processor do:
  - Find the physical processor with the minimum arithmetic load
  - Assign the next cluster to this processor.

The algorithm's complexity is $O(|V_c|\log|V_c|)$ for the necessary sorting of the clusters and $O((|V_c|-|V_p|)|V_p|)$ for the rest of the procedure, concluding in $O(|V_c|(\log|V_c| +|V_p|))$. In terms of load balancing, this algorithm can easily be proved to be *optimal*. Consequently, after this step, the resulting clusters obtain two essential properties: the communication with their neighbors is kept at low levels and their computational load does not accumulate in a particular location on the cluster graph. Both properties will eventually lead to an efficient exploitation of the target parallel architecture's resources.

2.3   Physical Mapping

The final stage of the multi-step approach consists of the physical mapping of clustered task graph onto the processor graph. As shown in the two previous stages, the number of clusters is now equal or smaller than the number of processors. Therefore, it is feasible to assign to each processor a single cluster. On the whole, the physical mapping problem is NP-Complete. The sole known algorithm that can provide optimal mapping is the exhaustive algorithm, which selects the most efficient mapping combination, after having tested all the possible ones. The extremely high complexity of this algorithm $O(|V|!)$ necessitated the elaboration of heuristic methods to tackle the problem of physical mapping. Many scheduling tools, such as Oregami, Pyrros and Parallax, use occasionally efficient heuristics or approximation algorithms. More specifically, Gerasoulis et Yang in Pyrros [7] have used Bokhari's heuristic which is based on simulated annealing. This algorithm starts from an initial assignment, and then, performs a series of pairwise interchanges by reducing the cost function. Furthermore, Oregami's [9] "MAPPER", a tool for task allocation on distributed architectures, uses a greedy heuristic, called the NN-Embed, a rather simple method, that lists all edges in ascending order of their weights and assigns them to the processor's network edges. However, "MAPPER" currently supports only mesh and hypercube processor networks, limiting its potential use. Finally, the PMAP algorithm, proposed by Koziris et al in [10] has proven to be efficient regardless of the processor topology. More specifically, this heuristic detects the most communication-intensive clusters, mapping them along with their neighbors on the closest possible processor-nodes.

Many of the heuristics' testing results fail to guarantee, in advance, the efficiency of a certain heuristic, when elaborating assorted types of task and processor graphs. In order to enhance the tool's efficiency and global use, TOPPER implements not only the exhaustive algorithm, which can be applied for graphs with a limited number of nodes, but also three heuristics, which are proven to be quite competent in the physical mapping problem of large cluster graphs on heavy-sized target architectures. These heuristics are the *NN-Embed*, the *PMAP* and the *PMAP-Exhaustive*, an alternative version of the PMAP, which combines this successive heuristic with the exhaustive algorithm. According to the size of the cluster graph, TOPPER selects the appropriate method and compares the results of the implemented heuristics. The final choice of a certain mapping is made with the help of a cost function, which objectively computes the theoretical execution time of a parallel application mapped on the available processor network.

2.4   Single Shell Command Execution

TOPPER can automatically execute all the above-mentioned operations. TOPPER's users are involved only with the drawing of two graphs, describing the application's tasks with computation/communication costs and precedence constraints and the processor network. The user must form the two graphs, following certain syntax constraints in order for the graphs to be recognizable by TOPPER. Additionally, these constraints should not deprive the user of the ability to exploit repetitive or flow-control syntax structures and predefined procedure definition, since they may provide valuable help with large and complicated graphs. A detailed description of these syntax limitations can be found in [8]. The two graph definitions are stored in two separate ASCII files and are inserted as the input in TOPPER.

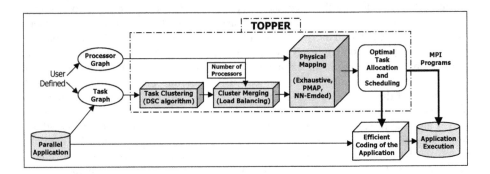

Fig. 1. An outline of TOPPER's flow chart is presented. The bold arrows on the right side of this figure demonstrate the two alternative paths that can be followed towards the application's execution on the target multiprocessing machine.

From this point onwards, only a single command issued in the command line is sufficient for the execution of the necessary operations of the multi-step approach. Before executing a certain parallel program, a TOPPER's user can consult the efficient task allocation results specified by the tool. The left hand path in Fig.1 is followed, and the application is programmed according to these task specifications. Furthermore, a TOPPER's user could even execute automatically, with the appropriate command switches, a parallel application, whose communication primitives are provided by MPI (Message Passing Interface). In this case, the right hand path towards the application execution is followed, and the user is certain that the application's tasks will be properly allocated on the multiprocessor system. Obviously, this facilitates the optimization of parallel applications, requiring a minimal effort by the programmer.

3   Experimental Results

In order to verify TOPPER's efficiency, two parallel applications have been developed and executed in a processor cluster of 16 Pentium III processors, working on 500 MHz and having 128 MB RAM each. The cluster works under LINUX 2.2.17 and supports MPICH (MPI CHameleon). These processors are connected on a Fast Ethernet network (100 Mbps).

The applications consist of several tasks, which follow the atomic task execution model. More specifically, every task is blocked until all messages are received and, after executing its computational load, sends sequentially the respective messages to its neighboring tasks. Serializing all sends is a reasonable assumption, since in most cases the network interface card does the same thing. Message passing between tasks was implemented using MPI primitives and several Mbytes of data were transferred through the processor network.

The only problem arising, during the applications' coding, was the inability of MPI to exchange faster messages between tasks running on the same processor. In fact, not only was the transmission of these messages slower, but it was measured to have no relevance whatsoever with the size of the message. This drawback cannot be alleviated with the use of other type of communication, such as system pipes, due to the MPI initialization of the tasks. In order to ensure the tasks' synchronization, semaphores were implemented, which are quite sufficient for simulation applications, but unsuitable for actual parallel programs. As a result of this problem, the programming of simulation applications became a strenuous procedure, degrading the tools most functional feature: single shell command execution of these applications. However, the potential elimination of this drawback in future versions of MPI will allow the development of an example generator and the thorough evaluation of the tool's efficiency. In this way, the testing of all of the multi-step procedures, which is operation.

Let us now demonstrate the generated results during the programming and execution of two simulation parallel applications.

3.1   Example 1

The first application is a fork-join graph, with slight variations, that consists of 10 tasks, and is mapped on a 4-ring processor network.

The application's task graph is presented in Fig.2, which shows an intensive communication path $\{0\rightarrow1\rightarrow5\rightarrow9\}$ on the left side of the task graph. The task and the processor graph were given to TOPPER, and the result was the following task allocation, defined as "MAP" implementation, as shown in Fig.2. The finding of an optimal physical mapping was not a very complicated procedure, since less than 4! different mappings could be made. It should be noted that all the heavy-weighted edges are zeroed, and that 30% acceleration was achieved, compared to the serial execution time (Fig.3). Next, three alternative ways of task allocation were implemented, which will be called Alt.1, 2 and 3.

In Alt.1 only 3 out of the 4 processors available are used. It is easy to observe that P_2 and P_3 exchange data only with P_1, and thus an optimal mapping of these three clusters is easy to find. As Fig.3 indicates, the acceleration achieved with Alt.1 is approximately 20%, significantly lower than MAP's acceleration.

In Alt.2 the clusters' content was the same as he ones in MAP, and a different mapping was chosen, in order to demonstrate physical mapping's importance. Indeed Alt.2 demonstrated lower acceleration, due to several two hop-needing messages but it still remains the best possible alternative implementation, achieving sufficient acceleration approximately 25%, even though the clusters are not allocated on processors the optimal way. This fact is a strong indication of the high efficiency of the DSC algorithm in decreasing the parallel execution time.

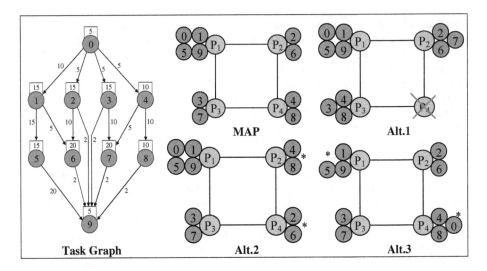

Fig. 2. In this figure the first application's task graph is presented, along with the different mapping implementations. The node tags and the figures beside the links of the task graph represent respectively the computational load and the communication cost between the tasks. The tasks are executed on the attached processors. The asterisks in Alt.2 mapping indicate the swapped clusters while the asterisk in Alt.3 shows that task 0 is moved from P_1 to P_4.

In Alt.3 the arithmetic load of the clusters is being equalized, by moving 0 from P_1 to P_4, and the physical mapping is the same as in MAP. Although the new clusters dispose a better-balanced load, the edge $0 \rightarrow 1$ is no longer a zero-valued edge. This feature renders Alt.3 inferior to all other implementations, exhibiting the worst execution time, speeding up the application only by 9%.

3.2   Example 2

The second application's graph is more complicated, consisting of 11 tasks (Fig.4), and is mapped on a hypercube of 8 processors. Once again the task and processor graph were given to TOPPER, and the results were 5 clusters, an indication that after task clustering no further merging was done. These clusters' synthesis is shown in Fig.4, and this is defined as the "MAP" implementation. The optimal physical mapping of the 5 clusters on the 8 available processors was obtained with an exhaustive search and is also presented in Fig.4. Furthermore, as a result of the DSC algorithm, P_2 is a significantly overweighed cluster with an aggregate arithmetic load of 100 units, which is more than 1/3 of the total arithmetic load of all the tasks. Nevertheless, the elimination of edges $1 \rightarrow 4$, $4 \rightarrow 9$ and $9 \rightarrow 10$ offers a parallel time speedup of 20% that cannot be achieved otherwise (Fig.3).

In Alt.1 the clusters dispose a more balanced computational load, while task 1 was moved from P_2 to P_4. All clusters possess an arithmetic load between 45 and 70, though the general scheme of the previous implementation has not been altered. The resulting execution time speedup is quite reasonable, remaining though lower than TOPPER's implementation.

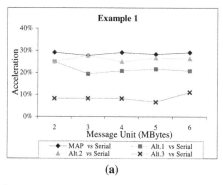 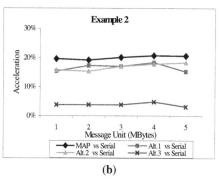

(a)   (b)

Fig. 3. Acceleration of the two applications' execution time achieved for every one of the four implementations, relatively to the message unit size. This increase of send/receive unit was, of course, followed by a proportional increase of computation unit, maintaining the task model's consistency.

In Alt.2, although the clusters used are identical to those in the MAP implementation, their allocation on the processor network is altered, with clusters P_1 and P_3 swapping places correspondingly with P_5 and P_4. As a result of this "misplacement" the parallel execution time increases (Fig.3).

In Alt.3 all the 8 processors are used. Firstly, the three most demanding communication edges are zeroed by the formation of three clusters $\{2, 5\}$, $\{4, 9\}$ and $\{3, 8\}$. Then, every one of the remaining tasks is assigned to a free processor, forming the remaining clusters. Finally the clusters are allocated on the processor network trying to avoid multiple hops for large messages (Fig.4). The result constitutes a quite well "handmade" implementation, in practice, however, its acceleration is minimal, and the parallel time almost equals to the serial time.

4.   Conclusions - Future work - Acknowledgments

Summarizing we should mention that TOPPER is an integrated software tool that can contribute to the improvement of parallel applications' effectiveness, thus taking one more step towards the expansion of parallel programming as a solution to computational need. TOPPER manages to overcome the task allocation problem, one of the main handicaps of parallel programming. Additionally, TOPPER is portable, as it is not specifically designed for a certain type of architecture, but it can be useful for executing parallel applications over all types of processor networks.

We are currently working on expanding TOPPER's capabilities as follows:

- A hybrid merging algorithm, which takes into account both the arithmetic load and the interprocessor communication, is currently being elaborated for use in future versions of the tool.
- A graphic user interface is being designed and aims to facilitate the construction of graphs and to give TOPPER's users a more accurate vision of the parallel application and the processor topology and to provide certain run-time information, in a windows environment.

- Automatic parallelization and MPI code generation for serial programs should be supported, in order to enhance the tool's functionality.

This research was supported in part by the Greek Secretariat of Research and Technology (GSRT) under a PENED 99/308 Project.

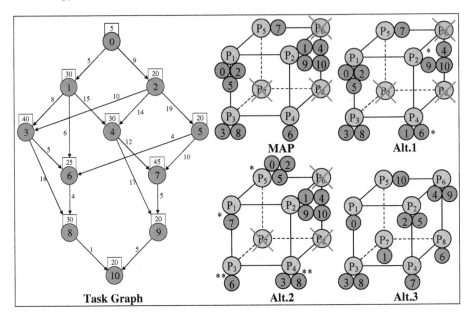

Fig. 4. In this figure the second application's task graph is presented, along with the alternative mapping implementations. Again the node tags and the figures beside the links of the task graph represent respectively the computational load and the communication cost between the tasks. The asterisk in Alt.1 shows that task 1 was moved from P_2 to P_4 and the asterisks in Alt.2 indicate the swapped clusters.

References

[1] H. Ali et H. El-Rewini. Task Allocation in Distributed Systems: A Split Graph Model. *Journal of Combinatorial Mathematics and Combinatorial Computing*, vol. 14, pp. 15-32, October 1993.

[2] H. El-Rewini, T. G. Lewis and H. Ali. *Task Scheduling in Parallel and Distributed Systems.* Prentice Hall, 1994.

[3] N. Koziris, G. Papakonstantinou and P. Tsanakas. Optimal Time and Efficient Space Free Scheduling for Nested Loops. *The Computer Journal, vol. 39, no 5, pp 439-448, 1996.*

[4] C. H. Papadimitriou and M. Yannakakis. Toward an Architecture-Independent Analysis of Parallel Algorithms. *SIAM J. Comput.*, vol. 19, pp. 322-328, 1990.

[5] V. Sarkar. *Partitioning and Scheduling Parallel Programs for Execution on Multiprocessors.* Cambridge, MA: MIT Press, 1989.

[6] A. Gerasoulis and T. Yang. On the Granularity and Clustering of Directed Acyclic Task Graphs. *IEEE Trans. Parallel Distrib. Syst.*, vol. 4, no. 6, pp. 686-701, Jan. 1993.

[7] T. Yang and A. Gerasoulis. PYRROS: Static Task Scheduling and Code Generation for Message Passing Multiprocessors. *Proc 6th Int'l Conf. Supercomputing (ICS92),* ACM Press, New York, N. Y., 1992, pp. 428-437.

[8] D. Konstantinou and A. Panagiotopoulos, T*hesis*, Dept. of Electrical Engineering, NTUA, Athens 2000

[9] V. Lo, S. Rajopadhye, S. Gupta, D. Keldsen, M. Mohamed, B. Nitzberg, J. Telle and X. Zhong. OREGAMI: Tools for Mapping Parallel Computations to Parallel Architectures. *Int'l Journal of Parallel Programming*, vol. 20, no. 3, 1991, pp. 237-270.

[10] N. Koziris, M. Romesis, G. Papakonstantinou and P. Tsanakas. An efficient Algorithm for the Physical Mapping of Clustered Task Graphs onto Multiprocessor Architectures, *(PDP2000), IEEE Press, pp. 406-413, Rhodes, Greece.*

[11] J.-C. Liou and Michael. A Palis. *A Comparison of General Approaches to Multiprocessor,* 11th International Parallel Processing Symposium (IPPS'97), Geneva, Switzerland, April 1997, pp. 152-156.

Programming Parallel Applications with LAMGAC in a LAN-WLAN Environment*

Elsa M. Macías, Alvaro Suárez, C.N. Ojeda-Guerra, and E. Robayna

Grupo de Arquitectura y Concurrencia (G.A.C.)
Departamento de Ingeniería Telemática
Las Palmas de Gran Canaria University (U.L.P.G.C.), Spain
{elsa,alvaro,cnieves}@cic.teleco.ulpgc.es

Abstract. Traditionally, the Local Area Network (LAN) has been used for parallel programming with PVM and MPI. The improvement of communications in Wireless Local Area Network (WLAN) achieving till 11 Mbps make them, according to some authors, candidates to be used as a resource for Grid Computing. Although CONDOR-MW tool can manage wireless communications using PVM (it is intended the use of MPI in future), in this paper we present a library based on LAM/MPI named LAMGAC in order to ease the programming of LAN-WLAN infrastructures. The novel of this research is that we can vary the parallel virtual machine in runtime, we generate algorithms in which computations and communications are efficiently overlapped and we include a web interface to offer our system as a Grid resource. At this moment we have measured the execution time of some algorithms and the functions of LAMGAC, obtaining interesting results.

1 Introduction

In the past a lot of experimentation in the execution of parallel programs using a LAN has been achieved [1], [2] using MPI and PVM. With the recent release of WLANs at 11 Mbps [3] some authors think that the future of these networks depend on the demonstration of its application to real problems. Moreover, there are authors [4] that consider the WLAN as a useful infrastructure (Grid Domain) for Grid Computing.

Although there are a lot of tools that use PVM or MPI, the most related work to our research is CONDOR-MW [5] because it can manage the execution of Master-Worker parallel programs taking wireless nodes into account. But it is also true that they use PVM for managing runtime availability in the parallel machine. At this moment we know they are planning to use MPI [6]. Due to MPI-2 is a globally accepted standard we are interested in using it for programming parallel algorithms in a LAN-WLAN Grid Domain. The aim for using this domain is to put into practice the ideas that other authors only mention. But

---

* Research partially supported by Spanish CICYT under Contract: TIC98-1115-C02-02.

Y. Cotronis and J. Dongarra (Eds.): Euro PVM/MPI 2001, LNCS 2131, pp. 158–165, 2001.

the most important is that experimenting [7] with WLANs can be very useful for proving the usefulness of these networks.

We have defined the LAN-WLAN architecture [8] and a preliminary programming model [9] and the interest of this paper is the presentation of a new library that manages the attachment and detachment of computers in runtime, changing the parallel virtual machine during the parallel algorithm execution. We implemented our library over LAM/MPI [10] because it is the only one free we know that implements the process creation and management funcionality defined in MPI-2. Our system can be also integrated with Grid tools like [4] using a web interface. In section 2 we present the current implementation of the library and the programming model. In section 3 a problem of nonlinear optimization using our library is measured. In sections 4 and 5 the attachment and detachment operations are explained. Finally we sum up some conclusions.

2   Current Implementation

LAMGAC is our new runtime library that helps the programmer in the development of parallel applications for a LAN-WLAN infrastructure. It handles the attachment and detachment of computers to/from parallel application, changing the parallel virtual machine in runtime. Before starting the presentation of the library, we first introduce the type of nodes we consider:

- *Access Node* (AN) that allows the remote access to the LAN-WLAN infrastructure, interconnects the remainder nodes of the system, communicates messages among nodes (of the same or different type), and manages the attachment and detachment of PNs.
- *Wired Node* (WN) that communicates with nodes of the same or different type (in the latter, through AN).
- *Portable Node* (PN) that communicates with nodes through AN. In the scope of this paper, a portable node is a wireless node that can be attached to or detached from the parallel virtual machine in runtime. On the other hand, a wired node can also be seen as a portable node if this node is attached to or detached from the parallel virtual machine during the program execution.

2.1   Internal Implementation

The library has been implemented in C language, using MPI library to communicate nodes and employs two UNIX IPCs mechanisms: shared memory and semaphores. The library includes the following three functions:

- **LAMGAC_Init.** It must be called before any other LAMGAC function is invoked. It is not necessary to call LAMGAC_Init function as the first executable statement in the program, or even in *main()*. The function must be called once by AN. It complies with attachments and detachments of nodes. The syntax for the function is

    *void LAMGAC_Init(char *argv[] /* in */);*

- **LAMGAC_Update.** It updates the parallel virtual machine: it fixes and broadcasts the number of WNs and PNs in the network and allocates a consecutive rank to the PNs. A rank equal to −1 is sent to a running process on a PN to convey it that it must finalize the program execution because this PN is going to be detached from virtual machine. The syntax for the function is

   LAMGAC_info_PN *LAMGAC_Update (int *rank /* in/out */,
   int *count_WN /* in/out */, int *count_PN /* in/out */,
   char *prog_wireless /* in */, LAMGAC_info_PN *info_PN /* in */,
   MPI_Comm COMM /* in */);

   The parameter *rank* refers to the rank of the nodes (0 for the AN, *1..count_WN* for the WNs and *count_WN+1..count_WN+count_PN* for the PNs). It's an input/output parameter for the PNs. The *count_WN* is the amount of WNs. The *count_PN* is the amount of PNs and its value can be changed dynamically during the application execution. The *prog_wireless* is the program name that is used by AN to spawn the processes in the PNs newly attached. The parameter *info_PN* is an array that stores information related to the PNs (IP address, intercommunicator with AN which is returned by the MPI_Comm_spawn function, and a flag). The *COMM* is a communicator (for the AN, it consists of AN; for the WNs, it is MPI_COMM_WORLD communicator; for the PNs, it is the intercommunicator between one PN and AN). The return value by LAMGAC_Update function is the *info_PN* array updated: removing the detached nodes, adding the newly attached nodes and keeping the remainder nodes. In this case, the *flag* has a special meaning (it is 1 for newly attached nodes and 0 for old ones). This function must be called by the nodes involved in the current application, at the time they are ready to update the parallel virtual machine (for instance when a computation task has been finished).
- **LAMGAC_Finalize().** This function is called by AN to free memory allocated by LAMGAC. After this routine call, no attachment and detachment of nodes are complied with. The sintax is

   void LAMGAC_Finalize();

2.2 Programming Model

In [9] we present the preliminary programming model in pseudocode for a LAN-WLAN infrastructure. In this section we explain the real implementation of our programming model using LAMGAC library. Our programming model consists of three algorithms (figure 1). Each squeleton must contain the preprocessor directive *#include "lamgac.h"*. This file includes the definitions and declarations necessary for compiling an MPI program that uses LAMGAC library. The squeletons for AN and WNs are rather similar but *LAMGAC_Init()* and *LAMGAC_Finalize()* calls. At the beginning, it's created one communicator (*COMM_0*) whose underlying group consists of the AN process (because of the

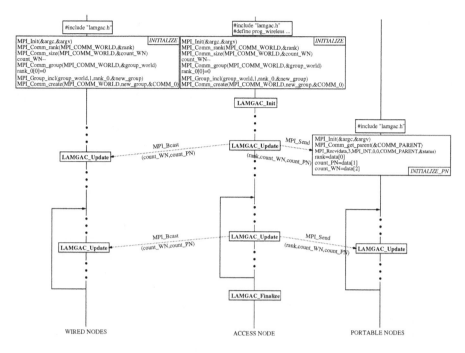

Fig. 1. Programming model in LAMGAC

process 0 is the only involved in the spawning process). After this, both nodes execute the application. Periodically, by calling LAMGAC_Update function, the parallel virtual machine is tested and updated. The checking frequency depends on the programmer.

The squeleton for a PN is a bit different because it's a spawned process by AN. The spawned process starts up like any MPI application and accesses to its intercommunicator (by using *MPI_Comm_get_parent()* call). After this, the PN receives its rank, the *count_PN* and *count_WN* from AN and begins to execute the application. Periodically, by calling LAMGAC_Update function, the parallel virtual machine is tested and updated.

3   Programming with LAMGAC

The applications that fit best in our environment are those that share the following characteristics: low communication requirement and weak synchronization. An example of this kind of application is the global nonlinear optimization, that is, searching the global minimizer of a given function defined in a multidimensional space. In this section, we present the algorithm that finds the global minimum of a nonlinear real valued continuous bidimensional function, using the LAMGAC library in the LAN-WLAN Grid Domain.

3.1   The Optimization Problem and Experimental Results

The algorithm uses the steepest descent to quickly locate the local minimum and an iterative process to reduce the search domain [11]. The algorithm uses an initial box as large as possible. Once the domain is defined, the next step is to mesh it into a finite set of boxes. Then we use the MRS algorithm [12] to calculate the multiple stationary points on each subdomain. A fully exploration of the objective function is made into each box. Once the algorithm has explored all the boxes (for a finite number of iterations), the box containing the global minimum so far is selected. The remainder boxes are deleted. The above process is repeated a finite number of iterations and hopefully the global minimum is reached. In figure 2 the tasks made by each process in order to solve the optimization problem are shown. In the computation/communication blocks, at the same time a minimum is searched (iteration i), the minimum of the previous iteration is communicated (iteration $i-1$) from each node to AN. Finally, the AN chooses the global minimum and broadcasts the box that contains the minimum to the current processes. The above algorithm was tested in our LAN-WLAN infrastructure that consists of two machines in the Ethernet LAN at 10 Mbps: one monoprocessor machine (633 Mhz Celeron processor) and one SMP with four Pentium Xeon processors. The access node is a 266 MHz dual Pentium Pro and the number of available portable PCs are three (two 550 Mhz and one 633 Mhz Celeron processors). All the machines run under Linux operating system. The wireless cards complies with IEEE 802.11 standard and they use the Direct Sequence Spread Spectrum. The link speed is 2 Mbps and they operate at 2.4 Ghz ISM band. We have made four kinds of execution time measurements for LAMGAC_Update, repeating them several times and then calculating the arithmetic mean (μ) and the standard deviation (σ) for the obtained values:

- No attachments (μ=0,158 ms and σ=1,121E-09).
- CPU attachment in the SMP. The cost to attach one CPU is around 31,116 ms (σ= 3,012E-04). The cost to attach one and two more CPUs sequentially is rather similar (about 33 ms). The cost to attach three CPUs simultaneously is less than the cost of three CPUs sequential attachments (73,359 ms).
- Wired nodes attachment. The average cost to attach one wired node is 22,129 ms (σ= 1,529E-06).
- Attachment of wireless nodes. To attach one wireless node located close to the AN is 1,07 s (σ=15,63). Far to the AN: 1,859 s (σ=1,567). Let us note the attachment time is considerably high and due to wireless channel characteristics it can be unpredictable. The cost to attach three portable nodes sequentially is 0,653 s (σ= 1,65E-02), 0,529 s (σ= 9,93E-04) and 1,663 s (σ= 1,282). The time spent to attach two portable nodes simultaneously is 1,219 s (σ= 3,382E-03). For three nodes we obtained an average time of 3,157 s (σ= 1,267). It is important to notice the irregular behaviour of the wireless channel in which the latency is high and the variation among measurements is also high.

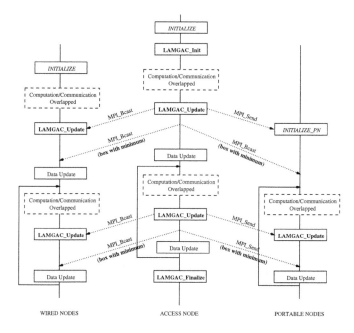

Fig. 2. Tasks made by each node in the optimization problem

4   Inside of Attachment and Detachment Processes

Using a browser, a user can attach or detach a node in our cluster. Via an easy web interface, a user has access to a web server in the AN ((1) in figure 3). When a user is connected to this server, it is shown a form with the name of the parallel applications currently in execution ((2) in figure 3). To avoid the AN be interrupted whenever an attachment (or detachment) takes place, the sequential program in the AN creates two LAM non-MPI processes when it is started, invoking the LAMGAC_Init function (one process is in charge of the attachments and the other one of the detachments. To clarify, we will call them as attachment and detachment processes). The send_LAM_SIGA program sends to the attachment process a LAM user signal reserved for application programmers (LAM_SIGA signal, (3) in figure 3). Initially, attachment process installs a signal handler for LAM_SIGA signal. When this process is signalled, execution control passes inmediately to the LAM_SIGA handler which instead returns the node identifier newly attached. This process writes the returned information in a shared memory segment ((4) in figure 3). The program in the AN reads this shared information whenever it checks new attachments ((5) in figure 3). If there is a new attachment the AN will make the spawning of a process on the portable PC (invoking MPI_Comm_spawn function) and it updates the virtual machine ((6) in figure 3). After that, the spawned process is ready to receive data from the AN and contribute to the calculation of the parallel algorithm. The detachment mechanism is rather similar.

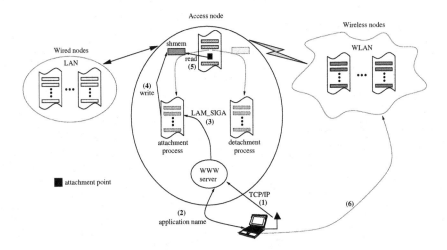

Fig. 3. Attachment management

5   WEB Interface

In the access node there is a web server to manage the access to the LAN-WLAN infrastructure (jobs submission and the attach/detach of computers). The users that can use our computing environment are two different ones: *guest user* and *computing user*. The former lends his/her machine to be attached to the cluster (notice that this user can detach the machine from the cluster whenever he/she wants via web). The latter submits jobs to be executed in the parallel virtual machine (via a client-server web interface with CGIs support).

The computing user must be an authenticated user in the web server. When this user accesses to the infrastructure successfully, he/she submits jobs via a web page. It is important to clarify that this user has not to attach his/her machine to the infrastructure so any authenticated user via Internet can execute jobs in our environment.

When the system is available, the access node transfers the sources to the remainder nodes, compiles them in the local and remote machines, edits the application schema, and orders the execution. Whatever happens in the execution, the results or the errors are noticed by e-mail to the computing user.

6   Conclusions

In this paper we presented a library based on LAM/MPI for programming parallel programs in a LAN-WLAN infrastructure. We overlap computations and communications efficiently. The interest of this work is to experiment with wireless communications that are slow but also non reliable. Experimental results show the short execution time of our LAMGAG library. We are planning to measure the application performance and integrate faults management in our library in order to the parallel program can restart its execution automatically.

References

1. V. Sunderam: PVM: A Framework for Parallel Distributed Computing. Concurrency: Practice and Experience. (1990)
2. M. Zhou, M. Mascagni: The Cycle Server: A Web Platform for Running Parallel Monte Carlo Applications on a Heterogeneous Condor Pool of Workstations. Proceedings of the 2000 International Workshop on Parallel Processing. (2000) 111–118.
3. IEEE Standards Products Catalog: Wireless (802.11): http://standards.ieee.org/catalog/IEEE802.11.html
4. K. Krauter and M. Maheswaran: Architecture for a Grid Operating System. Proceedings of the 1st IEEE/ACM International Workshop on Grid Computing. Springer Verlag, LNCS 1971. (2000). Also available in: http://www.csse.monash.edu.au/~rajkumar/Grid2000/grid2000/book/19710064.ps
5. E. Heymann, M. A. Senar, E. Luque, Miron Livny: Adaptive Scheduling for Master-Worker Applications on the Computational Grid. First IEEE/ACM International Workshop on Grid Computing. (2000)
6. Running MPICH jobs in Condor: http://www.cs.wisc.edu/condor/manual/v6.1/2_9 Running_MPICH.html
7. E. Macías, A. Suárez, C. N. Ojeda-Guerra, E. Robayna: Experimenting with the Implementation of Parallel Programs on a Communication Heterogeneous Cluster. International Conference on Parallel and Distributed Processing Techniques and Applications. Las Vegas (Nevada), U.S.A. To appear. (2001)
8. A. Suárez, E. Macías: Management of Portable Nodes in a WLAN-LAN Cluster. Proceedings of the World Multiconference on Systemics, Cybernetics and Informatics. Orlando (Florida), U.S.A. Vol. VII (2000) 151-155.
9. E. Macías, A. Suárez, C. N. Ojeda-Guerra, L. Gómez: A Novel Programming Strategy Considering Portable Wireless Nodes in a Heterogeneous Cluster. Distributed and Parallel Systems. From Instruction Parallelism to Cluster Computing. Kluwer Academic Publishers (2000) 185-194.
10. LAM/MPI Parallel Computing: http://www.mpi.nd.edu/lam/
11. J. M. Hernández, E. Macías, L. Gómez, A. Suárez: A Parallel Branch&Bound Algorithm for Non-smooth Unconstrained Optimization. Proceedings of the IASTED International Conference on Parallel and Distributed Computing and Systems. Boston, U.S.A. (1999)
12. R. S. Andersen, P. Bloomfield: Properties of the random search inglobal optimization, J. Optim. Theory Appl. Vol. 16 (1975) 383-398.

A Dynamic Load Balancing Architecture for PDES Using PVM on Clusters

Arnold N. Pears[1] and Nicola Thong[2]

[1] Department of Computer Systems, University of Uppsala,
Box 325, 751 05 Uppsala, Sweden
Ph: +46 18 4710000, Fax: +46 18 55 02 25
arnoldp@docs.uu.se
[2] Department of Computer Science and Software Engineering, University of
Melbourne, Victoria, Australia
nicola@cs.mu.oz.au

Abstract. Scalable simulation in a cluster computing environment depends on effective mechanisms for load balancing. This paper presents an object-oriented software architecture and specifies communication protocols which support efficient load balancing via object migration in cluster based PDES. A general decomposition of functionality is given in the context of a flexible adaptable architecture for migration and load adjustment.

The key contributions are the development of a general architecture into which specific migration policies and load balancing heuristics can be easily integrated and a specification of the communication protocols required. An implementation of the proposed architecture using PVM for communication indicates the approach is both flexible and well suited to customization. This enhances simulation scalability and usability in computing platforms including both traditional cluster computing and Web-computing.

1 Introduction

Several popular and well known approaches to both optimistic [10] and conservative [8] methods for implementing Parallel Discrete Event Simulation (PDES) have emerged.

Global Virtual Time (GVT) computation for PDES [2,4,9] has received a lot of attention in recent years. The other area that is critical to the development of scalable efficient simulation environments is load balancing. To improve simulation efficiency on clusters it is vital that the processing load on each station be tuned to the capabilities and resources available.

This paper presents the integration of OBJECTSIM [7], a generic object-oriented simulation environment, with the PVM cluster computing environment [3]. The problem of designing a load balancing approach for PDES executing on cluster computing environments is explored, and a general architecture based on an object-oriented framework is proposed. The framework supports enhanced

Y. Cotronis and J. Dongarra (Eds.): Euro PVM/MPI 2001, LNCS 2131, pp. 166–173, 2001.

modularity and localizes host specific hardware interface code in a few key objects. An object-oriented (OO) approach also makes it easy to tailor heuristics and migration methods to the heterogeneous environment, while retaining common protocols for the exchange of load metrics and the transmission of simulation entities between stations in the cluster.

2   Load Balancing and Entity Migration

2.1   DES and PDES

Continuous simulations advance time at a fixed rate and this can result in long time periods where nothing interesting is being simulated while time ticks by. Distributed Event Simulation (DES) systems eliminate the periods where an idle system is being simulated by advancing the simulation clock in irregular sized steps. The steps correspond to the intervals between consecutive "interesting events".

Applying object-oriented design principles to the implementation of DES is extremely intuitive. Abstract components of a system are represented by Objects, allowing a general behavioral specification of each entity in a process. The interface to that entity can be clearly defined, and its semantics specified as an automata which responds to events and generates events across the interface making simulation entities truly generic [7].

Extending the DES model to support simulations on parallel computers, clusters of workstations and internet computers introduces several new problems. The most important are **synchronizing the global clocks** in the simulation engines and **communicating events and objects** between kernels.

Static load balancing methods based on task graphs and other a priori information about the computational and communication requirements of simulation entities often provide a good starting point for load allocation. However, in addition an estimate of the load capacity of each workstation, and the ability to dynamically relocate resources between the workstations is needed to guarantee that allocation mechanisms function well.

2.2   Virtual Computers

In a shared workstation cluster or web computing environment the machines being used to execute the simulation are often being used for day to day computing tasks. Such a situation can be imagined when constructing a cluster using the machines in a postgraduate computer laboratory together with the personal workstations of faculty staff. In such circumstances it seems both natural and sensible to prioritize the owner's use of the machine at the expense of the cluster. However, not all machines are in use all the time and so the ability to migrate load over the cluster to machines that are currently idle is very attractive. How should such migration mechanisms be designed and implemented?

Related work in this area includes the shared cluster computing support suites PVM [3] and MPI [1]. The limitation in these systems is that work allocation is still mostly left under user control, and tends to be coarse grain. Processes can be placed on a named computer and instructed to execute there, but support for dynamic load optimization over the cluster is typically absent.

2.3   Supporting PDES

We propose a structured approach to cluster enabled PDES, defining an abstract architecture (supported by PVM) for a general object-oriented simulation kernel. This model defines the additional objects needed to handle the communication and the data marshaling methods necessary to implement dynamic load migration. The key elements are thus as follows.

- Specification of load monitor and migration policy entities.
- The ability to locate an object and deliver events to it.
- Reliable protocols for communication between kernels.

3   Kernel Architecture

An OSim-D simulation, a distributed version of OBJECTSIM, consists of a set of Objects (state machines which specify the reactions to specified input stimuli) and a mapping table which describes the event flows (flows of stimuli) between the simulation objects.

The kernel itself is composed of a number of interacting objects that provide the basic simulation functions discussed in Section 2. A summary of the kernel object relationships, and the simulation objects that are communicated between them is given in Figure 1. A full description of OSim's operation and modeling methodology has been outlined in a previous paper [6], while the application of the simulation method to several problem domains has also been reported elsewhere [5]. The functions of the major objects can be summarized as follows.

- ObjectStore: An indexed data structure in which simulation entities are stored while idle.
- EventStore: An indexed data structure storing events in increasing timestamp order.
- Scheduler: The active entity that drives the kernel forward. It selects the next event, matches it to an object in the store and then delivers the event to the object.
- Resolver: Maps Event (srcOid,Port#) to (dstOid,Port#) using the simulation interconnection mapping table provided by the simulation experiment. All events are made up of an address tuple of Object ID (Oid) and a Port Number (Port#) which tells it which Kernel and Object it belongs to.
- Mapper: Works out in which kernel (KernOid) a dstOid object is currently located, and delivers events there. The mapper also updates its database of Oid to KernOid mappings in response to observed object migration actions.

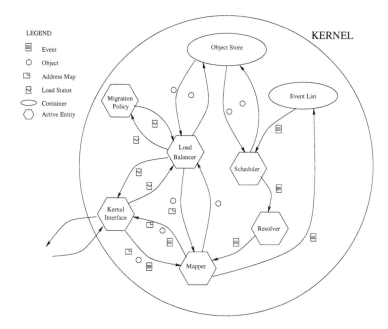

Fig. 1. Kernel Architecture

- KernelInterface: Kernel peer to peer protocol implementation that provides communication and connection between kernels.
- LoadBalancer: Maintains load estimates and implements load collection policies using interfaces to the native operating system load functions.
- MigrationPolicy: Decide when and where to move local objects.

Implementing dynamic load balancing requires added functionality which can be used to move objects between running engines and to resolve the addressing issues associated with event delivery to objects that are moving dynamically between the simulation engines.

The architecture localizes this activity in the last four kernel entities. Namely the KernelInterface, Mapper, LoadBalancer, and MigrationPolicy.

3.1   Resolver and Mapper

In a single OSim-D kernel the Resolver is used to determine the destination of an event (dstOid,Port#) given the source of that event (srcOid,Port#). In a PDES simulation the destination object can be located in a different kernel to the one in which the event source object is executing. How can we deliver events to the correct Kernel?

To solve this problem the Mapper object collaborates with its peers to maintain a consistent distributed database which maps object identifiers (which are globally unique) to kernel instances (which are also allocated unique identifiers at startup).

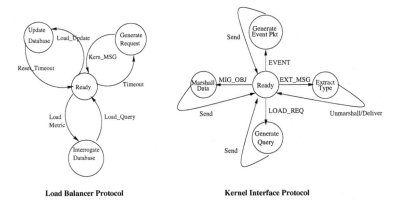

Load Balancer Protocol   **Kernel Interface Protocol**

Fig. 2. State diagrams for load balancer and kernel interface

When an object migration operation takes place the origin and destination are registered immediately with the mapper instances in the source and destination kernels. Propagation of this information to additional mappers is managed by an "on demand" protocol when objects resident in other kernels attempt to route events to the object at its original location.

3.2 Load Estimation and Migration

The Load Estimator maintains a database which attempts to record the current load on each machine in the cluster/internet computing platform. This information is stored on a per node basis, where each entry consists of a (nodeID, load, timestamp) triple.

The idea is to localize the interface to the local operating system functions used to provide the load measures and convert them to a higher level abstract measure that is understood by all kernels. Communication of these platform independent load measures between Load Estimators gives each Load Estimator an impression of the distribution of simulation effort over the cluster. Load on the local machine is sent to other kernels through the Kernel Interface.

Migration policy is separated from load estimation to allow flexible tailoring of migration policies to be achieved on a per-station basis. The migration policy object uses the data maintained by the load estimator to decide when it is desirable to relocate objects executing in its local kernel and, subsequently, to decide to which other kernel they should be sent.

3.3 Kernel Interface

Providing the mapping from internal kernel protocol data units (PDU's) onto the facilities provided by the transport protocol service access points (SAP's) is accomplished in the Kernel Interface object.

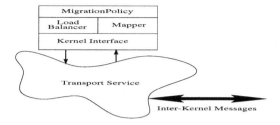

Fig. 3. Application communication layering

The Kernel Interface has the primary responsibility of managing the marshaling and unmarshaling of data exchanged between the kernels. Complex data structures such as simulation objects and events which utilize dynamic data structures are required to implement self marshaling and unmarshaling methods. These methods must be exported by the objects, since including knowledge of the internal structure of user defined simulation objects in the Kernel Interface would be a violation of object-oriented encapsulation principles.

This provides a consistent SAP interface to the internal kernel objects, encapsulating all transport and host specific implementation details in a single location. Implementing versions of OSim-D to function with any data transport service is consequently localized entirely within the Kernel Interface object implementation.

4   Implementation

OSim-D uses a transport layer service (provided in this case by PVM) as a basic carrier infrastructure and extends it with several application protocol layers within the PDES kernel. The relationships between the PDES kernel communication layers and standard carrier layers is shown in Figure 3.

Some implementation issues associated with the communicating elements arise in relation to the integration of PVM.

4.1   Mapper

A key mapper implementation issue is how to track the motion of objects between kernels and keep the location maps relatively well synchronized. The on demand protocol and message piggybacking used to achieve this are too complex to describe here given the page limitations. A forthcoming publication will deal with the implementation and operational evaluation of this protocol.

4.2   Load Balancer

The current load estimation only attempts to take into account the computational effort allocated to each kernel instance. Currently the length of the event queue is used to estimate load in each kernel.

An alternative for UNIX clusters is to use the UNIX system load figure as the kernel load metric. This would have the effect of taking into account overall load in the station rather than trying to evaluate the simulation effort allocated to the Kernel. This is desirable when the stations in the cluster are multi-purpose and some of them are on people's desk-tops.

Updates to the load estimates are required relatively frequently if we are to have reliable information upon which to base migration decisions. In the absence of events arriving from a desired node a timeout expiry event in the load manager generated by comparing the local GVT estimate with load entry timestamps is used to trigger a direct request for a new load value from the relevant nodes (see Figure 2).

4.3   Migration Policy

The problem of when and where to migrate objects can be quite personal. It is easy to foresee situations where the policies used to make this decision should be easily adjusted to local machine hardware and usage characteristics. For example, a cluster station which belongs to the organization's research director! He/She doesn't expect to have to compete with a simulation for cycles on his personal machine. In such a situation the kernel migration policy should start to migrate objects at a much lower load level than one might choose for the stations in the graduate student laboratory. Consequently this aspect of load balancing has been encapsulated separately to simplify tuning the kernels to the requirements of individual machines in a cluster.

Once migration has been decided upon an appropriate object is extracted from the Object Store and its events removed from the event list. All these are passed to the Mapper. The Mapper in turn updates its record of object to kernel allocations and passes everything to the Kernel Interface for marshaling and transmission to the destination (dstKid).

4.4   Kernel Interface

The integration of PVM with OSim-D allows multiple instances of kernels to be transparently and easily launched on the participating cluster station. Communication between the kernels is then managed using a mixture of PVM commands and direct TCP socket communication.

PVM provides support for transfer of marshaled data. However direct support for marshaling and unmarshaling of complex user data structures such as Objects, Events and Address Updates is lacking. Some direct TCP socket communication must also be used to avoid stalling for messages in some communication situations since PVM_send() blocks on a matching PVM_receive().

Underlying the process of marshaling and unmarshaling objects for transmission are implementation issues associated with memory management. Objects are deleted from the sending kernel space after marshaling is completed. New object instances are created and populated from arriving marshaled data during the unmarshaling process in the receiving kernel.

5   Conclusions and Further Work

This article describes a general purpose architecture for managing load migration and balancing in object-oriented simulation kernels. The target simulation platform is clusters of PVM workstations or internet computers.

A prototype implementation combines a generic OO simulation kernel OSim-D (which implements the OBJECTSIM modeling methodology) with PVM which provides a virtual programming environment for cluster computing. The result is a flexible engineered approach to implementing dynamic load migration.

Testing the efficiency of our approach is beyond the scope of this publication. Experimental studies comparing the average simulation duration time and average load over the cluster for benchmark simulations of multistage interconnection networks are the focus of current work.

References

1. MPI Forum. Mpi: A message-passing interface standard. *International Journal of Supercomputer Application*, 8(3/4):293–311, 1994.
2. R. M. Fujimoto. Parallel discrete event simulation. *Communications of the ACM*, 33(10):30–53, October 1990.
3. A. Geist, A. Berguelin, J. Dongarra, W. Jiang, R. Manchek, and V. Sunderam. *PVM: Parallel Virtual Machine*. MIT Press, 1994.
4. J. Misra. Distributed discrete-event simulation. *Computing Surveys*, 18(1):39–65, 1986.
5. A.N. Pears and M.L. Ojczyk. Performance analysis of a parallel target recognition architecture. volume 3371, pages 404–414, April 1998.
6. A.N. Pears and E. Pissaloux. Using hardware assisted geometric hashing for high speed target acquisition and guidance. In *SPIE AeroSense '97, Orlando Florida, USA, OR11*, pages 312–320, 22-25 April 1997.
7. A.N. Pears, S. Singh, and T.S. Dillon. A new approach to object oriented simulation of concurrent systems. In *SPIE AeroSense '97, Orlando Florida, USA, OR11*, 22-25 April 1997.
8. L. F Pollacia. A survey of discrete event simulation and state-of-the-art discrete event languages. *Simulation Digest*, pages 8–25, March 1985.
9. F. Quaglia, V. Courtellessa, and B. Ciciani. Trade-off between sequential and time warp-based parallel simulation. *IEEE Transactions on Parallel and Distributed Systems*, 10(8):781–794, 1999.
10. P.L. Reiher. Parallel simulation using the time warp operating system. In *Proceedings of the 1990 Winter Simulation Conference*, 1990.

Dynamic Partitioning of the Divide-and-Conquer Scheme with Migration in PVM Environment[*]

Pawel Czarnul[1], Karen Tomko[2], and Henryk Krawczyk[3]

[1] Electrical Engineering and Computer Science, University of Michigan, USA
pczarnul@eecs.umich.edu
[2] Dept. of Electrical and Computer Engineering and Computer Science
University of Cincinnati, USA
ktomko@ececs.uc.edu
[3] Faculty of Electronics, Telecommunications and Informatics
Technical University of Gdansk, Poland
hkrawk@pg.gda.pl

Abstract. We present a new C++ framework which enables writing of divide-and-conquer (DaC) applications very easily which are then automatically parallelized by dynamic partitioning of the DaC tree and process migration. The solution is based on DAMPVM – the extension of PVM. The proposed system handles irregular applications and dynamically adapts the allocation to minimize execution time which is shown for numerical adaptive quadrature integration examples of two different functions.

1 Introduction

The divide-and-conquer scheme is widely used since it can be used in many algorithms e.g. sorting, integration, n-body simulation ([1]). Mapping a DaC tree onto a distributed network is not an easy task since it may be significantly unbalanced and may change at runtime unpredictably as intermediate results are obtained. Moreover, available hardware may be heterogeneous in the aspect of the system architecture, topology and processor speeds. We have developed an easy-to-use object-oriented C++-based DaC framework which is adaptively mapped to a multi-user distributed memory system at runtime. At first, top-level branches of the DaC tree are executed as separate processes to keep all the processors busy. Then the proposed framework is automatically parallelized by DAMPVM ([2], [3], [4]) on which it is based. It partitions the DAC tree dynamically if some processors become underutilized, then migrates tasks from other processors to achieve the lowest possible execution time. The only development a user has to do is to derive their own C++ class from the supported DAC template class and override only a few virtual methods. The framework does not require any parallel programming specific code, is very general and thus very easy to

---

[*] This research was supported in part by the Army Research Office /CECOM under the project for "Efficient Numerical Solutions to Large Scale Tactical Communications Problems", (DAAD19-00-1-0173).

use. There are existing systems which facilitate parallel implementation of DaC based algorithms. APERITIF (Automatic Parallelization of Divide and Conquer Algorithms, formerly APRIL, [5]) translates C programs to be run on parallel computers with the use of PVM ([6]). REAPAR (REcursive programs Automatically PARallelized, [7], [8]) derives thread-based parallel programs from C recursive code to be executed on SMP machines. Cilk ([9]) is a similar thread-based approach. An extension of this language towards a more global domain with the use of the Java technology is presented in the ATLAS ([10]) system. Satin ([11]) is another Java-based approach targeted for distributed memory machines. Other framework-based approaches are Frames ([12]) and an object-oriented Beeblebox ([13]) system. An algebraic DaC model is described in [14]. The main contribution of this work is its capability of using heterogeneous process migration (Section 2.2) in mapping work to a system in addition to dynamic partitioning. Migration can tune the assignment at runtime and balance work without spawning more tasks if enough tasks are already in the system. This enables to handle unbalanced DaC applications (on the contrary to PVM-based APRIL where subtrees should be of the same size to achieve good performance) in a multi-user environment. Other works describe mapping of the DaC scheme to various system topologies ([15], [16]). Architecture-cognizant analysis in which different variants may be chosen at different levels is presented in [17].

2   Divide-and-Conquer Framework

Figure 1 presents a general DaC paradigm (as considered by all the other approaches) as a pseudocode with some extensions provided by the proposed framework. It is assumed that each node in the DAC tree receives a data vector delimited by left and right pointers `Object *vector_l` and `Object *vector_r`. `Object` is a template parameter for the abstract class `DAC` and a user-derived class from `DAC` should instantiate `Object` with a class/type suitable for its needs e.g. `double` for sorting vectors of `double` numbers. In general vector (`vector_l`, `vector_r`) is either a terminal node (if a user-defined method `DaC_Terminate(vector_l, vector_r)` returns true) or is divided further into some number of subvectors by method `DaC_Divide(vector_l, vector_r)` which returns a list of left and right pointers to the subvectors. In the first case method `DaC_LeafComputations (vector_l, vector_r)` should provide leaf computations and in the latter method `DaC_PreComputations(vector_l, vector_r)` is executed and then the recursive call takes place. The procedure is repeated at deeper recursion levels. Upon return method `DaC_PostComputations(new_vectors, vector_l, vector_r)` may provide code which merges subvectors into the parent vector. This scheme is general and allows different numbers of subvectors at each node and different depths (unbalanced trees) depending on an application's needs (e.g. computation accuracy may determine the depth).

2.1   Dynamic Partitioning

The above well-known ([1]) scheme has been extended with dynamic partitioning of the DAC tree as well as migration provided by DAMPVM. The main idea

```
   template <class Object> // main recursive method
 2 void DAC<Object>::DaC(Object * &vector_l,Object * &vector_r) {
      if (DAC has been requested for higher level than current) {
 4       spawn children; send data;
      }
 6    if (the highest depth) {
         receive data from parent; DaC_Initialize( vector_l , vector_r );
 8    }
      if (DaC_Terminate(vector_l,vector_r)) DaC_LeafComputations(vector_l,vector_r);
10    else {
         DaC_PreComputations(vector_l,vector_r);
12       nHowManyNewVectors=Dac_HowManyNodes(vector_l,vector_r);
         new_vectors=DaC_Divide(vector_l,vector_r);
14       if (more tasks needed) {
            spawn tasks; send data;
16          inform DAMPVM that my size=DaC_VectorSize(new_vectors[0],new_vectors[1]);
            DaC(new_vectors[0],new_vectors[1]);
18       } else {
            if (no tasks have been spawned) enable migration for this process;
20          for(int nTask=0;nTask<nHowManyNodesExecutedByThisTask;nTask++)
               DaC(new_vectors[2*nTask],new_vectors[2*nTask+1]);
22       }
         if (data has been sent to children) {
24          inform DAMPVM I may be idle waiting; receive data;
         }
26       DaC_PostComputations(new_vectors,vector_l,vector_r);
      }
28    if (the highest level and I am not the root) send data to parent;
   }
30 template <class Object>
   void DAC<Object>::Run(void) {// initialization method
32    Object *vector_l,* vector_r ;
      if ((PC_Parent()==PCNoParent) || (PC_How_Started()==migrated)) {
34       InitializeData( vector_l , vector_r ); // root process or migrated one
         DaC_Initialize( vector_l , vector_r ); // execute this
36    }
      DaC(vector_l,vector_r); // activate the recursive code
38                            // (children read data inside)
      if (PC_Parent()==PCNoParent) MasterReport(vector_l,vector_r);
40 }
```

Fig. 1. DAC Recursive Code and Initialization Method

is that if the tree is very unbalanced then the initial partitioning may provide very poor utilization of some processors resulting in low speed-up values and poor scalability. Since often computation times may not be known in advance static partitioning may not give good scalability and dynamic reassignment is necessary. Dynamic partitioning of the DAC tree is shown in Figure 2. Initially the whole tree shown is supposed to be executed by one process. Each process keeps variable nCurrentPartioningLevel which indicates the highest level (higher levels are closer to the root and are denoted by lower numbers) on which the tree may be partitioned. There are as many DAMPVM schedulers as the number of nodes each running on a different host. When a scheduler detects a load below a certain threshold on its machine it requests dynamic partitioning of the largest processes on more loaded neighboring nodes – a neighbor graph may be freely defined ([2]). New tasks are always spawned on underloaded nearest neighbors and migration is used to tune the allocation ([3]). A user-defined function DaC_VectorSize(vector_l, vector_r) returns predicted amount of work (in some units) for the given vector. Sometimes this may be determined by the

complexity function for the algorithm with some coefficients. Sometimes however it may not be known precisely but only estimated in advance. The purpose
of this method is to prompt DAMPVM which processes should be dynamically
partitioned and which migrated. When a process with the tree as shown in
Figure 2 receives dynamic partitioning request 0 it partitions the tree at the
nCurrentPartioningLevel level. It means if there are more iterations of the loop
(line 20 in Figure 1) at the nCurrentPartioningLevel level they are assigned
to other processes which are automatically created at runtime and the corresponding data is forwarded to the new processes. The process continues to work
on its part of the tree and then receives results from the dynamically spawned
processes at the nCurrentPartioningLevel level. Moreover, the proposed scheme
allows many requests to be received by a process and thus multiple partitioning.
If the same process receives request 1 it does not partition itself at level 2 since
there are no more iterations at this level available for other processes. However,
it partitions itself at level 3 at which there are 2 more iterations which are assigned to processes 2 and 3 respectively and corresponding tasks are spawned.
Again at level 3 data is collected from the spawned processes.

2.2   Process Migration

DAMPVM provides the ability of moving a running process from one machine
(stopping it there) to another (restarting) which we refer to as process migration. The state of a process is transferred at the code level (at the expense of
additional programming effort) not the system level but still provides the same
functionality. For spawn/migration details see [3] and [4]. As shown in Figure
1 migration is enabled for the current process ([2]) if the system does not need
more tasks to be spawned (because all the processors are already busy) and no
tasks have been spawned before by the process. As described in [2] and [3] migration can be triggered by dynamic process size changes including spawns/ter-
minations (irregular applications) and other users' time-consuming processes.
If a DAMPVM scheduler wants to migrate a task its execution is interrupted
by calling a PVM message handler which activates function PackState(). This
function packs all the necessary data which describes the current process state.
A new copy is spawned in a special mode on another machine which unpacks
its state in function UnPackState(). Both of them need to be supported by a
programmer. Moreover, some programs require a special programming style to
recover the process state. In return for that a user is provided with very fast,
flexible and heterogeneous migration.

3   Numerical Adaptive Quadrature Integration Example

As an example we have implemented a DAC-based numerical integration example which integrates any given function. The idea is proposed in [1]. In general
a certain function $f(x)$ and range $[a, b]$ are given. As shown in Figure 3a if area
C is small enough then we can terminate the DAC strategy and compute $\int_a^b$
as a sum of areas A and B. Otherwise range $[a, b]$ is divided into two $[a, \frac{a+b}{2}]$

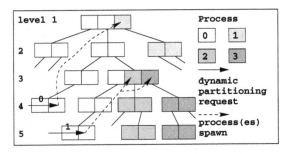
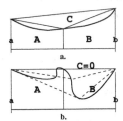

Fig. 2. Dynamic Partitioning of DAC Tree   **Fig. 3.** Integration Algorithm

and $[\frac{a+b}{2}, b]$ and the operation is repeated. To prevent the algorithm from termination for functions as shown in Figure 3b we pick out ten different points inside the $[a, b]$ range instead of only one $\frac{a+b}{2}$ and perform the C area check ten times before going deeper. Intuitively, such an algorithm will give similar execution times for the same size subranges for some classes of functions e.g. periodic functions with the period much smaller than the initial range. However, execution times may vary greatly for irregular functions. Since the algorithm does not know the function in advance it can assume that two ranges $[a, b]$ and $[c, d]$ take the same time if $(b - a) = (d - c)$. If only static partitioning is used it will result in some processors finishing their work much sooner than the others. The proposed scheme is to create a sufficient number of processes at runtime thanks to dynamic partitioning and balance them using migration.

The above example has almost a trivial implementation in the proposed DAC scheme which is presented in Figure 4. In this implementation a vector has always three elements: the left and right contain the range extreme left and right coordinates and the middle one contains 0 when the integration of the range starts and the computed value upon return. Thus functions for migration (Figure 4, [2] and [3]) can contain only these three values. If a process computes a vector it always does this from left to right as it is implied by the DAC scheme. If its execution is interrupted i.e. process migration occurs it simply delivers the value for the left computed subrange as the initial value and the right not yet computed range for the migrated process (its new copy).

4   Experimental Results

The initial implementation of the DaC scheme described above has been tested on Linux/Sun workstations. The first four nodes run Linux and the fifth is a Sun workstation. The Linux workstations have a performance of 25 relative to the Sun with a performance of 16.1. The experiments were performed on two different integration examples, the functions are given below:

1. $f(x) = sin^2(x)$: $\int_0^{400} f(x)dx$; this function is periodic and thus the execution times for same-size ranges should be similar,
2. $g(x) = \begin{cases} sin^2(x) & 0 \leq x \leq 120\pi \\ x - 120\pi & 120\pi \leq x \leq 240\pi \end{cases}$: $\int_0^{240\pi} g(x)dx$

```
   double fInitialLeft, fInitialRight ; // top level range for this process
 2 double fRight,fVal=0; // current right coordinate and value
                         // for range [ fInitialLeft ,fRight]
 4 double fIntegrationRange[3];// initial left coordinate, value, right coordinate

 6 void PackState() {// pack the values (left coordinate, computed value and the
      //  right coordinate) for the uncomputed range [fRight,fInitialRight]
 8   PC_PkDouble(&fRight); PC_PkDouble(&fVal); PC_PkDouble(&fInitialRight);
   }
10 void UnPackState() {// unpack the coordinates and initial value
      PC_UPkDouble(&(fIntegrationRange[0])); PC_UPkDouble(&(fIntegrationRange[1]));
12    PC_UPkDouble(&(fIntegrationRange[2])); fVal=fIntegrationRange[1];
   }
14 int MyDAC::Dac_HowManyNodes(double *vector_l,double *vector_r) { return 2; }
   long MyDAC::DaC_VectorSize(double *vector_l,double *vector_r) {
16   return (long)(100*(*vector_r-*vector_l)); // predicted process size
   }
18 void MyDAC::DaC_Initialize(double *vector_l,double *vector_r) {
      // remember top level range and set the initial  right coordinate
20    fInitialLeft =*vector_l;  fInitialRight =*vector_r; fRight= fInitialLeft ;
   }
22 double **MyDAC::DaC_Divide(double *tab_l,double *tab_r) {
      double fPivot=(*tab_l+*tab_r)/2;
24    // these are new subvectors -- allocate memory for them
      new_vectors[0]=*tab_l; new_vectors[1]=0; new_vectors[2]=fPivot;
26    new_vectors[3]=fPivot; new_vectors[4]=0; new_vectors[5]=*tab_r;
      // return pointers to subvectors' coordinates
28    tab_pointers[0]=new_vectors; tab_pointers[1]=new_vectors+2;
      tab_pointers[2]=new_vectors+3; tab_pointers[3]=new_vectors+5;
30    return tab_pointers;
   }
32 void MyDAC::DaC_PostComputations(double **new_tab,double*&tab_l,double*&tab_r){
      *(tab_l+1)+=(*new_tab)[1]+(*new_tab)[4]; // add results from children
34    // deallocate memory associated with the subvectors
   }
36 bool MyDAC::DaC_Terminate(double *tab_l,double *tab_r) {
      if (ComputeCArea(tab_l,tab_r,&fPivot)<0.000000001 for 10 pivots)
38      return true; else return false;
   }
40 void MyDAC::DaC_LeafComputations(double *tab_l,double *tab_r) {
      float fArea=ComputeTrapezoidArea(tab_l,tab_r); // add (A+B) areas and remember
42    *(tab_l+1)+=fArea; fVal+=fArea; fRight=*tab_r; // the current right coordinate
   }
44 void MyDAC::InitializeData(double * &tab_l,double * &tab_r) {
      // executed for every new process - initialize data
46    tab_l=(double *)fIntegrationRange; tab_r=((double *)fIntegrationRange+2);
      fInitialRight =*tab_r;  fInitialLeft =*tab_l; fRight= fInitialLeft ;
48 }
   void MyDAC::MasterReport(double *vector_l,double *vector_r) {
50   cout << (*(vector_l+1)); // print the final  result
   }
52
   main(int argc,char **argv) {
54   DaC_Init(&argc,&argv); // DaC intialization
      if (PC_How_Started()!=migrated) { // if this is not a migrated task
56    // set initial  left coordinate, result and right coordinate
      fIntegrationRange[0]=0; fIntegrationRange[1]=0; fIntegrationRange[2]=400;
58   }
      MyDAC mdcIntDAC(&PC_PkDouble,&PC_UPkDouble); // create a DaC object, pass data
60   // packing and unpacking functions for the double type and activate the object
      mdcIntDAC.Run(); // simply run the object and wait for results
62   DaC_Finish(); // DaC termination
   }
```

Fig. 4. Numerical Adaptive Quadrature Integration – Complete Source Code

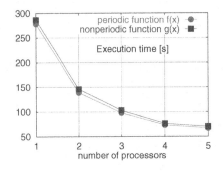

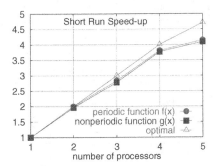

Fig. 5. Execution Time   Fig. 6. Speed-up

Notice that in the latter case range $\int_{120\pi}^{240\pi} g(x)dx$ will be integrated almost immediately since area C will always be 0 regardless a pivot point chosen. On the other hand the integration of $\int_0^{120\pi} g(x)dx$ will take long time. If there's only static partitioning and two available nodes, two processes would be attached to these processors to integrate ranges $[0, 120\pi]$ and $[120\pi, 240\pi]$. After a while the second node becomes idle which results in the total execution time practically the same as for one processor. On the other hand, DAMPVM detects load imbalance and activates our dynamic DaC scheme which partitions range $[0, 120\pi]$ into two $[0, 60\pi]$ and $[60\pi, 120\pi]$ and places one process on the idle processor which results in almost the best speed-up. Obviously, the $\int_0^{400} sin^2(x)dx$ case should give better performance as there are no such dynamic load imbalances. The execution times and speed-ups for short runs are shown in Figures 5 and 6. Heterogeneous migration between Linux and Solaris workstations is extremely fast for this example. Benefits from migration vs. the dynamic DaC only were observed for configurations with other users disturbing the load balance when migration can balance the load before dynamic partitioning is invoked if enough processes are available.

5   Conclusions and Future Work

We presented a dynamic divide-and-conquer scheme which aims at partitioning load dynamically and dynamic load balancing with the use of migration procedures supported by DAMPVM. The proposed implementation is able to detect load imbalance in a parallel environment at runtime, partition data to keep all the processors busy and balance their workloads. Such a scheme can partition and map a binary tree to a 3-processor system quite well which is shown in our experiments. The proposed DaC software will be available at the DAMPVM Web site ([4]). Future work will focus on closer integration of the DaC and migration schemes. A better load balancing algorithm ([18]) is currently being incorporated into the code which will give better performance for larger networks. We plan to implement many different examples and possibly enhance the proposed scheme for efficient parallel execution of different applications as well as test it on larger heterogeneous LAN networks. Also, a direct performance comparison with other existing approaches will be made including Java-based ones.

References

1. B. Wilkinson and M. Allen, *Parallel Programming: Techniques and Applications Using Networked Workstations and Parallel Computers.* Prentice Hall, 1999.
2. P. Czarnul and H. Krawczyk, "Dynamic Assignment with Process Migration in Distributed Environments," in *Recent Advances in Parallel Virtual Machine and Message Passing Interface*, Vol. 1697 of *Lecture Notes in Computer Science*, pp. 509–516, 1999.
3. P. Czarnul and H. Krawczyk, "Parallel Program Execution with Process Migration," in *Proceedings of the International Conference on Parallel Computing in Electrical Engineering*, (Trois-Rivieres, Canada), IEEE Computer Society, August 2000.
4. DAMPVM Web Site: http://www.ask.eti.pg.gda.pl/~pczarnul/DAMPVM.html.
5. T. Erlebach, *APRIL 1.0 User Manual, Automatic Parallelization of Divide and Conquer Algorithms.* Technische Universitat Munchen, Germany, http://wwwmayr.informatik.tu-muenchen.de/personen/erlebach/aperitif.html, 1995.
6. A. Geist, A. Beguelin, J. J. Dongarra, W. Jiang, R. Manchek, and V. S. Sunderam, "PVM 3 user's guide and reference manual," Tech. Rep. ORNL/TM-12187, Oak Ridge National Laboratory, May 1993. http://www.epm.ornl.gov/pvm.
7. L. Prechelt and S. Hngen, "Efficient parallel execution of irregular recursive programs," *Submitted to IEEE Transactions on Parallel and Distributed Systems*, December 2000. http://wwwipd.ira.uka.de/~prechelt/Biblio/Biblio/reapar_tpds2001.ps.gz.
8. S. Hngen, "REAPAR User Manual and Reference: Automatic Parallelization of Irregular Recursive Programs," Tech. Rep. 8/98, Universitat Karlsruhe, 1998. http://wwwipd.ira.uka.de/~haensgen/reapar/reapar.html.
9. R.D.Blumofe, C.F.Joerg, B.C.Kuszmaul, C.E.Leiserson, K.H.Randall, and Y.Zhou, "Cilk: An efficient multithreaded runtime system," in *Proceedings of the 5th ACM SIGPLAN Symposium on Principles and Practice of Parallel Programming*, pp. 207–216, July 1995.
10. J.Baldeschwieler, R.Blumofe, and E.Brewer, "ATLAS: An Infrastructure for Global Computing," in *Proceedings of the Seventh ACM SIGOPS European Workshop on System Support for Worldwide Applications*, 1996.
11. R. V. van Nieuwpoort, T. Kielmann, and H. E. Bal, "Satin: Efficient Parallel Divide-and-Conquer in Java," in *Euro-Par 2000 Parallel Processing, Proceedings of the 6th International Euro-Par Conference*, No. 1900 in LNCS, pp. 690–699, 2000.
12. J. P. i Silvestre and T. Romke, "Programming Frames for the efficient use of parallel systems," Tech. Rep. TR-183-97, Paderborn Center for Parallel Computing, January 1997.
13. A. J. Piper and R. W. Prager, "Generalized Parallel Programming with Divide-and-Conquer: The Beeblebrox System," Tech. Rep. CUED/F-INFENG/TR132, Cambridge University Engeneering Department, 1993. ftp://svr-ftp.eng.cam.ac.uk/pub/reports/piper_tr132.ps.Z.
14. Z. G. Mou and P. Hudak, "An algebraic model for divide-and-conquer and its parallelism," *The Journal of Supercomputing*, Vol. 2, pp. 257–278, Nov. 1988.

15. V. Lo, S. Rajopadhye, J. Telle, and X. Zhong, "Parallel Divide and Conquer on Meshes," *IEEE Transactions on Parallel and Distributed Systems*, Vol. 7, No. 10, pp. 1049–1057, 1996.

16. I. Wu, "Efficient parallel divide-and-conquer for a class of interconnection topologies," in *Proceedings of the 2nd International Symposium on Algorithms*, No. 557 in Lecture Notes in Computer Science, (Taipei, Republic of China), pp. 229–240, Springer-Verlag, Dec. 1991.

17. K. S. Gatlin and L. Carter, "Architecture-Cognizant Divide and Conquer Algorithms," *SuperComputing '99*, Nov. 1999.

18. P. Czarnul, K. Tomko, and H. Krawczyk, "A Heuristic Dynamic Load Balancing Algorithm for Meshes," (Anaheim, CA, USA), 2001. accepted for presentation in PDCS'2001.

Using Monitoring Techniques to Support the Cooperation of Software Components[*]

Roland Wismüller

Lehrstuhl für Rechnertechnik und Rechnerorganisation (LRR-TUM)
Technische Universität München, D-80290 München
wismuell@in.tum.de

Abstract. Many applications in computing could be implemented by re-using existing programs, if only it were possible to easily make them interact with each other. MPI or PVM do not support these applications, as inserting a direct communication makes the components depend on each other. Even middleware for distributed computing still requires that the components already implement the necessary interfaces and that these are sufficient for the required interactions. The paper shows how on-line monitoring techniques can be used to improve this situation. An extended version of the monitoring system OCM allows both to incorporate the necessary interfaces and to define interactions between components at run-time, without having to modify their source code.

1   Introduction

In parallel applications, several components (usually processes) tightly interact in order to perform a common computation. In distributed applications, on the other hand, client components use common services that are provided by usually remote server components. In both scenarios, the components need an infrastructure supporting an efficient communication. Such infrastructures have been available for quite some time with communication libraries like MPI and PVM or distributed middleware, such as Corba.

There is, however, an important scenario, which is not adequately supported by these infrastructures. Imagine a set of independently developed software components, which normally are used as stand-alone applications. In some cases, however, it may be desirable to use them in a cooperative manner in order to exploit a synergy. Examples for this situation are so called multi-physics codes [3]. E.g. to simulate air pollution, a climate code, which models air and water transport, can be loosely coupled to an atmospheric chemistry code, which accounts for chemical reactions. Other examples, which have been the motivation for this research, are tool environments to monitor and control program executions. E.g. a visualizer shows a coarse grained overview of an execution, while a debugger can provide ample details about a specific program state. However,

---

[*] This work was partly funded by *Deutsche Forschungsgemeinschaft*, Special Research Grant SFB 342, Subproject A1, and Research Grant WI-1346/1-1.

Y. Cotronis and J. Dongarra (Eds.): Euro PVM/MPI 2001, LNCS 2131, pp. 183–191, 2001.

the states shown by the visualizer are already in the past, while the debugger can only examine a program's current state. Combining these tools with a checkpointing system allows to transparently restore the execution to a state selected in the visualizer, which then can be examined in detail [9].

The distinctive properties of the scenario outlined in the previous paragraph are the following:

1. The components have been developed independently, usually by different institutes or companies.
2. The components are designed to be used stand-alone. They are not prepared for any cooperation with other components.
3. The source code of these components often is not available for modification. Even if it is, modifications are often undesirable for reasons discussed later.
4. The intended cooperation requires only infrequent communication with a comparatively low data volume.

When these properties are met, we will speak of *interoperating components* in the remainder of this paper.

In the following, we outline the requirements of applications based on interoperating components and show that they are not met by existing communication libraries and middleware. Section 4 presents a solution that emerged from our research in the area of monitoring tools. We propose to use on-line monitoring techniques in order to implement the cooperation between software components. In this approach, the monitoring system plays the role of an extended, event oriented middleware. Section 4 also provides some details on the implementation of the approach in the framework of the OCM project [10].

2   Example: An Environment for Efficient Debugging

To provide the following discussions with a concrete background, we will use the tool environment already mentioned in the introduction. The environment consists of three independently developed run-time tools for parallel programs based on PVM: a program flow visualizer, a debugger, and a checkpointing system[1] [9]. The visualizer VISTOP animates the program's execution behavior by displaying a sequence of snapshots. Each snapshot shows a global view of the program's processes and their interactions, i.e. communication and synchronization, as well as other global state information, e.g. the contents of PVM groups. The debugger DETOP allows to examine and modify the current state of all or a selectable subset of processes. The checkpointing system CoCheck finally can save and restore the complete global state of the program.

By using all three tools concurrently, a feature strongly required by users could easily be implemented: When a suspicious state is found in the high level view of the program execution presented by the visualizer, users want to be

---

[1] The real tool environment actually contains a fourth tool – a deterministic execution controller – which is not discussed here for simplicity.

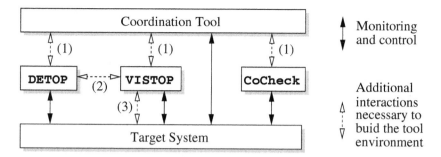

Fig. 1. Component architecture of the example environment

able to examine this state more closely with the debugger. This is not possible without a cooperation between all three tools, since the visualizer shows the history of the program's execution, i.e. the state in question has already passed. In order to implement this feature in the tool environment, the target program is checkpointed periodically. When the user wants to inspect a selected state more closely, the program is first reset to the most recent checkpoint preceding this state. Next, it is automatically re-executed from this point on until it reaches the target state, which then can be examined using the debugger.

Such a concurrent, cooperative use of run-time tools is only possible if (1) there are no conflicts between these tools, e.g. due to resources that can only be used exclusively [9], and (2) the tools can interact with each other and can be controlled by a higher-level coordination tool. We will only discuss the second issue in this paper.

Figure 1 indicates the interactions necessary to realize the feature outlined above:

(1) Certain actions of DETOP, VISTOP, and CoCheck need to be invoked by the coordination tool in order to reset the target program to the desired state. E.g. the coordination tool must be able to get information on the state currently displayed by VISTOP, to restore the proper checkpoint, or to print a notice in DETOP's output window.

(2) When certain event occur in a tool, actions in other tools must be invoked. For example, when VISTOP shows a global state that is in the past, DETOP should inactivate its state display, since it shows information on the current state, which is inconsistent with the one presented by VISTOP.

(3) Finally, the tools need to execute additional actions when certain events occur in the target program. E.g. when a target process is stopped by DETOP, VISTOP must update its displays in order to provide consistent information.

These classes of interactions are also representative for other applications based on interoperating components. While for the first two classes this is rather obvious, the third one seems to be specific to interoperating run-time tools. However, even in other applications the components often should react on events in their execution environment, e.g. node failure, low memory, etc.

3   Requirements of Interoperating Components

In the scenario of interoperating components, each component is implemented by
a separate processes or a set of processes. Thus, the required interactions between
these components must be implemented by some kind of message exchange. The
attempt to use MPI or PVM for this message passing, however, results in severe
problems:

- First of all, explicit send and receive calls must be inserted into the components. This means that the source code must be modified and the components need to be recompiled. Furthermore, the developers of the components
  should be convinced to incorporate these changes into their 'official' version.
  Otherwise, the changes must be re-applied to each new version.
  Developers are, however, rather reluctant to include such modifications, especially, since there may be different, possibly conflicting requests from users
  applying the component in different environments.
- Adding explicit communication also makes the components depend on each
  other [8]. For example, in the environment outlined in Section 2, VISTOP
  could send a message to DETOP asking it to inactivate its state display.
  This would lead to the following consequences: VISTOP needs to know that
  a DETOP process is running and must find its task ID or rank in order to
  send a message to it. When used in a different environment, VISTOP needs
  to locate and notify different components.
  Furthermore, VISTOP must know the message format expected by these
  components (or vice versa, they must know the format defined by VISTOP).
  If this interface is modified in one of the components, VISTOP has to be
  adapted accordingly.
  Without special measures, the components even are no longer usable in a
  stand-alone fashion, since e.g. the sending component will be blocked or
  buffers will overflow when there is no receiver. Such dependencies are usually
  unacceptable to component developers.

The situation is not much better when using object oriented middleware, like e.g.
Corba. The components still have to be modified, unless they already provide
Corba interfaces for *all* required interactions. In addition, the component acting
as a client still depends on the component acting as a server.

A general principle in computer science to break dependencies is the use of
indirection. In our example, instead of letting VISTOP send the notification directly to DETOP, it could always send it to a special *broadcast message server*
(BMS). Components interested in this type of notification have to register with
the BMS, which then forwards the proper messages to them. Thus, the components no longer need to know which other components exist in the current
environment. This scheme might be implemented using PVM broadcasts and
dynamic groups. A more specific BMS implementation is ToolTalk [5].

There is, however, still a dependency, as all the components need to agree on
common message types and formats. This dependency can be removed by using
mediators [8]. A mediator is an object that monitors events in some components.

When an event is detected, the mediator executes an associated piece of code that typically invokes actions in other components. The mediator is free to adjust the parameters when invoking the action; it can also call a sequence of actions in potentially different components. In general, a mediator implements an arbitrary behavioral relationship between two ore more components. While the mediator – of course – depends on all these components, the components themselves now are fully independent.

All current implementations of mediators, like e.g. the Corba-based MediatorService [7], still require the components to explicitly provide interfaces for events and actions. As a consequence, all components depend on the concrete implementation of the mediator approach and can no longer used purely stand-alone. E.g. components using the MediatorService cannot be used outside a Corba environment any more. Another drawback of current implementations is that we still need to modify the components' source code, if the interfaces are not yet available or are incomplete w.r.t. our needs. Sullivan, who originally invented the idea of mediators, states that "[anticipating] *the events needed in an object's interface [...] is unreasonably hard*" ([8], Section 8.10). This problem, however, only occurs if we view components as static, unchangeable entities. Modern monitoring techniques allow to dynamically modify programs during their execution. Thus, the idea is to insert the required component interfaces dynamically during run-time, as needed. In this way, we avoid source code modifications and can even use legacy applications as components. Stand-alone usage is no problem at all, since the static executable remains unchanged. Dependencies to other components or the infrastructure (like ToolTalk or Corba) are completely avoided.

4   The Monitoring System as Communication Middleware

If we examine the operation of a modern, distributed monitoring system like the OCM [10] more closely, we even see that its operation is already very similar to a mediator: the monitoring system waits for events in the target program and executes a set of actions when the event occurs. In addition, the OCM supports run-time programmable relations between the events and lists of actions to be executed. It transparently accounts for the distributedness of the target system by automatically routing event information and requests for actions to the proper nodes in the system. Thus, the OCM not only enables the dynamic insertion of component interfaces, but also supports the definition of behavioral relations between components that can be adapted on demand.

These features lead to the following idea: if the OCM attaches to all interoperating components in the same way as it attaches to its regular target processes, it can realize the mediator concept without a need to specifically prepare or adapt the components. There are, however, two requirements, which must be met before this idea can be realized: (1) the OCM must be able to detect all events of interest in the components, and (2) it must be able to invoke actions, which are executed inside other components.

4.1   Detection of Events

The OCM already can detect certain classes of events in its target processes without a need for an explicit source code instrumentation. These classes include the beginning and end of function or subroutine invocations in a target process. Currently, two different implementations are available: The first one is based on binary wrapping [2]. It requires to relink the target program, since it modifies the object code of programming libraries. Thus, this technique is mainly suited to monitor library calls. A recently finished second implementation uses the dyninstAPI [4] that allows to dynamically insert code into running processes.

In both cases, the inserted code notifies the OCM about the event. It first writes the relevant event information into a shared memory segment and then executes a trap instruction. This results in a signal being sent to the OCM. The event information usually consists of the values of parameters or the result of the monitored subroutine, but it can also include the values of global variables or other state information. The event information to be acquired for different subroutines is specified in a definition file – in analogy to an interface definition file used e.g. with Corba – and thus can easily be adapted if necessary.

When the OCM receives the trap signal, it reads the current event information from the shared memory segment and examines the currently defined event/action relations. If necessary, it sends messages to remote OCM processes requesting the execution of actions on their respective nodes.

4.2   Execution of Actions

In the OCM's normal use, actions are executed directly by the OCM processes. In order to support interoperating components, we have extended the OCM with means allowing to execute actions within a target process. Like in the regular case, the interface of such an action is specified in a special interface definition file, while the action itself is implemented as a piece of C or C++ code. Similar to standard middleware environments, an IDL compiler generates a stub routine from this file, which adapts the internal generic action interface of the OCM to the specific C/C++ interface of the action's implementation.

When such an action needs to be invoked in the context of a target process, two steps must be executed:

1. If not already done before, the stub routine and the code of the action's implementation must be loaded into the target process's address space.
2. The target process must be forced to call this stub routine and, thus, the action.

Step 1 is achieved by using dynamic linking techniques. The stub and the action's code are compiled into a dynamically linkable library. The target process then is forced to execute a call to the **dlopen** routine in the UNIX run-time system, using the same technique as described for step 2 below. This routine loads the library into the target process. Since the dynamic linker resolves symbol references in both directions, the action's implementation can access any public symbol in the target process, i.e. it can read and write global variables and can call subroutines in the target process's code. Thus, the action's implementation code usually just calls such an already existing subroutine, passing on its own parameters to it.

Step 2 is realized by manipulating the stack and the program counter of the target process using operating system services. The dyninstAPI already offers a routine for this so-called inferior RPC. A technical problem is that the inferior RPC is performed completely asynchronously w.r.t. the target process's execution. This can result in problems, if e.g. the inferior RPC interrupts the target process in a routine that is later called again by the invoked action. Most routines behave incorrectly when reentered in such a way. Thus, in the OCM the simple inferior RPC has been extended towards a synchronized scheme: when an action has to be invoked, the OCM installs a local event/action relation in the target process, where the event signals that the execution reached a point where it is safe to call the action.

4.3 Support for Interoperating Components

Due to the dynamic instrumentation of subroutine invocations supported by the OCM, events need not be explicitly announced by components, as long as they are related to the beginning or end of such an invocation. The experience with the tool environment presented in Section 2 suggests that this regularly is the case with reasonable well structured codes. While technically, dynamic instrumentation also enables to monitor the execution of arbitrary points in a program's code, the problem is to identify and to name them in a meaningful way.

The OCM's implementation of actions allows to invoke any subroutine in a component from the outside, possibly adapting its interface as needed. In addition, completely new actions can be added to a component. The component's source code does not need to be modified in any of these two cases. Thus, neither a dependency to other components nor to the OCM infrastructure is created.

Since the OCM loads the information on available events and actions at run-time, it can be adapted to different applications without a need to alter its code. Via the OCM's monitoring interface, a coordination "tool" can define the necessary relations between events and actions in the components of a specific application at run-time. In addition, the coordination tool can directly invoke actions in the components via this interface.

4.4 Performance

The OCM's enhanced flexibility asks for some price to be paid in terms of performance. On a Sun Sparc Ultra-10 running Solaris 2.6, the time between the occurrence of an event in a component and the execution of the associated action in another one[2] currently amounts to $1.7ms$. This compares to $0.33ms$ needed

---

[2] In order to enable the use of high resolution timers, both components have been located on the same machine.

by an equivalent standard PVM communication between two processes on the same node. The higher overhead mainly comes from two sources:

1. While with PVM a communication only requires two hops when the processes are located on the same node, the OCM always needs four: (1) The event in the component is notified to the local OCM process via a signal. (2) A message is sent to the OCM's routing process. (3) The message is forwarded to the local OCM process on the destination node. (4) The action is invoked in the destination component using an inferior RPC. This communication accounts for about half of the total time.
2. The other half is caused by the management of event/action relations and the marshaling and unmarshaling of the communicated data structures, which are more complex than in an equivalent PVM communication.

We are still in the process of optimizing the OCM. Since in the scenario of inter-operating components, communication is typically of low-volume and infrequent, this is, however, not of crucial importance in practice.

5   Related Work

We have already discussed the relation of our work to communication libraries, middleware and broadcast message servers like ToolTalk in Section 3. The approach most closely related to our work is the idea of *mediators* [8]. We have extended this idea with dynamic code modification techniques, which allow to completely avoid dependencies between components. This extension is the distinctive feature of our work.

The *interactor* approach [6] is a different extension of the mediator idea, which addresses the case where components can also call external services rather than just creating events. In our OCM-based environment, this is possible, too. Just like the coordination tool, any other component can use the OCM's monitoring interface to call remote actions. However, since we are concerned about components that normally are used stand-alone, we do not need this feature.

The goal of coupling independent components finally is addressed by a large number of *coordination languages* [1], which allow to uniformly specify the interfaces and interactions of distributed components. These approaches are orthogonal to the one proposed here: while current implementations of coordination languages still require the interfaces to be statically compiled into the components, the OCM currently requires three different specifications for events, actions and their relationships, respectively. Thus, the integration of our approach with a coordination language will be an important issue addressed in future work.

6   Conclusions

The paper has shown that the interoperability between independently developed components is not adequately supported by existing middleware. In order to avoid undesired component dependencies, it is necessary to define the required interfaces and interactions separate from the components' code and to link them to the components only at run-time. We have shown that this is possible by using on-line monitoring techniques. Furthermore, the monitoring system can play the role of the communication middleware, since it already implements the necessary features. We have demonstrated the usability of this approach by composing a tool environment for advanced parallel debugging from existing tools. However, the approach can support any kind of loosely coupled components, like e.g. multi-physics applications or distributed simulations.

References

1. F. Arbab, P. Ciancarini, and C. Hankin. Coordination Languages for Parallel Programming. *Parallel Computing*, 24(7):989–1004, 1998.
2. J. Cargille and B. P. Miller. Binary Wrapping: A Technique for Instrumenting Object Code. *ACM SIGPLAN Notices*, 27(6):17–18, June 1992.
3. T. J. Downar, R. Eigenmann, J. A. B. Fortes, and N. H. Kapadia. Issues and Approaches in Parallel Multi-Component and Multi-Physics Simulations. In H. R. Arabnia, editor, *Intl. Conf. on Parallel and Distributed Processing Techniques*, volume II, pages 916–922, Las Vegas, NV, USA, June 1999. CSREA Press.
4. J. K. Hollingsworth and B. Buck. *DyninstAPI Programmers' Guide, Release 1.2*. Univ. of Maryland, College Park, MD, USA, Sept. 1998.
   http://www.cs.umd.edu/projects/dyninstAPI/dyninstProgGuide.pdf.
5. A. M. Julienne and B. Holtz. *ToolTalk & Open Protocols: Inter-Application Communication*. SunSoft Press, 1994. ISBN 0-13-031055-7.
6. G. Paul, K. Sattler, and M. Endig. An Integration Framework for Open Tool Environments. In H. König, K. Geihs, and T. Preuß, editors, *Distributed Applications and Interoperable Systems*, pages 193–200, Cottbus, Germany, Sept. 1997. IFIP, Chapman & Hall.
   http://wwwiti.cs.uni-magdeburg.de/~sattler/papers/dais97.ps.
7. G. Paul and K.-U. Sattler. MediatorService - Integration von verteilten Objekten durch Beschreibung der Interaktionen. In *Proc. Int. Workshop Trends in Distributed Systems*, Aachen, Oct. 1996.
   http://wwwiti.cs.uni-magdeburg.de/~sattler/papers/trevs.ps.
8. K. J. Sullivan. *Mediators: Easing the Design and Evolution of Integrated Systems*. PhD thesis, Dept. of Computer Sciences and Engineering, Univ. of Washington, USA, 1994. Techn. Bericht 94-08-01.
   ftp://ftp.cs.washington.edu/tr/1994/08/UW-CSE-94-08-01.PS.Z.
9. R. Wismüller and T. Ludwig. Interoperable Run-Time Tools for Distributed Systems - A Case Study. In H. R. Arabnia, editor, *Proc. of the International Conference on Parallel and Distributed Processing Techniques and Application, PDPTA'99, Volume IV*, pages 1763–1769, Las Vegas, NV, USA, June 1999. CSREA Press.
10. R. Wismüller, J. Trinitis, and T. Ludwig. OCM — A Monitoring System for Interoperable Tools. In *Proc. 2nd SIGMETRICS Symposium on Parallel and Distributed Tools SPDT'98*, pages 1–9, Welches, OR, USA, Aug. 1998. ACM Press.
    http://wwwbode.in.tum.de/~wismuell/pub/spdt98.ps.gz.

An Integrated Record&Replay Mechanism for Nondeterministic Message Passing Programs

Dieter Kranzlmüller, Christian Schaubschläger, and Jens Volkert

GUP Linz, Johannes Kepler University Linz,
Altenbergerstr. 69, A-4040 Linz, Austria/Europe
kranzlmueller@gup.uni-linz.ac.at
http://www.gup.uni-linz.ac.at/

Abstract. Nondeterminism is a characteristic of many parallel programs that needs dedicated support from analysis tools and programming environments. In order to allow cyclic debugging of such programs, record&replay mechanisms are used most frequently. Such techniques operate in two phases, where the record phase traces a program's execution that can be arbitrarily repeated during subsequent replay phases. In contrast to most existing approaches, this paper describes a mechanism that is transparently integrated in the underlying message passing interface. The main advantage of this approach is its omnipresence, such that a program's execution can be repeated immediately after it has been observed. Other benefits are the lack of instrumentation and a corresponding simplification of the whole technique for inexperienced users. The difficulties addressed by this approach are concerned with the amount of monitor overhead, which must neither perturb the program's execution nor generate huge amounts of trace data.

1   Introduction

Parallel programming is a difficult activity that requires a certain degree of attention from program developers. This is true for every stage of the software lifecycle, leading to corresponding tool support for program design, implementation, and analysis. One characteristic that complicates parallel programming even more is nondeterminism. Simplified, a program is nondeterministic, if it yields different results in subsequent executions even if the same input data is provided. While this can be achieved in sequential programs due to dedicated functionality (e.g. the random number generator), it occurs in parallel programs due to the nature of concurrently executing and communicating tasks. Examples in this area include race conditions, which are observed whenever a receive operation selects one of multiple incoming messages.

Problems with nondeterministic behavior are experienced during debugging, when an application is analyzed for detecting incorrect behavior. In this case, several anomalies may be observed [6]: On the one hand, a user may notice a program's error during one execution, but is unable to repeat this error afterwards, because the program delivers completely different results. Furthermore,

Y. Cotronis and J. Dongarra (Eds.): Euro PVM/MPI 2001, LNCS 2131, pp. 192–200, 2001.

the program's behavior may be seriously perturbed by the overhead induced by the debugger, such that debugging is almost useless. On the other hand, errors may occur only sporadically and some executions may never be executed at all.

Corresponding solutions are offered by two kinds of techniques, record&replay and controlled execution, which provide functionality for equivalent program executions [6]. Record&replay techniques achieve reproducibility with two phases. An initial record phase traces a program's execution, whereas subsequent replay phases use the obtained data to enforce the same program behavior. Controlled execution techniques follow the idea of pre-defining the behavior at nondeterministic choices, such that a program can be arbitrarily re-executed if the same rules are applied. Both techniques are equally applicable for the problems stated above. However, record&replay techniques are more common in practice, because they do not limit a program's freedom during the initial record phase.

The solution described in this paper implements a record&replay technique. Yet, in contrast to other existing techniques, our approach is integrated into the functionality of the underlying message passing interface. This means, that recording takes place whenever a program is executed, which allows to repeat any observed execution without requiring the otherwise mandatory instrumentation step. Consequently, an integrated mechanism simplifies the analysis activities of nondeterministic parallel programs, especially for in-experienced users.

This paper is organized as follows: The next section introduces definitions and basics about nondeterminism in messages passing programs, as well as related work in this area. Section 3 describes our strategy in connection with the underlying message passing interface, with details about optimizations in Section 4. Afterwards, some measurements of our current implementation are given.

2   Nondeterminism in Message Passing Programs

Nondeterminism is observed in programs, if there may be situations for a given input, where an arbitrary programming statement is succeeded by one of two or more follow-up states. This freedom of choice may be determined by pure chance or unawareness of the complete state of the execution environment. In sequential programs, users introduce nondeterminism with dedicated functions, such as the random number generator or system calls like `gettimeofday`. The same functionality can be applied to introduce nondeterminism in parallel programs. However, in most cases, parallel programs will be nondeterministic due to process interaction. Simplified this means, that a parallel program is nondeterministic whenever it does not fully specify all possible execution sequences, but instead allows a degree of freedom in selecting from possible choices at communication and synchronization operations.

In terms of message passing programs, nondeterminism resulting from process interaction occurs most frequently at receive operations. Receive operations that permit more than one message to be accepted establish race conditions, whose result determines the subsequent program states of the receiving process. In concrete, two kinds of race conditions can be distinguished based on the type of receive operation.

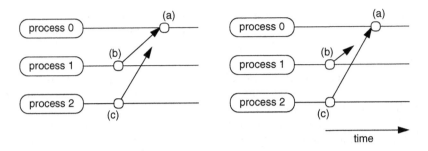

Fig. 1. Nondeterministic blocking receive operation on process 0.

On the one hand, race conditions occur at blocking receives (e.g. MPI_Recv or pvm_recv), when a process is waiting for a message and more than one message may fulfil the criteria given by the receive statement.This is achieved by using a wild card constant (e.g. "MPI_ANY_SOURCE" for MPI or "-1" for PVM) for a message's origin, which means that the next available message from any process will be accepted.

An example for this situation is given in Figure 1, where two possible results of a parallel program running on 3 processes are given. This event graph display describes a parallel program's execution and the interaction of its processes [6]. The time is given on the horizontal axis, while the processes are arranged vertically. Process interaction is displayed by directed arcs connecting corresponding send and receive operations, with the tip of the arc pointing to the receiver. The race condition in this example stems from receive operation (a), which specifies a wild card for the message's origin. In the left diagram, the receive accepts the message submitted by process 1 at send operation (b), while in the right diagram the message originating from send operation (c) on process 2 is accepted. Depending on which scenario takes place, the results of process 0 after receive operation (a) may be completely different.

On the other hand, race conditions may also occur at nonblocking receive operations (e.g. MPI_Irecv/MPI_Test or pvm_probe). Firstly, such a nonblocking receive operation may also specify a wild card for a message's origin. In this case, the nondeterministic behavior is equivalent to the race conditions described for blocking receives above. Secondly, nonblocking receive operations may also influence a processes behavior, whether the message's origin is specified or not. An example for this situation is given in Figure 2, which describes a parallel program's execution on 2 processes. In this case, the program may reveal different results depending on the number of calls to the nonblocking receive operation. For example, in the left diagram of Figure 2 the message from send operation (c) on process 1 is available when process 0 executes the nonblocking receive operation (b). Yet, in the right diagram, the same message is already available at nonblocking receive operation (a). Consequently, if a program's behavior depends on the number of nonblocking receive operations that occurred before a message is accepted, nondeterministic behavior may be observed.

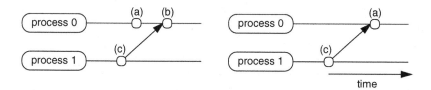

Fig. 2. Nondeterministic nonblocking receive/test operations on process 0.

Corresponding to the reasons for nondeterministic behavior, dedicated functionality must be provided by record&replay techniques. The history of record and replay techniques dates back to an approach described in [3], which records the contents of each message as it is received by the target process (compare also with [16,9]). The biggest drawbacks of this data-driven approach are its significant monitor overhead and storage requirements.

This disadvantage is reduced by control-driven approaches, which preserve only the order of occurring operations, and generate the transferred data again during each replay execution. Probably the first control-driven approach was Instant Replay [8]. Based on this work, several other approaches have been implemented as described in [11,12,2]. Of special interest is the approach described in [14], which tries to minimize the amount of required trace data. Other optimizations concerning space and time efficiency are described in [4,10,15,1].

3   Integrated Record&Replay for MPI Programs

In addition to all existing record&replay techniques, the approach described in this paper tries to accomplish the following requirements:

- Perform functionality transparent and invisible for the user.
- Permit replay on demand by recording program executions all the time.
- Generate as little overhead as possible in order to limit the probe effect.
- Minimize the storage requirements for the hidden trace files.

These requirements are achieved by integrating a record&replay mechanism directly into the message passing environment. In our case, modifications to the source codes have been applied to MPICH, a freely available, portable implementation of MPI, version 1.2.1 (see ftp://ftp.mcs.anl.gov/pub/mpi). In addition, we have studied the source code of the LAM MPI implementation, and we are confident that similar changes are applicable. The same is expected for PVM, which was not investigated until now.

After adding the necessary changes to the code of the chosen MPI implementation, the following functionality is available for the user. Firstly, record&replay functionality is present without the user's knowledge. There is no need to perform any instrumentation or manage any tracefiles, because all these steps are transparently performed within MPI. Secondly, tracedata is only obtained, when a program is actually nondeterministic. In this case, an observed program execution can be arbitrarily re-executed, by providing a corresponding parameter to the mpirun command (e.g. "-replay"). If the user decides not to replay a program's execution, the MPI implementation itself dumps the obsolete tracefiles before the next record phase is executed.

At present, the following functions have been modified for record&replay activities: MPI_Init, MPI_Finalize, MPI_Recv, MPI_Test. The function MPI_Init has been modified to process the "-replay" parameter at the mpirun command. In addition, it handles the tracefile management (open, close) together with the MPI_Finalize routine. The functions MPI_Recv and MPI_Test have been modified to record the choices occurring during a program's execution. In both cases, optimizations have been applied in order to reduce the runtime overhead and the size of the generated tracefiles.

4   Optimization Concerning Non-overtaking Message Transfer

The biggest advances of our approach compared to other existing record&replay mechanisms are its optimizations towards message passing interfaces, that prohibit overtaking of messages. Overtaking of messages is a feature that imposes a high potential for erroneous behavior [6]. For that reason, most message-passing implementations guarantee non-overtaking of messages between processes. For example, the standard Message Passing Interface [13] clearly states, that every valid MPI implementation must offer non-overtaking message transfer (see MPI Standard 1.1, Section 3.5. "Semantics of point-to-point communication", http://www.mpi-forum.org/docs/mpi-11-html/node41.html). A similar prerequisite is provided by the Parallel Virtual Machine PVM [5], which also defines non-overtaking message order of point-to-point communication functions (see PVM 3.3 Manual pages, pvm_send - Immediately sends the data in the active message buffer, http://www.epm.ornl.gov/pvm/man/pvm_send.3PVM.html).

Based on this knowledge, every existing record&replay technique can be optimized, if it is applied to a message passing system, that guarantees non-overtaking message transfer. If this is the case, the record&replay mechanism does not need to establish a temporal relation between the communication events on different processes [7]. Instead, the implicit characteristic of the message transfer strategy determines the set of messages that may be accepted at a particular receive event. This feature allows to define the choice taken at non-deterministic receive events solely by the identification of a message's sending process. Consequently, it is not necessary to trace any data at the send event, nor to generate a temporal relation with some kind of logical clock mechanism [6]. Even message tags are not needed. Instead, only the origin of a message (as stored in the MPI_Status structure or returned by pvm_bufinfo) must by obtained at a corresponding wild card receive event.

An example for this optimization is given in Figure 3. Receive operation (a) is assumed a wild card receive, while send operations (b) through (e) specify parameters as provided by Table 1. For these operations, two cases can be distinguished. In both cases, the receive specifies a wild card for the source process of a message. In case (1) the message tag is also a wild card parameter, while

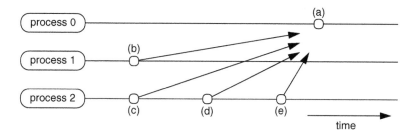

Fig. 3. Race condition candidates at receive event (a).

Table 1. Race condition candidates for example shown in Figure 3.

Operation	Type	process ID	case (1)		case (2)	
(a)	**receive**	source = ?	tag = ?		tag = x	
(b)	send	dest = 0	tag = x	yes	tag = x	yes
(c)	send	dest = 0	tag = y	yes	tag = y	no
(d)	send	dest = 0	tag = x	no	tag = x	yes
(e)	send	dest = 0	tag = x	no	tag = x	no

in case (2) the message tag is fixed. While this specification of the wild card at the message tag enforces different choices to be taken, their acceptance order at receive operation (a) is defined due to the non-overtaking characteristic by their order on the corresponding sender process.

5   Implementation Details and Measurements

Corresponding to the optimization based on non-overtaking message transmission, the functionality given in Table 2 is performed during the record phase, whenever a nondeterministic receive operation is executed. The first column of this table specifies the kind of operation, while the second column defines the value of the source parameter as provided at the function call. The third column provides the value of the variable, that is obtained during the record phase and stored in the tracefiles for the replay phase. This determines the size of the tracefiles, since only the values of the third column of Table 2 need to be stored. With simple encoding techniques, a few bits are sufficient for every occurrence of a nondeterministic receive operation. Please note, that this value is only available after the actual MPI function call has been performed. Thus, the changes to the code have been applied immediately after `MPID_RecvDatatype` in the file `mpich-1.2.1/src/pt2pt/recv.c`.

The data obtained during the record phase is used during subsequent replay phases as follows: Whenever a nondeterministic blocking receive operation is encountered, the recorded parameters of the original function call are read from the traces and used as replacement for the replayed function call. In case of replaying nonblocking receives, two cases need to be distinguished. If the return

Table 2. Data obtained during record phase.

operation	source	record
blocking receive	MPI_ANY_SOURCE	status->MPI_SOURCE
	fixed	
nonblocking receive	MPI_ANY_SOURCE	status->MPI_SOURCE return code
	fixed	return code

Table 3. MPI_Receive Measurements.

	constant source		wild card source	
mode	measurement	percentage	measurement	percentage
original	9 ms	100 %	9 ms	100 %
record (with buffer)	9 ms	100 %	14 ms	155 %
record (with flush)	9 ms	100 %	22 ms	244 %

code during the record phase was false, the replay mode can skip the call to the original nonblocking receive operation and simply return false. If the return code during the record phase was true, which indicates that a message has actually been available, the replay mode needs to delay the program's execution until the equivalent message is available. This is sufficient to provide equivalent replay of nondeterministic parallel programs [6].

In order to assess the impact of the integrated record&replay mechanisms, a few measurements have been conducted. Firstly, we counted the operations added to the code at blocking receive operations. For the C-code, we added less than 9 lines of code (3 if, 2 else, 1 fprintf, 1 fscanf, 2 MPID function calls). This code was translated into additional 81 lines of assembler code on an x86 compatible processor.

In addition, we measured the execution time at the blocking and nonblocking receive operations. The results of these measurements for the original function call, the function call in record mode, and the function call in replay mode are given in Tables 3 and 4, respectively. The measurements have been performed on a dual processor x86 compatible workstation. For the record mode we distinguish between tracing with buffered tracefile access, and with immediate tracefile access (= flush). The former uses the standard file I/O buffer provided by the operating system to delay write accesses to the harddisk, which is the default behavior during record mode. The latter ensures that all data is stored to disk as soon as it is generated, which is needed whenever a program's execution may be terminated before the buffers have been written to the disk. Consequently, record mode with flush represents the worst case with the largest observable monitor overhead.

For Table 3 it is important to notice that the integrated record&replay mechanism does not raise the execution time of blocking receive operations, when constant source code parameters are provided. This can be attributed to the fact, that these operations perform only one additional if-statement for checking the source parameter. Please note, that no timings are given for the replay mode of

Table 4. MPI_Test Measurements.

mode	measurement	percentage
original	23 ms	100 %
record (with buffer)	25 ms	109 %
record (with flush)	34 ms	148 %
replay	5 ms	22 %

the blocking receive operation, because during replay its behavior is equivalent to the receive operation with constant source parameter.

The interesting part of Table 4 is the measurement of the replay mode at the nonblocking receive operation, which can even be faster than the original function call. The reason for this "strange" measurement is, that the original function call must actually check for the availability of a message, while during replay we only need to return the results provided in the tracefile.

6 Conclusions

The record&replay mechanism described in this paper offer equivalent execution for arbitrary nondeterministic message passing programs. By integrating the system directly into the MPI library, the record phase is performed all the time. This allows to replay a program's execution on demand without instrumentation of the original code. Based on the possibility of perturbations due to the probe effect and limitations in terms of available trace memory, certain optimizations have been included in the mechanisms functionality. The biggest part of these optimizations can be attributed to the non-overtaking message transfer characteristic, which is guaranteed by the message passing interface.

The resulting overhead of the integrated record&replay mechanism is very small, which is also shown with the minimal deviations of the measurements from the original functionality. Additionally, the overhead occurs only at blocking receive operations, that specify a wild card, and at nonblocking receive operations. Thus, the concrete overhead for a given application can be calculated based on the corresponding number of function calls. Some experiments with real-world applications are currently conducted, which prove the small influence of the record functionality.

In terms of future work, several ideas are investigated: on the one hand, nondeterminism introduced by other functions needs to be studied and possibly included in the approach. Some examples in this area are the MPI functions MPI_Cancel and MPI_Waitany, which are explicitly defined as potential sources of nondeterminism in the MPI Standard [13]. Besides that, we are intending to extend the basic record&replay functionality with our event manipulation technique [6], which allows to modify the data observed during the record phase in order to enforce these modifications during subsequent replay phases.

Remark: The necessary modifications for integrating the record&replay functionality in MPICH or LAM can be requested from the authors via email.

References

1. Chassin de Kergommeaux, J., Ronsse, M., De Bosschere, K.: MPL*: Efficient Record/Replay of Nondeterministic Features in Message Passing Libraries. Proc. 6th EuroPVM/MPI Users' Group Meeting, Barcelona, Spain, 141–148 (Sept. 1999).
2. Clemencon, C., Fritscher, J., Rühl, R.: Visualization, Execution Control and Replay of Massively Parallel Programs within Annai's Debugging Tool. Proc. High Performance Computing Symposium, HPCS '95, Montreal, Canada, 393–404 (July 1995).
3. Curtis, R.S., Wittie, L.D.: BugNet: A Debugging System for Parallel Programming Environments. Proc. 3rd Intl. Conf. on Distr. Computing Systems, Miami, FL, USA, 394–399 (October 1982).
4. Fagot, A., Chassin de Kergommeaux, J.: Systematic Assessment of the Overhead of Tracing Parallel Programs. Proc. EUROMICRO PDP '96, 4th EUROMICRO Workshop on Parallel and Distributed Processing, IEEE Computer Society Press, Braga, Portugal, 179–186 (January 1996).
5. Geist, G.A., Sunderam, V.S.: Network-based Concurrent Computing on the PVM System. in: Concurrency - Practice & Experience, **4**, No. 4, 293–311 (1992).
6. Kranzlmüller, D.: Event Graph Analysis for Debugging Massively Parallel Programs. PhD Thesis, GUP Linz, Joh. Kepler Univ. Linz, Austria, (September 2000) http://www.gup.uni-linz.ac.at/~dk/thesis.
7. Lamport, L.: Time, Clocks, and the Ordering of Events in a Distributed System. Communications of the ACM, 558–565 (July 1978).
8. LeBlanc, T.J., Mellor-Crummey, J.M.: Debugging Parallel Programs with Instant Replay. IEEE Transactions on Computers, **C-36**, No. 4, 471–481 (April 1987).
9. LeBlanc, T.J., Robbins, A.D.: Event Driven Monitoring of Distributed Programs. Proc. 5th Intl. Conference on Distributed Computing Systems, IEEE Computer Society Press, Denver, CO, USA, 515–522 (May 1985).
10. Leu, E., Schiper, A.: Execution Replay: A Mechanism for Integrating a Visualization Tool with a Symbolic Debugger. in: Roberts, Y., Bouge, L., Cosnard, M., Trystram, D., (Eds.), Proc. CONPAR 92 - VAPP V, Lecture Notes in Computer Science, **634**, Springer-Verlag (1992).
11. Mackey, M.: Program Replay in PVM. Technical Report, Concurrent Computing Department, Hewlett-Packard Laboratories (May 1993).
12. May, J., Berman, F.: Panorama: A Portable, Extensible Parallel Debugger. Proc. 3rd ACM/ONR Workshop on Parallel and Distributed Debugging, San Diego, CA, USA (May 1993).
13. Message Passing Interface Forum: MPI: A Message-Passing Interface Standard - Version 1.1. (June 1995) http://www.mcs.anl.gov/mpi/.
14. Netzer, R.H.B., Miller, B.P.: Optimal Tracing and Replay for Debugging Message-Passing Parallel Program. Proc. Supercomputing 92, Minneapolis, MN, USA, 502–511 (November 1992).
15. Ronsse, M.A., Kranzlmüller, D.: RoltMP - Replay of Lamport Timestamps for Message-Passing Parallel Systems. Proc. EUROMICRO PDP '98, 6th EUROMICRO Workshop on Par. and Distr. Processing, Madrid, Spain, 87–93 (January 1998).
16. Smith, E.T.: Debugging Tools for Message-Based, Communicating Processes. Proc. 4th Intl. Conference on Distributed Computing Systems, San Francisco, CA, 303–310 (May 1984).

Fast and Scalable Real-Time Monitoring System for Beowulf Clusters

Putchong Uthayopas and Sugree Phatanapherom

Parallel Research Group, CONSYL
Department of Computer Engineering
Faculty of Engineering, Kasetsart University,
Bangkok, Thailand 10400
{pu,b40sup}@ku.ac.th

Abstract. Fast real-time monitoring of system information is important to the understanding of parallel system especially for a large cluster system that appeared recently. Making the system fast and scalable at the same time is still a challenging task. This paper presents the design and implementation of a fast and real time monitoring system called SCMS/RMS. This system is a part of more comprehensive cluster management tool called SCMS. SCMS/RMS is designed to be flexible, highly scalable, and efficient. Many techniques that are used to increase the monitoring speed and to achieve high scalability have been described in this paper. The experiment has been conducted on a 72 nodes Beowulf Cluster and the results show that SCMS/RMS is very fast and highly scalable.

1 Introduction and Related Works

System performance monitoring is important since it allows system administrator to understand system behavior or, in many cases, predict the malfunction earlier. Moreover, performance information can help many important subsystems such as batch scheduler to make a better decision about system resource allocation. Nevertheless, the timelines of the information is crucial for its usefulness. The implementation of fast and efficient real time monitoring is a challenging task especially for the recent large-scale clusters with thousands of node.

Many works related to system monitoring appeared in the literatures. Many tools such as CIS [1], ClusterProbe [2], GARDMON [3], PARMON [4], Co-Pilot [5], are built specifically for system monitoring. Moreover, many monitoring systems appear as a part of system management tools such as SCMS [6] and VACM [7]. The design of each tools is rather different due to the goal and complexity of the monitoring subsystem itself. For example, many tools rely on usual system interface such as /proc in Linux to access operating system performance data. But some tools such as CIS develops their own kernel probe to increase the speed of access and to reduce the level of intrusiveness. Scalability is also a major issue being addressed by many works. Many monitoring subsystems [1][3][4][8][5] are still based on centralized daemon or applications that collect the information from the distributed agents running on every node in the system. This obviously limits the scalability of the

Y. Cotronis and J. Dongarra (Eds.): Euro PVM/MPI 2001, LNCS 2131, pp. 201–208, 2001.
© Springer-Verlag Berlin Heidelberg 2001

system. Hierarchical monitoring is employed to enhance the scalability in ClusterProbe. A performance study of this is presented in [9]. ClusterProbe uses advanced techniques such as data filtering and merging to further reduce the monitoring traffic. The tool called CIS introduces the technique of adjusting the monitoring frequency to reduce the impact. This technique and more are also supported by our implementation. Most monitoring is based on their own protocol over TCP/UDP link except in [10] which builds a large scale monitoring based on SNMP protocol. Extension to the large system such as grid is discussed in [11]. The use of information obtained to predict the behavior of large wide area system is presented by the work on Network Weather Service (NWS)[12].

This paper presents the work on real-time monitoring subsystem called SCMS/RMS. This subsystem is a part of a more comprehensive cluster management tool called SCMS. The goals of SCMS/RMS are to provide a monitoring subsystem for Beowulf cluster that is fast, efficient, low intrusive, and scalable to a very large cluster. Many design techniques and experiences are presented.

The remainder of this paper is organized as follows. Section 2 gives a brief overview of SCMS/RMS system architecture and implementation. Section 3 gives the detail of performance improvement techniques used. The experiments performed to evaluate the performance and results obtained are described in Section 4. Finally, the conclusion and future work are discussed in Section 5.

2   SCMS/RMS System Architecture

SCMS/RMS consists of two kinds of agent processes, namely, CMA (Control and Monitoring Agent) and SMA (System Management Agent). CMA is an agent that runs on every node in the system. CMA main function is to collect the information requested by SMA.

SMA has several functions in the system. Firstly, SMA is a contact point for application. Applications under SCMS/RMS contact SMA through a set of API called RMI (Resource Monitoring Interface). API bindings are currently supported C, C++, Java, and Python. Secondly, SMA also acts as a message router so that the system can be configured to form tree structure as described in the following section.

The organization of SCMS/RMS monitoring subsystem is as illustrated in **Fig. 1**. To make system scalable and efficient, SCMS/RMS monitoring system is organized into *Partition*. Partition is a set of one SMA and multiple CMA, or a set of SMA. In **Fig. 1**, three partitions are shown. Partition 2 and 3 forms a basic structure of monitoring system. They consist of one SMA and a number of CMAs. In this partition, SMA knows about the underlining CMA. SMA and SMA can be connected together to form a partition as illustrated by partition 1. This connection allows us to connect the basic partition into a tree structure. This technique allows SCMS/RMS to easily scale to a very large number of nodes. Using this design also allows user to freely configure the monitoring subsystem into many forms. One example is the partitioning of monitoring system to match the underlining physical sub networks and localize the communication traffic.

Clients are free to connect to any of the SMA. But the scope of information is limited to the information gather from that SMA and the partition below it. Hence, only clients that connect to top-level SMA can access all the information provided by the cluster. In practice, the system can be designed such that client that connectd to any SMA can request any information. But doing so will increase the number of control message substantially.

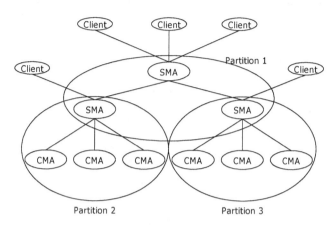

Fig. 1. SCMS/RMS System Structure

In SCMS/RMS, one decision is to clearly separate the information retrieval mechanism from data interpretation. SCMS/RMS only provides mechanism to name any object in cluster system and gather the information about named objected efficiently. The interpretation of data gathered is left to application that uses API. The purpose of this design is to allow SCMS/RMS to be employed in a much broader scope rather than limited to resource monitoring task only. For instance, one can write a probe module that attaches to MPI parallel program. Then gather the usage information of MPI via the MPI profiling interface. Then, a real-time visualization application can be developed easily to gather such information using SCMS/RMS as a delivering mechanism. This is, in fact, being target as one of our future work.

To allow extreme flexibility and extensibility, SCMS/ RMS allows users to build a new plug-in module and load it dynamically into CMA. This feature is based on shared library feature supported by many operating system including Linux. Each plug-in module is identified by 32-bit plugging id. In each plug-in, user can define many interface functions. This set of function is an agreement between end application and plug-in. SCMS/RMS only delivers the information between these two endpoints. Currently, there are two standard plug-in: Hardware Plug-in and Process Plug-in. These two plug-ins provide the interface to local operating system and get system information for SCMS. Using plug-in makes SCMS/RMS very modular and very portable.

In SCMS/RMS, applications can easily access the monitoring services using a set of API called Resource Management Interface (RMI). These API allows users to get node id and name of all live nodes in the cluster, get any information from any object in the system known to SCMS/RMS, and manage plug-in module attached to CMA.

3   Performance Improvement Techniques

The key concept in achieving high scalability and performance is to reduce amount of information passed on the network and the overlapping of processing and communicating activities. The following techniques are employed in SCMS/RMS to reduce network traffic and optimize the performance of monitoring subsystem.

3.1   Dual Mode Operation

SCMS/RMS separates its operation into two modes: Asynchronous and Synchronous mode. By default, SCMS/RMS will operate in Asynchronous mode. In this mode, the information will be sent back along the tree by each CMA and collected centrally in SMA. The frequency of information gathering can be adjusted by users to a suitable one depending on the level of intrusiveness required. Application that requests the information will get the one that are cached. So, the information is only the estimation of the system state at that point but not a precise one. The purpose of this mode is to support the system monitoring tools.

In contrast, synchronous mode is designed for the application that requires more precise and more updated system state. When application requests the information in synchronous mode, all the cached data is bypassed for that session and the request is passed through the system to destination CMA. The information obtained is delivered to requested application through SMA agent that application attachs to. This mode is intended to support systems such as batch scheduling where the correctness of information has more impact.

3.2   Compact Name Space

One important issue is the naming of resources in cluster systems. One approach is to use SNMP style of hierarchical object named space. Although very flexible, this scheme results in a very complex and slow implementation.

In SCMS/RMS, each resource is given a unique integer id. So, the resource can be addressed with a tuple of *<node id, resource id>*. Multiple resources information can be accessed simultaneously by forming a *Resources vector*. Resource vector is a list of integer values that represent a set of resources or objects. Since most API requires users to supply node id, multiple resource addresses on one node can be derived from the resources vector. By using resource vector, the naming can be done very efficiently.

3.3   Message Pipelining

Since SCMS/RMS is structured as a tree of process, it is natural choice to exploit this to speed up the system. All requests are pipelined along the tree structure of the system. The effect is the overlapping between processing time and communicating time. Moreover, the tree structure guarantees that the time complexity of the information gathering scale as $O(\log n)$ where n is number of nodes in the system.

3.4   Filtering and Merging

In SCMS/RMS, the requests from several client applications will be merged together and sent down the tree to destination CMA. Once CMA gets the required data, it will be sent back to destination SMA. Intermediate SMA will merge the data along the way up to the top level SMA. Then the data will be splitted and distributed to the client application that requests them. This method reduces the number of message exchanged substantially.

4   Performance Evaluation

SCMS/RMS has been evaluated on PIRUN Beowulf Cluster system, the main computing facility at Kasetsart University. The experimental system has 72 diskless nodes of Pentium III 500 MHz with 128 Mbytes RAM, 3 Xeon 500 MHz fileserver with 1 Gigabytes of memory. The interconnection used is 3COM Superstack II Fast Ethernet switch. The tests had been carefully conducted when the system had no task to interfere with the communication and computation time. Due to node availability, we were able to test the configuration up to 64 nodes only. Each experiment has been conducted 3-5 times and the average results are used.

The first experiment simulates the task of finding least loaded nodes in the cluster. This is done by getting load information from all nodes and comparing them to find the minimum one. This operation is usually performed by a batch scheduler to locate a set of best node in the cluster. The elapse time from the start of the request to the end of request is measured. The results are as shown in Table 1.

Table 1. Time used to determine the least loaded node

Nodes	Time Used (msec)
1	86
2	46
4	38
8	33
16	45
32	68
64	116

Table 1 shows that SCMS/RMS monitoring system is very fast. The time used to find this information is less that 116 millisecond for 64 nodes. One fact that can be observed from the graph in Fig. 2 is that the time used is quite high for a small number of nodes and then decreases to a minimum at about 10 nodes, then increases again. This is caused by the overhead of monitoring system such as the hierarchical structure organization. So for a small number of nodes, the system will not benefit from the optimization techniques used. Only medium to large cluster can take the advantages of the technique.

In the second experiment, the test is conducted by measuring the average response time when requesting large number of performance data from multiple nodes. This test simulates the case when system administrator wants to monitor the whole system in very detail. The size of information from each node is about 11 Kbytes. Number of nodes used in the experiment is 8, 16, 32, and 64. The comparison has also been made between SCMS/RMS and an open source monitoring software from SGI Incorporate called performance co-pilot (PCP). This software is available on Internet at http://oss.sgi.com/projects/pcp/. The performance co-pilot version 2.2.0 has been used in this experiment. The time used is measured as presented in **Table 2.** and the graph are as illustrated in Fig. 3.

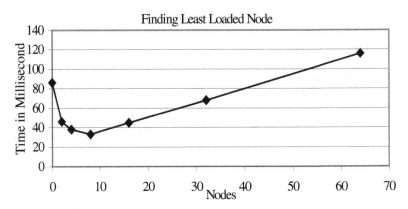

Fig. 2. Timed used to find least loaded node as number of nodes increases

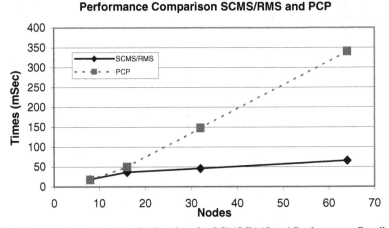

Fig. 3. Comparison of the Monitoring time for SCMS/RMS and Performance Co-pilot

Table 2. Time used to request large amount of information from a set of nodes

Nodes	SCMS/RMS (millisecond)	PCP (millisecond)
8	18.53	17.20
16	36.35	49.39
32	45.69	147.71
64	65.76	340.57

As can be seen from Fig. 3, the time used increases as number of node increases. In all cases, SCMS/RMS is much faster than performance co-pilot. As the number of nodes increase, the monitoring time of SCMS/RMS increases much slower than performance co-pilot. This means that SCMS/RMS can achieve a better scalability.

5   Conclusion

Good real-time performance system is an important component in cluster software tools. Apart from using them to obtain more insight about large system behavior, this subsystem can be applied in a much wider context. For example, real-time monitoring can be used to feed the resource information to batch scheduler. This allows us to build a smart scheduling policy that better utilizes system resources. Another application is the monitoring of running parallel program for debugging or optimizing purposes. This is the area that will be explored as a future work for SCMS/RMS.

Acknowledgement

This work was partially funded by Kasetsart University Research and Development Institute under SRU Grant and Faculty of Engineering Research Grant. Equipments used for the development is supported in part by AMD Far East Inc.

References

1. J. Astalos, "CIS - Cluster Information Service". http://ups.savba.sk/parcom/cluster/
2. Z. Liang, Y. Sun, and C. Wang. "ClusterProbe: An Open, Flexible and Scalable Cluster Monitoring Tool", Proceedings of the First International Workshop on Cluster Computing, pp. 261-268, Dec. 2-3, 1999
3. R. Buyya, B. Koshy, and R. Mudlapur, "Gardmon: A javabased monitoring tool for gardens non-dedicated cluster computing", In Proceedings of Workshop on Cluster Computing Technologies, Environments, and Applications, PDPTA 99, Monte Carlo Resort, Las Vegas, Nevada, USA, 1999
4. R. Buyya, "PARMON: A Portable and Scalable Monitoring System for Clusters", International Journal on Software: Practice & Experience (SPE), John Wiley & Sons, Inc, USA, pp. 723-739, June 2000
5. Silicon Graphics, Inc, "Performance Co-Pilot". http://oss.sgi.com/projects/pcp/
6. Putchong Uthayopas, Jullawadee Maneesilp, Paricha Ingongnam,"SCMS: An Integrated Cluster Management Tool for Beowulf Cluster System", Proceedings of the International Conference on Parallel and Distributed Proceeding Techniques and Applications 2000 (PDPTA'2000) , Las Vegas, Nevada , USA , 26-28 June 2000
7. VA Linux Systems, "VACM Cluster Management". http://www.valinux.com/projects/vacm/
8. R. Flanery, A. Geist, B. Luethke, and S. Scott, "Cluster Command & Control (C3) Tools Suite", http://www.epm.ornl.gov/~sscott/
9. K. Meyer, M. Erlinger, J. Betser, C. Sunshine, G. Goldzmidt, and Y. Yemini, "Decentralizing Control and Intelligence in Network Management", Proceedings of 4th International Symposium on Integrated Network Management, Integrated Network Management IV, Ed. Sethi et.al., pp. 4-15, ISBN 0412715708, Chapman &Hall, 1995

10.Rajesh Subramanyan, Jose Miguel-Alonso, and Jose A.B Fortes, "A scalable SNMP-Based distributed monitoring system for heterogenous network computing", Proceedings of SC2000, Dallas, Texas, 2000
11.B. Tierney, R. Aydt, D. Gunter, W. Smith, V. Taylor, R. Wolski, M. Swany, and the Grid Performance Working Group, "White Paper: A Grid Monitoring Service Architecture (DRAFT)", Global Grid Forum, February 2001.
   http://www-didc.lbl.gov/papers/Grid.Monitoring.wp.pdf
12.R. Wolski, N. T. Spring, and J. Hayes, "The Network Weather Service: A Distributed Resource Performance Forecasting Service for Metacomputing", Proceedings of 6th High-Performance Distributed Computing, 1998

Dynamic Process Management
in KSIX Cluster Middleware

Thara Angskun, Putchong Uthayopas, and Arnon Rungsawang

Parallel Research Group, Computer and Network System Research Laboratory,
Department of Computer Engineering, Faculty of Engineering,
Kasetsart University, Bangkok, Thailand 10900
Phone: (662) 942-8555 Ext. 1416
{g4265087,pu,fenganr}@ku.ac.th

Abstract. Dynamic process management is a much-needed feature for applications and tools development in Beowulf cluster environment. A well-defined and efficient dynamic process management in cluster middleware layer can simplify the programming task of parallel tool developers. This paper presents the design and implementation of dynamic process management in a cluster middleware called KSIX. KSIX provides a rich set of system call that handles rapid process creation, termination, and remote signal delivery. Moreover, KSIX also handles the correct redirection of standard input, standard output and standard error of process or group of processes. The experiment has been conducted and the results are presented to illustrate the performance of our implementation.

1 Introduction

The development of applications and tools under Beowulf cluster [1] environment is still a challenging task because this platform provides only a traditional single processor UNIX programming environment. Proper extension of OS services can simplify the programming task for parallel software tools developer tremendously. The use of middleware seems to be a promising approach since it can create a very portable and efficient implementation. This reason motivates us to develop a middleware called KSIX for our SCE (Scalable Cluster Environment) [2] software system. Currently, KSIX is used to support our SQMS batch scheduling system, and SCMS cluster management tool [3]. In this paper, we will focus only on the fast dynamic process management in KSIX middleware, which is the extension from our previous work described in [4]. The detail discussion focuses on the implementation of fast process control using dynamically formed tree algorithm. This results in a process creation on multiple hosts that is much faster than the traditional implementation.

The organization of this paper is as follows. In Section 2, we describe KSIX overview followed by KSIX implementation issue in Section 3. Section 4 presents the experimental results. Section 5 presents the discussion about some related works. Finally, we present the conclusion and future work in section 6

Y. Cotronis and J. Dongarra (Eds.): Euro PVM/MPI 2001, LNCS 2131, pp. 209–216, 2001.
© Springer-Verlag Berlin Heidelberg 2001

2   KSIX Overview

KSIX system consists of a set of daemons running in user space, no kernel modification is required for the current version. An application can use KSIX services by calling a set of API provided with KSIX library. An application can use KSIX to spawn a new task or a set of tasks on any node in the cluster. KSIX automatically provides a unified process space without the machine boundary by allocating a global process id and global process group to these newly created processes. KSIX task id is used for task identification in the subsequent operations. KSIX also supports the operations of sending signal to task, getting process status and more. In KSIX, tasks are in one of the three mode of operations. In *Normal* Mode, a task acts the same way as normal Unix process. In *Restart Mode*, the task will automatically be restarted by KSIX when terminated. Finally, in *Migration Mode,* KSIX starts the task on different node after the task termination is detected. The purpose of this design is to provide a future support for process migration and check pointing.

KSIX also offers a mechanism for distributed event services, lightweight directory services, and membership management. There are more than 30 KSIX system calls available now for tool developers.

For process management, KSIX APIs for C and C++ are as listed in Table 1. KSIX process management APIs can be divided into three groups.

Table 1. Process management API

KSIX Process Management API		Description
C++	C	
Process::kxSpawn(...)	kx_spawn(...)	Spawn process
Process::kxSpawnIO(...)	kx_spawnIO(...)	Spawn process with input/output
Process::kxWaitPid(...)	kx_waitpid(...)	Wait for process terminate
Process::kxPkill(...)	kx_pkill(...)	Send signal to process
Process::kxGkill(...)	Kx_gkill(...)	Send signal to group of process
Process::kxAllPs(...)	kx_alllps(...)	Report all process status
Process::kxUserPs(...)	kx_userps(...)	Report status of user process
Process::kxSetPmode(...)	kx_setpmode(...)	Change class of process
Process::kxSetGmode(...)	kx_setgmode(...)	Change class of a process group

Process creation and control: These groups of functions control the creation and termination of KSIX processes. The function kx_spawn() is used to   spawn the processes with default I/O redirected to standard input and standard out of the caller process. But in kx_spawnIO, the I/O redirection can be   fully controlled by programmer. Both APIs are spawned in non-blocking mode. Parent process can wait for the successful return of child processes by calling kx_waitpid().

Signal delivery: programmer can send signal to process using kx_pkill(). Signal is delivered to process across machines transparently. Moreover, signal can be sendt to a group of processes using an function called kx_gkill(). One restriction is that a signal can only be sent to processes that belong to the current user.

Process status handling: Programmer can query process status of whole system using kx_allps(), and from particular user using kx_userps(). User can change process mode or a single process by calling kx_setpmode() and can change process mode of the whole group by calling kx_setgmode(). Both APIs intend to support batch scheduling, which allow user to submit process with checkpointing and process migration.

There are many possible applications for KSIX process management possible. For example, in cluster environment, users usually rely on slow rsh and ssh mechanisms for remote command execution. This method cannot scale well for very large system. There is an effort to define a parallel extension to Unix command by parallel tool consortium. This SUT (Scalable Unix Tools) effort is well explained in the literature. Using KSIX fast and collective process management, a powerful SUT implementation can be done by replacing rsh with KSIX based remote execution command. Remote process can be started simultaneously on the remote machines to execute local Unix command. Then KXIO can be used to relay back the result efficiently.

KSIX dynamic process management is designed such that process creation, process termination, process group, and signal delivery can be extended to support dynamic process management of MPI-2 standard with ease. MPI_COMM_SPAWN can be mapped to KSIX spawn. Processes in KSIX always form into a group or context. This organization can easily be mapped to the context-based concept of MPI communicator. Parent and child group can be created by first create KSIX group, then using KX_Spawn to create child group. Group id and intercommunicator can be built and kept track later. Finally, efficient dynamic process creation and control provided by KSIX can be mapped directly to MPI2 approach. It will help ease the development effort greatly.

3   KSIX Implementation

Dynamic process management subsystem in KSIX consists of two main components. The first one is KSIX daemon (KXD) that keeps track processes status, maintain global process ID, create processes, terminate processes, send signal to processes and membership management. KXD uses 2 table called *process table* and *node table* to keep track of all these information. Another component is KXIO, which is responsible for input and output management.

KSIX system can be started using a shell-script called *ksixboot*. This script starts KXD and KXIO on a first node called *boot node*. Then, the script starts KXD and KXIO on the rest of cluster node. New KXD will register itself with KXD on a booted node (called proxy KXD) using information given in the startup script. After the registration, newly created KXD obtains node table from proxy KXD and announce it existence to proxy KXD. When proxy KXD detects a new KXD from the registration process, it sends the information of a new KXD to the first node in node table, called root node. Root node will help synchronize node table of every node using a broadcast mechanism in KSIX, which will be discussed in detail later. After all KSIX boot processes finish, the structure of the system will be as illustrated in Fig. 1.

Each KXD detects the absence of other KXD daemon by sending a heartbeat message around the logical ring that constructed from host ID. KXD also identifies the exit status of process to prevent forever restart or migrate. If user process terminate abnormally, KSIX will automatic change process type of that process to normal.

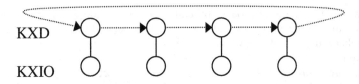

Fig. 1. KSIX structure after the end of boot process

KXD communicates with each other using 3 message types. First, a message called UNICAST is used for point-to-point communication between daemons. For the group communication, there are 2   modes of operation available, namely, MCAST and BCAST modes. Both of them are based on tree algorithm to increase speed and scalability. The difference is that MCAST mode will send message to a set of adjacent node while BCAST mode send message to every node.

MCAST and BCAST modes are used mainly for group operations. This is useful for operation such as sending process startup message to a set of daemons to request them to start a group of new processes. The message exchanges operation occurred according to the following algorithm.

In BCAST mode, suppose a node with node id M wants to send messages from a node with id R to cluster size S where $0 \leq M < S$, $0 \leq R < S$ and $S > 2$. We can define child ID from following steps.

Step 1: Build node table
Table $_K = (K + R)$ mod S
Step 2: Define child ID
Child $_{Left}$ = Table $_{2M+1}$ where $2M + 1 < S$
Child $_{Right}$ = Table $_{2M+2}$ where $2M + 2 < S$

MCAST uses similar algorithm except that the node set is limited to the set of adjacent nodes only. In the other cases, MCAST are simulated using UNICAST messages instead. With this communication structure, process creation and remote command submission can be done very quickly.

Although distributed approach is more of scalable, the important problem is race condition. To avoid this problem, all information updated such as the membership change must be originated from the root node (HOSTID=0) only. The rest of the operations can be done from any location.

KSIX uses KXIO to manage standard input, standard output, and standard error of KSIX processes. When KXD spawn a process, it redirects the standard input, standard output, and standard error of that process to KXIO. For standard output handling, KXIO on each node collects all output from local KSIX process and send them to KXIO of the first process. The output data gathered is then sent the to the device that user specified in KX_Spawn and KX_spawnIO. For the standard input handling, the standard input of first process will be delivered to every process in the group by KXIO. KXIO uses special message header to multiplex the data stream received from KSIX processes.

All of the major components in KSIX are implemented with C++ . C++ STL is used to implement all data handling such as message queue and process table, and global attribute. Using STL helps maintain a good code structure and reduce the bug in handling complex data management. Multithread programming is heavily used in KXD code to gain the performance. The structure of thread is as depicted in Fig. 2.

The daemon consists of 3 kinds of thread. The first one is called master thread. This thread receives an incoming message and pushes it into common message queue. The second one is called worker thread whose function is to get the message from message queue and then, execute the operation as specified in the message header. The third thread is called keep-alive thread. This thread periodically checks the status of neighboring KXD and notifies the system when some KXD is missing.

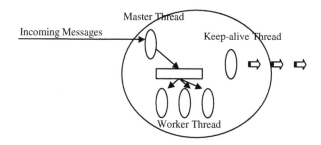

Fig. 2. KXD Architecture

4   Experiment

The experiment has been conducted to test KSIX implementation by comparing its process startup time for large number of tasks with LAM/MPI and rsh. The purpose of this test is to study two things. First, rsh is the most used way to start remote execution of scalable Unix tool on multiple hosts. By comparing KSIX fast process startup time with rsh mechanism, the potential improvement in term of speed can be seen. Second, the comparison with LAM can show the potential speed gained if KSIX has been used as a fast process startup for MPI implementation.

The experiment has been performed using a program, which is referred to as master task, to start a set of remote tasks on multiple hosts and wait for the reply messages from all slave tasks. Once started, each slave task sends a UDP message back to master task. After received all UDP messages from all slave tasks, the master task will terminate. The elapse time from the start of all the slave tasks until   master task received all the reply has been measured. Three sets of experiment have been conducted. The first one uses RSH mechanism to start the remote tasks. The second one use LAM/MPI 6.5.1 and the last one use KSIX. The experiment has been repeated for five times on 2,4,8, and 16 nodes. The results shown are the average value of the data obtained.

The experimental system is a Beowulf Cluster System called PIRUN at Kasetsart University, Thailand. This system consists of 72 Pentium III 500 MHz with a memory of 128 Mbytes/nodes and 3 Pentium Xeon 500 MHz file server nodes with 1 Gbytes of memory. A Fast Ethernet Switch is used as an interconnection network. The results of the experiment are as shown in Figure 3, Figure 4.

As illustrated in Fig.3, KSIX can start 16 processes on 16 nodes using only about 100 milliseconds. KSIX is faster than LAM/MPI and much faster than rsh. The reason is that by combining a quick broadcasting to process creation, more concurrency has been achieved on the process startup. The increase in process startup time is much less in KSIX than in RSH based and LAM/MPI implementation. Hence, this means that KSIX process creation can potentially scale much better.

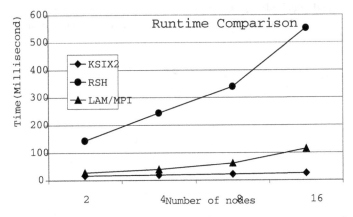

Fig. 3. Comparison of KSIX, LAM/MPI and RSH based remote process creation time

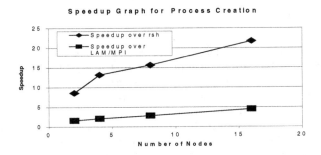

Fig. 4. Speedup of KSIX process creation over RSH and LAM/MPI

Figure 4 shows speed up of KSIX implementation of process startup over RSH based implementation. As number of machine increase, KSIX speed up also increases. At 16 nodes, the speedup increases over 20 times. Again, this clearly show that using KSIX to implement scalable UNIX command will be much faster in a large cluster.

5   Related Works

There are many ongoing projects related to the cluster middleware. Basically, these projects can be classified into two approaches. The first approach is to extend operating system functionality at kernel level. Some example works are Mosix[5], Nomad [6] ,and Bproc [7]. Mosix offers a kernel-level, adaptive resource sharing algorithms and low-overhead process migration. Nomad effort is quite similar to

Mosix but has a wider scope of building a scalable cluster OS that provides a single-system image, resources management, and distributed file system. Bproc attempts to provide a transparent global process control through the use of proxy process on central node. In effect, Bproc gives programmers a loosely view of SMP system on top of a cluster. Although this approach is very efficient, Kernel extension is not likely to be portable and hard to keep pace with rapid kernel changes.

The second approach is to use a middleware to provide a virtual machine view in user space with APIs to access that its services. Although, the performance and transparency are less than the first approach, the advantage is the higher portability. This is crucial factor considering a rapid changing nature of an operating system such as Linux. The reason is the reduction of the needs to frequently release new patches. The examples of this approach are PVM [8] and MPI [9], Glunix [10], and Score-D [11]. In a larger scope is the recent interest in Grid technology. Some of MPI implementations, which support MPI-2 standard has already had dynamic process management features including LAM/MPI [12], WMPI [13], and Fujitsu 's MPI [14]. Finally, there are many efforts that focus on middleware for wide area grid computing such as Harness [15], and Globus[16], and Legion[17] project.

6   Conclusion and Future Works

KSIX is a very important part of our fully integrated cluster environment called SCE. Many new features is planned such as extend the APIs to meet more needs in cluster development, providing the better support for MPI2, adding file services that allows users to move and manipulate remote file object easily and transparently, and adding a certain level of high-availability support. All these features will create a rich cluster environment for users and developers of Beowulf cluster.

References

1.   T. Sterling, D. J. Becker, D. Savarese, J. E. Dorband, U. A. Ranawake, and C. E. Packer, "Beowulf: A Paral;lel Workstation for Scientific Computation", in Proceedings of International Conference on Parallel Processing 95,1995
2.   Putchong Uthayopas, Sugree Phatanapherom, Thara Angskun, Somsak Sriprayoonsakul, "SCE: A Fully Integrated Software Tool for Beowulf Cluster System", to be appeared in Linux Cluster: The HPC Revolution, A conference for high-performance Linux cluster users and system administrators, University of Illinois, Ubana, Illinois, USA, June 2001
3.   Putchong Uthayopas, Jullawadee Maneesilp, Paricha Ingongnam, "SCMS: An Integrated Cluster Management Tool for Beowulf Cluster System", Proceedings of the International Conference on Parallel and Distributed Proceeding Techniques and Applications 2000 (PDPTA'2000), Las Vegas, Nevada, USA, 26-28 June 2000
4.   Thara Angskun, Putchong Uthayopas, Choopan Ratanpocha, "KSIX parallel programming environment for Beowulf Cluster", Technical Session Cluster Computing Technologies, Environments and Applications (CC-TEA), International Conference on Parallel and Distributed Proceeding Techniques and Applications 2000 (PDPTA'2000), Las Vegas, Nevada, USA, June 2000
5.   A. Barak, O. La'adan, and A. Shiloh, "Scalable Cluster Computing with MOSIX for LINUX", in Linux Expo 99, pp95-100, Raleigh, N.C., USA, May1999

6. E. Pinheiro and R. Bianchini, "Nomad: A scalable operating system for clusters of uni and multiprocessors", in Proceeding of IWCC99, Melbourne, Australia, December 1999

7. E. Hendriks, "BPROC: A distributed PID space for Beowulf clusters", in Proceeding of Linux Expo 99, Raleigh N.C, May 1999

8. V. Sunderam and J. Dongarra, "PVM:A Framework for Parallel Distributed Computing", Concurrency: Practice and Experience, pp. 315-339, 1990

9. W. Gropp, E. Lusk and A. Skjellum, "Using MPI: Portable Parallel Programming with the Message-Passing Interface", MIT Press, 1994

10. Douglas P. Ghormley, David Petrou, Steven H. Rodrigues, Amin M Vahdat, and Thomas E. Anderson, "GLUnix: a Global Layer Unix for a Network of Workstations", Software-Practice and Experience Volume 28, 1998

11. A. Hori, H. Tezuka, and Y. Ishikawa, "An Implementation of Parallel Operating System for Clustered Commodity Computers", in Proceedings of USENIX 99, 1999.

12. Greg Burns, Raja Daoud, James Vaigl, "LAM: An open cluster environment for MPI", Proceeding of Supercomputing Symposium 94, pp 379-386, University of Toronto, 1994

13. Marinho, j. and Silva, J.G., "WMPI: Message Passing Interface for Win32 Clusters", Proceeding of 5th European PVM/MPI User's Group Meeting, pp. 113-120, September 1998

14. Asai, N., Kentemich, T., Lagier, P., "MPI-2 Implementation on Fujitsu generic message passing kernel", Proceeding of Supercomputing'99, Portland, Oregon, US, November 1999

15. Micah Beck, Jack J. Dongarra, Graham E. Fagg, G. AL Geist, Paul Gray, James Kohl, Mauro Migliaridi, Keith Moore, Terry Moore, Philip Papadopoulous, Stephen L. Scott and Vaidy Sunderam, "HARNESS: A Next Generation Distributed Virtual Machine", International journal on future generation computer system Elsevier Publ, Volume 15, 1999.

16. I. Foster, C. Kesselman, "Globus: A Metacomputing Infrastructure Toolkit", International journal on Supercomputer Applications, 11(2): 115-128,1997

17. Andrew S. Grimsaw, Wm. A. Wulf, "Legion—A View From 5,000 Feet", in Proceedings of the Fifth IEEE International Symposium on High Performance Distributed Computing, IEEE Computer Society Press, Los Alamitos, California, August 1996.

Adaptive Execution of Pipelines*

Luz Marina Moreno, Francisco Almeida, Daniel González, and Casiano Rodríguez

Dpto. Estadística, I.O. y Computación, Universidad de La Laguna,
La Laguna, Spain
{falmeida,dgonmor,casiano}@ull.es

Abstract. Given an algorithm and architecture a tuning parameter is an input parameter that has consequences in the performance but not in the output. The list of tuning parameters in parallel computing is extensive: some depending on the architecture, as the number of processors and the size of the buffers used during data exchange and some depending on the application. We formalize the General Tuning Problem and propose a generic methodology to solve it. The technique is applied to the special case of pipeline algorithms. A tool that automatically solves the prediction of the tuning parameters is presented. The accuracy is tested on a CRAY T3E. The results obtained suggest that the technique could be successfully ported to other paradigms.

1   Introduction

Most of the approaches to performance analysis fall into two categories: Analytical Modeling and Performance Profiling.

Analytical methods use models of the architecture and the algorithm to predict the program runtime. The degree to which the results approximate reality is determined by the extents of the assumptions and simplifications built in the model. The analytical model provides a structured way for understanding performance

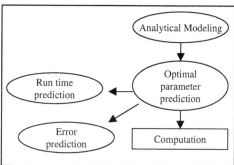

Fig. 1. Analytical Modeling. Squared Blocks represent automated steps

problems, and has predictive ability (figure 1). However analytical modeling is not a trivial task. Abstractions of parallel systems may be difficult to specify because they must be simple enough to be tractable, and sufficiently detailed to be accurate. Many Analytical Models have been proposed (PRAM [5], LogP [2], BSP [13], BSPWB [11], etc.) assuming different computation/communication models.

---

* The work described in this paper has been partially supported by the Spanish Ministry of Science and Technology (CICYT) TIC1999-0754-C03.

Y. Cotronis and J. Dongarra (Eds.): Euro PVM/MPI 2001, LNCS 2131, pp. 217–224, 2001.

Profiling may be conducted on an existing parallel system to recognize current performance bottlenecks, correct them, and identify and prevent potential future performance problems (figure 2). Performing measurements requires special purpose hardware and software and, since the target parallel machine is used, the measurement method can be highly accurate. In [4] was presented a prediction tool that is part of the Vienna Fortran Compilation System. The goal of this tool is to rank the performance of different parallel program versions without running

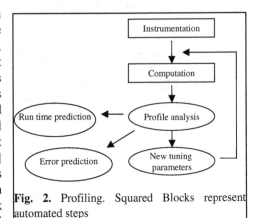

Fig. 2. Profiling. Squared Blocks represent automated steps

them. The tool assumes that parallelizations are based on loops, and worst-case scenarios of traffic and contention that can never occur, what constitutes a restriction on the approach.

Currently, the majority of performance metrics and tools devised for performance evaluation and tuning reflect their orientation towards the measurement-modify paradigm (figure 2). Many projects have been developed to create trace files of events with associated time stamps and then examine them in post-mortem fashion by interpreting them graphically on a workstation. The ability to generate trace files automatically is an important component of many tools like PICL [6], Dimemas [8], Kpi [3]. Trace file presentation programs like ParaGraph [7], Vampir [14], Paraver [8], etc. helps to understand the information provided.

Although much work has been developed in Analytical Modeling and in Parallel Profiling, sometimes seems to be a divorce between them. Analytical modeling use to be considered too theoretical to be effective in practical cases and profiling analysis sometimes is criticized of loss of generality. We claim that to obtain automatic and effective practical tools with predictive ability, both fields should be integrated (figure 3). When executing an algorithm, the user should know the analytical model and provide the complexity analytical formula of the algorithm implemented.

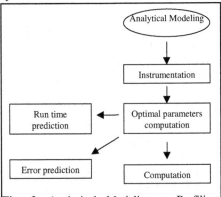

Fig. 3. Analytical Modeling + Profiling. Squared Blocks represent automated steps

According to this formula, the profiler could compute the parameters needed by this formula for the performance prediction and use them on the optimal execution of the algorithm.

In section 2 we formalize the General Tuning Problem and propose a generic methodology to approach it. In section 3 we restrict to the Tuning Pipeline Problem and apply the technique presented in the previous section to the Pipeline Paradigm.

Theoretical solutions for some of the steps required for the methodology have been devised. However, the inclusion of these methodologies in a software tool still remains a formidable challenge.

We have extended the La Laguna Pipeline tool with automatic profiling facilities. The La Laguna Pipeline *llp* [9] is a general purpose tool for the pipeline programming paradigm. The new system measures the relevant code sections, find the constants involved, minimizes the complexity function and finds the optimal parameters for the execution on the current input and architecture. These parameters may include the optimal number of processors, the buffer sizes and the number of processes per processor (granularity). The feasibility of the technique and its accuracy has been contrasted over pipeline algorithms for combinatorial optimization problems.

2   The Problem and the Methodology

As is common in Computability Theory, any algorithm A determines a function F_A from an input domain D to an output domain. Usually the input domain D is a cartesian product $D = D_1 x...xD_n$, where D_i is the domain for the *i-th* parameter

$$F_A : D = D_1 x...xD_n \subset \Sigma^* \rightarrow \Sigma^*$$

such that $F_A(z)$ is the output value for the entry z belonging to D. This algorithm A, when executed with entry z on a machine M, spends a given *execution time*, denoted $Time_M(A(z))$. In most cases this $Time_M(A(z))$ function can be approximated by an analytical Complexity Time formula $CTime_M(A(z))$. We assume that $CTime_M(A(z))$ represents with enough accuracy the actual function time $Time_M(A(z))$.

We will classify the parameters $D_1 x...xD_n$ of F_A into two categories $\mathbb{T}$ ($\mathbb{T}$ comes for tuning parameters) and I (for true Input parameters).

We define that $x \in D_i$ is a "*tuning parameter*" , $x \in \mathbb{T}$ if and only if, occurs that x has only impact in the performance of the algorithm but not in its output. We can always reorder the tuning parameters of A, to be the first ones in the algorithm:

$$\mathbb{T} = D_1 x...xD_k \text{ and } I = D_{k+1} x...xD_n$$

With this convention is true that:

$$F_A(x, z) = F_A(y, z) \text{ for any } x \text{ and } y \in \mathbb{T}$$

But, in general, $Time_M(A(x, z)) \neq Time_M(A(y, z))$.

The "*Tuning Problem*" is to find $x_0 \in \mathbb{T}$ such that

$$CTime_M(A(x_0, z)) = min \{ CTime_M(A(x, z)) \ /x \in \mathbb{T}\}$$

The list of tuning parameters in parallel computing is extensive: the most obvious tuning parameter is the Number of Processors. Another is the size of the buffers used during data exchange. Under the Master-Slave paradigm, the size and the number of data item generated by the master must be tuned. In the parallel Divide and Conquer technique, the tuning parameters not only control when subproblems are considered trivial but also govern the processor assignment policy. On regular numerical HPF-like algorithms, the block size allocation is a tuning parameter.

The general approach that we propose to solve the tuning problem is:

1. Profiling the execution to compute the parameters needed for the Complexity Time function $CTime_M(A(x, z))$.

2. Compute $x_0 \in T$ such that minimizes the Complexity Time function $CTime_M(A(x, z))$.

$$CTime_M(A(x_0, z)) = min \{ CTime_M(A(x, z)) / x \in T\} \quad (1)$$

3. At this point, the predictive ability of the Complexity Time function can be used to predict the execution time $Time_M(A(z))$ of an optimal execution or to execute the algorithm according to the tuning parameter T.

The success of the technique will depend on the accuracy of the three factors, the analytical model, the profiling and the minimization algorithm. Errors in any of these factors are cumulative and may imply the failure of the method.

3   The Pipeline Tuning Problem

As an example to illustrate the technique presented we will consider the pipeline paradigm. The implementation of pipeline algorithms with N stages on a target architecture is strongly conditioned by the actual assignment of virtual processes to the physical processors and their simulation, the granularity of the architecture, and the instance of the problem to be executed.

We will restrict our study to the case where the code executed by every processor of the pipeline is the M iteration loop of figure 4. In the loop that we consider, *body0* and *body1* take

```
void f() {
    Compute(body0);
    i = 0;
    While(i < M) {
        Receive();
        Compute(body1);
        Send();
        Compute(body2, i);
        i++
    }
}
```

Fig. 4.

constant time, while *body2* depends on the iteration of the loop. This loop represents a wide range of situations, as is the case of many parallel Dynamic Programming algorithms [12].

The virtual processes running this code must be assigned among the p available processors. To achieve this assignment the set of processes is mapped on a one way ring topology following a mixed block-cyclic mapping. The grain G of processes assigned to each processor is the second tuning parameter to ponder. Buffering data reduces the overhead in communications but can introduce delays between processors increasing the startup of the pipeline. The size B of the buffer is our third tuning parameter.

The Analytical Model

The optimal tuning parameters $(p_o, G_o, B_o) \in \mathbb{T} = \{ (p, G, B) / p \in \mathbb{N}, 1 \leq G \leq N/p, 1 \leq B \leq M \}$ must be calculated assumed that the constants characterizing the architecture and the constants associated to the algorithm have been provided. Usually the architectural dependent parameters can be computed just once, but the constants depending on the problem are not known in advance and should be computed for each instance of the problem. Although several Analytical Models have been provided for this kind of algorithms for particular cases [1], we will follow the more general model presented in [10].

In this model, the time to transfer B words between two processors is given by $\beta + \tau B$, where β is the message startup time (including the operating system call, allocating message buffers and setting up DMA channels) and τ represents the per-word transfer time. In the internal communications τ is neglected and only the time to access the data is considered. An external reception is represented by (β^E) and an internal reception by (β^I). The last constant includes the time spent in context switching. The variables t_0, t_1, t_{2i} respectively denote the times to compute $body0$, $body1$ and $body2$ at iteration i.

The startup time between two processors T_s includes the time needed to produce and communicate a packet of size B

$$T_s = t_0*(G - 1) + t_1 * G * B + G*\Sigma_{i = 1, (B-1)}\, t_{2i} + 2*\beta^I * (G - 1)* B + \beta^E * B + \beta + \tau *B$$

T_c denotes the whole evaluation of G processes, including the time to send M/B packets of size B:

$$T_c = t_0*(G - 1) + t_1*G*M + G*\Sigma_{i = 1, M}\, t_{2i} + 2*\beta^I *(G - 1)*M + \beta^E*M + (\beta + \tau*B)* M/B$$

According to the parameters G, B and p, two situations arise. After a processor finishes the work in one band it goes to compute the next band. At this point, data from the former processor may be available or not. If data are not available, the processor spends idle time waiting for data. This situation appears when the startup time of processor p (the first processor of the ring in the second band) is larger than the time to evaluate G virtual processors, i. e., when $T_s *p \geq T_c$. For a problem with N stages on the pipeline (N virtual processors) and a loop of size M (M iterations on the loop), the execution time is determined by:

$$T(p, G, B) = max \{ T_s * (p - 1) + T_c * N/(G*p), T_s * (N/G - 1) + T_c \}$$

with $1 \leq G \leq N/p$ and $1 \leq B \leq M$, where,

$T_s * p$ holds the time to startup processor p and $T_c * N/(G*p)$ is the time invested in computations after the startup.

According to the notation expressed in section 2 the tuning parameter is $T = (p,$ $G, B)$ and the input parameter is $I = (N, M, t_0, t_1, t_2)$. In this case, I depends not only on the problem but on the particular instance of the problem too and therefore the solution to problem (1) should be computed before the execution of the parallel algorithm. An important observation is that $T(p, G, B)$ first decreases and then increases if we keep G, B or p fixed and move along the other parameters. Since, for practical purposes, all we need is to give values for (p, G, B) leading us to the valley of the surface, a few numerical evaluations of the function $T(p, G, B)$ will be sufficient to find the optimal.

Embedding the Pipeline Tuning Solver in *llp*

The La Laguna Pipeline tool, *llp*, enrolls a virtual pipeline into a simulation loop according to the mapping policy specified by the user. This policy is determined by the grain parameter, G. *Llp* also provides a directive to pack the data produced on the external communications. The directive establishes the number of elements B to be buffered. Former work [9] proves that the performances obtained with *llp* are similar to those obtained by an experienced programmer using standard message passing libraries.

We have instrumented *llp* to solve automatically the Pipeline Tuning Problem. The profiling step runs sequentially just one stage of the pipeline so that the whole set of input parameters is known in advance. The minimization function for the analytical model supplying the parameters for an optimal execution is then applied.

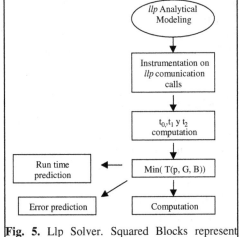

Fig. 5. Llp Solver. Squared Blocks represent automated steps

To solve the tuning problem the input parameters $I = (N, M, t_0, t_1, t_2)$ must be known before the minimization function be called. Given that N and M are provided by the user, (t_0, t_1, t_2) must be computed for each instance. Since the computations on the pipeline code (fig 4) are embedded into two *llp-communications* calls, during the profiling phase, these *llp* routines are empty and just introduce timers.

Computational Results

To contrast the accuracy of the model we have applied it to estimate (p_0, G_0, B_0) for the pipeline approach on the dynamic programming formulation of the knapsack problem (KP) and the resource allocation problem (RAP). The machine used is a CRAY T3E providing 16 processors. For the problems considered, we have developed a broad computational experience using *llp*. The computational experience has been focused to finding experimentally the values (p_0, G_0, B_0) on each problem. The tables denote the optimal experimental parameters as *G-Real*, *B-Real*. These were found by an exhaustive exploration of the *GxB* search space. *Best Real Time* denotes the corresponding optimal running time. The running time of the parallel algorithm for parameters (G_0, B_0) automatically calculated solving equation (1) is presented in column *Real Time*. The tables also show the error made ((*Best Real Time - Real Time*) / *Best Real Time*) by considering the parameters automatically provided by the tool. The very low error made with the prediction makes the technique suitable to be considered for other parallel paradigms.

Table 1. (G_0, B_0) prediction

P	G_0	B_0	Real Time	G-Real	B-Real	Best Real Time	Error
			Knapsack Problem (KP12800)				
2	10	3072	138.61	20	5120	138.24	0.0026
4	10	1536	70.69	20	1792	69.47	0.017
8	10	768	35.69	20	768	35.08	0.017
16	10	256	18.14	10	768	17.69	0.025
			Resource Allocation Problem (RAP1000)				
2	2	10	74.62	5	480	70.87	0.053
4	2	10	37.74	5	160	36.01	0.048
8	2	10	19.26	5	40	18.45	0.044
16	2	10	10.06	5	40	9.76	0.031

4   Conclusions

We have presented a formal definition of the General Tuning Problem and proposed a generic methodology to approach it. A special case for Pipelines has been approached. We have extended the La Laguna Pipeline tool with automatic profiling facilities. The new system measures the relevant code sections, find the constants involved, minimizes the complexity function and finally determines the optimal number of processors, the buffer sizes and the granularity for the execution on the current input and architecture. The steps involved use previous theoretical work, whose optimality has been soundly proved. The feasibility of the technique and its accuracy has been contrasted on pipeline algorithms for Knapsack Problems and for Resource Allocation Problems on a CRAY T3E.

Acknowledgments

We thank to the CIEMAT for allowing us the access to their machines.

References

1. Andonov R., Rajopadhye S.. Optimal Orthogonal Tiling of 2D Iterations. Journal of Parallel and Distributed computing, 45 (2), (1997) 159-165.
2. Culler D., Karp R., Patterson D., Sahay A., Schauser K., Santos E., Subramonian R., von Eicken T.. LogP: Towards a realistic model of parallel computaiotn. Proceeings of the 4th ACM SIGPLAN. Sym. Principles nad Practice of Parallel Programming. May 1993.
3. Espinosa A., Margalef T., Luque E.. Automatic Performance Evaluation of Parallel Programs. Proc. Of the 6th EUROMICRO Workshop on Parallel and Distributed Processing. IEEE CS. 1998. 43-49.
4. Fahringer T., Zima H.. Static Parameter Based Performance Prediction Tool for Parallel Programs. Proc. of ACM International Conference of Supercomputing. ACM Press. 1993. 207-219.
5. Fortune S., Wyllie J. Parallelism in Randomized Machines. Proceedings of STOC. 1978. 114-118.
6. Geist A., Heath M., Peyton B., Worley P.. PICL: Aportable Instrumented Communications Lybrary, C Reference Manual. Technical Report TM-11130. Oak Ridge National Laboratory. 1990.
7. Heath M. Etheridge J.. Visualizing the Performance of Parallel Programs. IEEE Software. 8 (5). September 1991. 29-39.
8. Labarta J., Girona S., Pillet V., Cortes T., Gregoris L.. Dip: A Parallel Program Development Environment. Europar 96. Lyon. August 1996.
9. Morales D., Almeida F., García F., González J., Roda J., Rodríguez C.. A Skeleton for Parallel Dynamic Programming. Euro-Par'99 Parallel Processing Lecture Notes in Computer Science, Vol. 1685. Springer-Verlag, (1999) 877–887.
10. Morales D., Almeida F., Moreno L. M., Rodríguez C.. Optimal Mapping of Pipeline Algorithms. EuroPar 2000. Munich. Sept. 2000. 320-324
11. Roda J., Rodriguez C., Gonzalez D., Almeida F..Title: Predicting the Execution Time of Message Passing Models. Concurrency: Practice & Experience. Addison Wesley. Vol 11(9). 1999. 461-447
12. Rodriguez C., Roda J., Garcia F., Almeida F., Gonzalez D.. Paradigms for Parallel Dynamic Programming. Proceedings of the 22nd Euromicro Conference. Beyond 2000: Hardware and Software Strategies. IEEE. (1996) 553-563.
13. Valiant L. A Bridging Model for Parallel Computation. Commun. ACM, 33 (8). 1990. 103-111.
14. Vampir 2.0 Visualization and Analysis of MPI Programs. http://www.pallas.com

MemTo: A Memory Monitoring Tool for a Linux Cluster*

Francesc Giné[1], Francesc Solsona[1], Xavi Navarro[1],
Porfidio Hernández[2], and Emilio Luque[2]

[1] Departamento de Informática e Ingeniería Industrial, Universitat de Lleida, Spain
{sisco,francesc,jnavarro3}@eup.udl.es
[2] Departamento de Informática, Universitat Autònoma de Barcelona, Spain
{p.hernandez,e.luque}@cc.uab.es

Abstract. Studies dealing with tuning, performance debugging and diagnosis in cluster environments largely benefit from in-depth knowledge of memory system information. In this paper a tool (called *MemTo*) for monitoring the behavior of the memory system through a Linux cluster is presented. *MemTo* has been designed to have as low intrusiveness as possible while keeping a high detail of monitoring data. The good behavior and usefulness of this tool are proved experimentally.

1 Introduction

Many cluster monitoring tools for obtaining information about resource availability and utilization [1,2,3,4] have been developed. Usually, these tools gather information provided by the *"/proc"* file system of every node making up the cluster. However, in some situations the collection of such information causes an unacceptable intrusiveness. Moreover, this information can be insufficient to obtain an accurate behavior of the distributed and local (user) applications (i.e. characterizing the paging activity) or state (i.e. over/under loading, resource occupation, etc ...) of a particular node (the overall system).

A tool (called *MemTo*) for obtaining accurate on-time information about memory and communication system performance of a Linux cluster is presented in this paper. Some properties of *MemTo* are its low intrusiveness, reported data visualization and/or saving facilities and its easy user utilization, provided by a web-based GUI (Graphical User Interface). This GUI integrates the facilities provided by *Monito* [5], a tool for measuring the state of all the communication message queues in a Linux cluster.

New facilities to collect the memory resources availability and the memory requirements of every parallel/local application have been added into *Memto*. Moreover, in contrast to other monitoring tools [1,2,3,4], *MemTo* provides an accurate memory pattern access (in terms of locality) and memory behavior (as page fault rating) of distributed (or local) applications. The task *locality* is defined as the set of different referenced memory pages in a sampling window

---

* This work was supported by the CICYT under contract TIC98-0433

Y. Cotronis and J. Dongarra (Eds.): Euro PVM/MPI 2001, LNCS 2131, pp. 225–232, 2001.
© Springer-Verlag Berlin Heidelberg 2001

extending from the current time backwards into the past[11]. MemTo can be very useful to study how efficient is the memory management when the process *locality* fits into the physical memory.

MemTo can be very useful in various research fields and this way its applicability is justified. For example: mapping and scheduling policies can take benefit of an in-depth knowledge of communication and memory availability resources. *MemTo* can locate system malfunctions, as thrashing or communication bottlenecks that can be detected in fault tolerance disciplines. Also, performance of synchronization techniques as coscheduling [8] (which deal with minimizing synchronization/communication waiting time between remote processes) can decrease drastically if memory requirements are not kept in mind [10].

The paper is organized as follows. Section 2 describes the Linux memory system. The *Memto* architecture and main features are presented in section 3 together with its GUI interface. In section 4, many examples of *MemTo* utilities are shown. Finally, conclusions and future works are detailed.

2   Linux Memory Management Subsystem

The Linux kernel (v.2.2) provides demand-paging virtual memory[1] based on a global replacement policy. Its main features are explained as follows:

Physical memory management. All the physical system pages are described by the *mem_map* struct. Linux uses the *Buddy Algorithm* to allocate and deallocate pages. In doing so, it uses the *free_area* vector, which includes information about free physical pages. Periodically (once per second) the *swap kernel daemon* (*swapd*) checks if the number of free pages on the *free_area* vector falls below a system constant, *free_pages_low*. If so, *swapd* will try to increase the number of free pages by applying *the clock replacement* algorithm. If pages to be discarded were modified (*dirty page*), Linux preserves their content into the *swap* file.

Data caches. Linux maintains three different *data caches* into physical memory. The *buffer cache* contains data buffers that are used by the block device drivers. As file pages are read into memory from disk or whatever device (i.e. the network), they are cached in the *page cache*. The *swap cache* tracks the swapped out pages that have not been modified.

Virtual memory management. Every process virtual memory is represented by a *mm_struct*, which can be accessed from the process *task_struct* (the Linux Process Control Block). When a process accesses a virtual address that does not have a valid page table entry, the processor will report a *page fault* to Linux. Linux looks for the missing page in the swap file or on the secondary memory by using the page table entry for this faulting virtual address.

---

[1] Page size = 4KB.

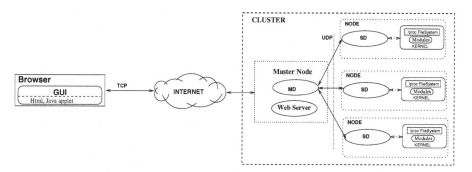

Fig. 1. MemTo architecture

All the above explained concepts are very useful for understanding how tasks memory requirements and memory resources availability are obtained by MemTo (as explained in the following section).

3   MemTo

Taking into account the Linux memory management described in the previous section, the *MemTo* architecture and facilities are presented next.

The *MemTo* architecture (shown in fig. 1) follows a master-slave paradigm. In each sampling period (sp) and during the monitoring interval (mi), every slave daemon (SD) collects the information gathered by various new local kernel modules. This samples are packed and sent to the master node by means of UDP protocol in order to reduce the network traffic. In addition, the master daemon (MD) manages the *SD* operation by means of UDP control packets too.

With the aim of providing remote access to *MemTo*, a web-based GUI for visualizing the system activity has been developed. Fig. 2 (left) shows the configuration window. This window allows the user to initialize the configuration parameters: *sp*, *mi*, events and name of the distributed/local application to be monitored, as well as monitoring mode (off/on-line). In off-line mode, the samples are saved in each monitored node in a *gnuplot*-format file. In on-line mode, collected information is sent to the *MD*, which will be visualized by the user's browser. Furthermore, the selection of the different cluster nodes to be monitored can be made from this configuration window (left frame). Fig. 2 (right) shows the browser window corresponding to memory information gathered from a specific node in on-line mode.

Limited *MemTo* intrusiveness is guaranteed because only the events required by the user are reported. Also, the I/O operation frequency, such as writing trace records into trace files, is reduced by means of buffering trace data. Experimentally, a 24kB buffer size has been chosen. Moreover, physical clock synchronization is performed by means of the NTP protocol [7], reaching a deviation below a millisecond. Thus, no event time stamp must be reported (the samples are collected simultaneously in every *sp* in all the cluster).

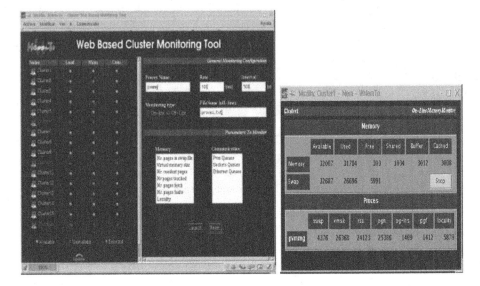

Fig. 2. Configuration window (left) and node memory information (right)

MemTo gathers information from module-associated files residing in each instrumented node. Several kernel modules have been implemented to monitor three different subsystems:

1. Communication: [5] describes three different modules to collect information from the communication subsystem (PVM, socket, protocol, logical and physical layer).
2. Workload: the ready to run task queue average length during the last 1, 5 and 15 minutes are collected from the */proc/loadavg* file.
3. Memory: the *statmem* module gathers memory information and maintains the */proc/net/statmem* file. The information provided by the *statmem* module in each node is the following:
   - Node Memory Information: physical and swapped memory available, used, free and shared, collected by means of the *si_meminfo* and *si_swapinfo* kernel functions respectively.
   - Node Cache Information: number of pages in the page, swap and buffer caches.
   - Process Memory Information (shown in table 1) gathered from *task_struct* and *mm_struct*.

4   Experimentation

The trials were performed in a homogeneous PVM-Linux cluster made up of a 100 Mbps Fast Ethernet network and eight PCs with the same characteristics: a 350Mhz Pentium II processor, 128 MB of RAM, 512 KB of cache, Linux

Table 1. Memory process information

Field	Explanation
swap	# of pgs in the swap file
vmsk	process virtual memory size in bytes
rss	# of resident pgs: distinguish between total, code, shared, date and dirty pgs
pg-in	# of pgs fetched from swap memory
pg-wrt	# of copy-on-write faults
pf	# of pgs faults
locality	# of different pgs touched during the sampling period

Table 2. Overhead introduced by MemTo

	Nominal Size	Ov (sp = 100ms)		Ov (sp = 500ms)		Ov (sp = 1s)	
		On	Off	On	Off	On	Off
IS	$2^{23}x2^{19}$	6.7%	5.1%	6.3%	4.1%	4.5%	2.4%
MG	256x256x256	6.2%	5.8%	4.9%	4.6%	3.9%	3.2%

o.s. (kernel v. 2.2.14) and PVM 3.4.0. In order to reflect the use of *MemTo* for evaluating the behavior of the memory system, two NAS [6] parallel benchmarks (*MG* and *IS* class A) were used.

4.1   Intrusiveness Evaluation

The overhead Ov introduced by *Memto*, in both *off-line* mode and *on-line* mode, has been defined as $Ov = \frac{T_{e\_mon} - T_{e\_pvm}}{T_{e\_pvm}} \times 100$, where T_{e_mon} (T_{e_pvm}) is the execution time of the distributed application when it was executed under monitoring (without monitoring).

Table 2 shows the added overhead in obtaining the *locality* parameter for three different *sp* values. In general, the added overhead is not excessive. The overhead introduced in on-line mode is bigger than off-line mode because monitoring network traffic is increased, disturbing the communication between remote tasks benchmarks. This perturbation is emphasized in intensive communication applications (IS). On the other hand, as it was expected, the off-line mode introduces more overhead in compute-bound applications (MG).

4.2   Application Examples

The IS and MG memory access pattern in terms of its *locality* is analyzed first. Fig. 3 shows the locality behavior and the total number of referenced pages (Total) for both benchmarks with an *sp* value of 100 ms. Both programs exhibit a phase-transition behavior: periodically and during a locality-dependent interval, all the referenced pages correspond to the same set. Every locality transition fits in a new benchmark iteration (10 in IS and 4 in MG). In both cases, the locality size in each transition grows rapidly, even near the total number of pages referenced. For this reason, the locality does not vary significantly in these

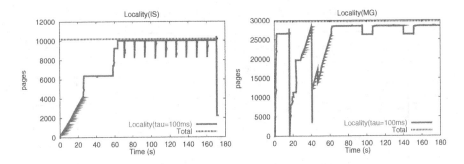

Fig. 3. Locality for IS (left) and MG (right) benchmarks

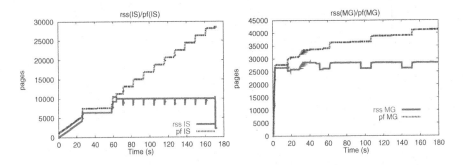

Fig. 4. (left) rss(IS)/pf(IS) and (right) rss(MG)/pf(MG)

specific benchmarks although the window size (sp) has been increased. *Locality* accuracy versus monitoring intrusion must be taken into account in the selection of *sp*.

Note as *MemTo* can help to take more efficient decisions in scheduling or mapping disciplines when the (distributed or local) application localities are taken into account. For example, no more distributed tasks should be assigned to a node when the total number of referenced pages overcomes the physical memory size.

Fig. 4 shows the resident set size (rss) and page fault number (pf) as a function of time for both benchmarks. We can see that the memory requirements of both benchmarks are completly differents: whereas *rss(MG)* nearly reaches the total number of physical memory pages (32007), the *rss(IS)* reaches a maximum of 8230pgs. It should be noted that the rss behavior follows the same way as locality in fig. 3 (right): each phase corresponds to one benchmark iteration. Also, note that in each iteration, due to a locality change, the number of free pages reaches the minimum kernel threshold (*free_pages_low*), so that the *Linux swapd* liberates pages applying the clock replacement algorithm. These page streams are reflected in a new step of the *pf* parameter. As it was expected and as consequence of the demand paging mechanism used by Linux, the number of page faults are quickly increased when processes are loaded in memory.

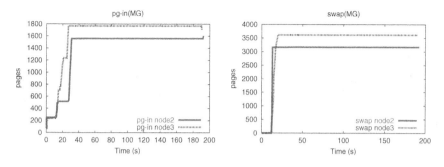

Fig. 5. (left) pg-in(MG) and (right) swap(MG)

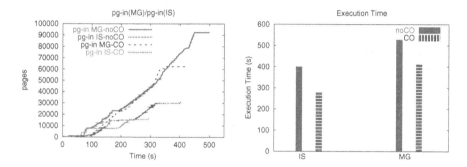

Fig. 6. (left) pg-in(MG)/pg-in(IS) and (right) coscheduling vs. non-coscheduling

Fig. 5 shows how the NTP (physical clock synchronization protocol) used by *MemTo* allows to correlate the samples obtained in different nodes. Fig. 5 (left) displays the MG *pg-in* parameter as a function of time obtained in a dedicated environment for two different nodes (2 and 3). These graphs prove that there is a substantial dissimilarity among both nodes. This is plausible because the sequence and number of instructions executed by different nodes in SPMD programs -such as our benchmarks- can be different due to data dependencies. Fig. 5 (right) shows the number of MG pages in the swap file *(swap(MG))*. As it was expected, *swap(MG)* grows according to the *pg-in* (see fig. 5 (left)).

Next, the performance obtained applying a coscheduling policy is analyzed by means of *MemTo*. Fig. 6 shows the *pg-in* and execution time obtained by MG and IS when they are executed together in two different environments (CO and noCO). In CO, a coscheduling policy is implemented; it is a policy based on simultaneously scheduling the most recently communicated processes (see [9] for further information). In noCO, no modifications have been performed in the base cluster. Fig. 6 (left) shows the *pg-in(MG)* and *pg-in(IS)* for both cases in node 2. The number of page faults in the IS benchmark is considerably lower in the CO case. The coscheduling policy gives a higher execution priority to the tasks with intensive communication (IS benchmark) and consequently, the page replacement algorithm liberates pages belonging to other applications. When IS finishes its execution, all the released memory will be available for MG. So, the requested MG memory fits now in the main memory, and for this reason *pg-in(MG)* does not increase anymore. Fig. 6 (right) displays the execution time obtained for every benchmark in both cases. As it was expected, a significant gain is obtained by applying the CO policy, even in environments where the main memory is overloaded. Note that the obtained gain is reflected in the previous explanation of fig. 6(left).

5   Conclusions and Future Work

MemTo, a tool for measuring the memory subsystem behavior in a Linux cluster, has been presented. *MemTo* incorporates memory and communication monitoring facilities. A web-based GUI allows the user to visualize the collected information. The usefulness of this tool has been shown by means of several examples.

MemTo will allow to analyze in-depth the interaction between the communication and memory subsystems with the aim of improving the performance either of distributed applications and local tasks through a cluster. Also, the scalability problem associated to the master-slave architecture of MemTo should be taken into account in a future work.

References

1. J. Astalos. "CIS - Cluster Information Service". *7th European PVM/MPI User's Group Meeting*, Lecture Notes in Computer Science 1908, September 2000.
2. R. Buyya. "PARMON: a Portable and Scalable Monitoring System for Clusters". *Int. Journ. on Software: Practice&Experience*, J. Wiley & Sons, USA, June 2000.
3. P. Uthayopas, S. Phaisithbenchapol and K. Chongbarirux. "Building a Resources Monitoring System for SMILE Beowulf Cluster". *Proc. of the Third International Conference on High Performance Computing in Asia-Pacific*, 1999.
4. Open Source Cluster Application Resources (OSCAR), http://www.csm.ornl.gov/oscar.
5. F. Solsona, F. Giné, J.L. Lérida, P. Hernández and E. Luque. "Monito: A Communication Monitoring Tool for a PVM-Linux Environment". *7th European PVM/MPI User's Group Meeting*, Lecture Notes in Computer Science 1908, September 2000.
6. Parkbench Committee. Parkbench 2.0. http://www.netlib.org/parkbench, 1996.
7. Network Time Protocol (NTP): http://www.ntp.org
8. J.K. Ousterhout. "Scheduling Techniques for Concurrent Systems." In *3rd. Intl. Conf. Distributed Computing Systems*, pp.22-30, 1982.
9. F. Solsona, F. Giné, P. Hernández and E. Luque. "Predictive Coscheduling Implementation in a non-dedicated Linux Cluster". *To appear in Europar'2001*.
10. D. Burger, R. Hyder, B. Miller and D. Wood. "Paging Tradeoffs in Distributed Shared-Memory Multiprocessors". *Journ. of Supercomputing*, vol.10, pp.87-104, 1996.
11. P.J. Denning. "Working Sets Past and Present". *IEEE Transactions on Software Engineering*, vol. SE-6, No 1, January 1980.

A Community Databank for Performance Tracefiles

Ken Ferschweiler[1], Mariacarla Calzarossa[2], Cherri Pancake[1],
Daniele Tessera[2] and Dylan Keon[1]

[1]Northwest Alliance for Computational Science & Engineering, Oregon State University
[2]Dipartimento di Informatica e Sistemistica, Università di Pavia

Tracefiles provide a convenient record of the behavior of HPC programs, but are not generally archived because of their storage requirements. This has hindered the developers of performance analysis tools, who must create their own tracefile collections in order to test tool functionality and usability. This paper describes a shared databank where members of the HPC community can deposit tracefiles for use in studying the performance characteristics of HPC platforms as well as in tool development activities. We describe how the Tracefile Testbed was designed and implemented to facilitate flexible searching and retrieval of tracefiles. A Web-based interface provides a convenient mechanism for browsing and downloading collections of tracefiles and tracefile segments based on a variety of characteristics. The paper discusses the key implementation challenges.

1 The Tracefile Testbed

Tracefiles are a valuable source of information about the properties and behavior both of applications and of the systems on which they are executed. They are typically generated by the application programmer as part of the performance tuning process. Our field studies of HPC programmers indicate that many experienced programmers also create suites of simple pseudo-benchmark codes and generate tracefiles to help establish basic performance characteristics when they move to new HPC platforms. The intent in both cases is to help the user better understand and tune his/her applications.

The developers of trace-based performance analysis and performance prediction tools (cf. [7, 8, 10, 9, 3]) also generate suites of tracefiles. In this case, the objective is to assist in the process of testing and fine-tuning tool functionality. According to the subjects interviewed in our field studies, tool developers do not often have access to "real" applications for these activities; rather, they construct artificial codes designed to generate tracefiles that will stress the tool's boundary conditions or generate demonstration visualizations.

Tool users and developers alike have indicated in several public forums (e.g., Parallel Tools Consortium meetings, BOF sessions at the SC conference, community workshops on parallel debugging and performance tuning tools) that it would be useful to construct a generally accessible testbed for tracefile data. This would make it possible for users to see if tracefiles from related applications can be of use in the design and tuning of their own application. It would also provide a more realistic foundation for testing new performance tools. Further, since tracefiles are typically large and unwieldy to store (the recording of key program events during one application run can generate

Y. Cotronis and J. Dongarra (Eds.): Euro PVM/MPI 2001, LNCS 2131, pp. 233–240, 2001.
© Springer-Verlag Berlin Heidelberg 2001

gigabytes of data), a centralized repository could encourage programmers to archive their tracefiles rather than deleting them when they are no longer of immediate use.

In response to this need, we created the Tracefile Testbed. The objective was to develop a database that not only supports convenient and flexible searching of tracefile data generated on HPC systems, but maximizes the benefit to others of performance data that was collected by a particular programmer or tool developer for his/her own purposes.

The Tracefile Testbed was implemented as a joint project of NACSE and the Università di Pavia. The work was supported by the NAVO (Naval Oceanographic Office) MSRC, through the PET program of the High Performance Computing Modernization Office, and will be maintained with support of the Parallel Tools Consortium. It was structured according to a data model that describes both the static and dynamic behavior of parallel applications, as captured in tracefiles. The tracefiles are maintained as separate file units. The source code that generated the tracefiles is also available (unless that code is proprietary). Metadata encapsulating the performance behavior and run-time environment characteristics associated with the tracefiles are maintained in a relational database using Oracle 8i.

A key aspect of tracefile storage is their size. This can pose difficulties for prospective users, who may find that storing many downloaded copies is quite resource-intensive. To reduce that burden, all file locations are maintained in the Tracefile Testbed's metadata database as URLs. This will allow users to "maintain" their own subsets of tracefiles by simply storing links or shortcuts to the files, rather than the files themselves. A secondary advantage of this approach is that it allows us to distribute the repository itself. That is, the actual tracefiles may be located on multiple servers, which can be different from the server(s) hosting the tool interface and the metadata database. The initial implementation involves three servers: a Web server maintains the interface, a relational database server hosts the metadata, and the tracefiles are stored on a separate file server. This architecture is illustrated in Figure 1.

A Web-based interface allows users to navigate through the repository, select tracefiles and segments from one or more applications, browse their characteristics, and download the data. The interface makes use of QML (Query Markup Language), a middleware product developed and distributed by NACSE. Performance data can be

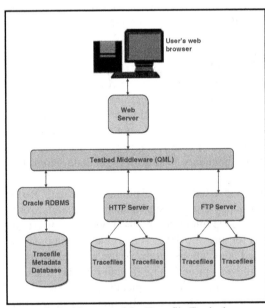

Figure 1. Architecture of the Tracefile Testbed

identified and extracted based on various selection criteria, such as "all data related to a given application," "data related to a class of applications," "data from programs executed on a particular system architecture," etc. The Tracefile Testbed provides performance summaries of selected trace data; alternatively, the tracefile data may be downloaded for analysis using available tools in order to derive detailed performance figures.

2 Data Model

In order to categorize and maintain tracefile data, we require a data model with the power to describe the characteristics of parallel applications and the performance measurements collected during execution. In large part, the framework we have chosen to describe tracefiles is based on user needs in searching the tracefile collection. Based on previous usability studies, we determined that users wish to select entire tracefiles or segments thereof, on the basis of machine architecture types and parameters, information related to the tracefile itself, and information related to the tracefile segments. Users should also be able to perform searches based on arbitrary keywords reflecting system platforms, problem types, and user-defined events.

2.1 Structure of the Data Model

The model must capture not just parallel machine characteristics, but also the design strategies and implementation details of the application. For this purpose, the information describing a parallel application has been grouped into three layers:

The *system layer* provides a coarse-grained description of the parallel machine on which the application is executed. The other two layers comprise information derived from the application itself; the *application layer* describes its static characteristics, whereas the *execution layer* deals with the dynamic characteristics directly related to measurements collected at run time. Most of the information comprising the system and application layers is not available in the tracefile, but must be supplied by the application programmer in the form of metadata. Execution layer information can be harvested directly from the tracefiles.

The system layer description includes machine architecture (e.g., shared memory, virtual shared memory, distributed memory, cluster of SMPs), number of processors, clock frequency, amount of physical memory, cache size, communication subsystem, I/O subsystem, communication and numeric libraries, and parallelization tools.

The static characteristics of the application layer range from the disciplinary domain (e.g., computational fluid dynamics, weather forecasting, simulation of physical and chemical phenomena) to the algorithms (e.g., partial differential equation solvers, spectral methods, Monte Carlo simulations) and programming languages employed. They also include information about the application program interface (e.g., MPI, OpenMP, PVM) and links to the source code. Problem size, number of allocated processors, and work and data distributions are further examples of static characteristics.

The execution layer provides a description of the behavior of a parallel application in terms of measurements generated at run time. These measurements are typically timestamped descriptions which correspond to specific events (I/O operation, cache miss, page fault, etc.) or to instrumentation of the source code (e.g., beginning or end of an arbitrary section of code, such as a subroutine or loop). The type and number of measurements associated with each event depend on the event type and on the monitoring methods used to collect the measurements. Application behavior might be described by the time to execute a particular program section or the number of events recorded in a particular time span.

2.2 Describing Tracefile Content

To maintain the system, application, and execution information describing the tracefile repository, we implemented a database of descriptive metadata. These exist at multiple levels: they include descriptions of individual tracefiles, sets of tracefiles, and segments of tracefiles. The use of off-the-shelf rDBMS software allows us to maintain and search these metadata with a great deal of power, flexibility, and robustness, and with a minimum of investment in software development.

As discussed previously, the choice of which metadata to maintain – the data model – was based on our assessment of user needs in searching the tracefile collection. The Tracefile Testbed provides the ability to search on machine, application, or execution parameters. The versatility of the database allows us to search based on flexible combinations of these parameters, but careful database design was required to make full use of the power of the rDBMS. Figure 2 presents a conceptual view of the database schema supporting user searches.

Note that tracefiles do not typically stand alone; they are usually generated in sets

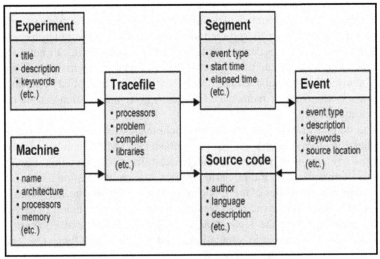

Figure 2. General structure of tracefile metadata

of related files pertaining to a larger project, or *experiment*. The metadata database allows us to maintain this information about the origin of tracefiles. In other cases, a number of tracefiles that were not generated together may still form a naturally cohesive set (e.g., they may demonstrate a common computational approach, or illustrate the effects of varying a particular parameter). Since cohesion of such sets would not always be apparent from the metadata described above, the system allows specification of *virtual experiments* – groups of tracefiles which, though not related in origin, have an *ex post facto* relationship which is useful to some researcher. This structure allows tracefiles to belong to multiple sets which cut across each other, allowing individual users to superimpose organizational schemes which fit their particular needs.

A key requirement for the Tracefile Testbed is that it be easy for members of the HPC community to add new tracefiles. We were fortunate in having access to a sizeable collection of files from a variety of machine and problem types to use as the initial population of the repository. We gathered almost 200 files in our benchmarking work with the SPEC suite [1]. Given the number of files we anticipate gathering from the APART (Automated Performance Analysis: Resources and Tools) working group and other members of the HPC community, it was important to be able to parse the files in batch mode, and our initial parser reflects this bias. A Web-based tool for uploading tracefiles will be implemented in the next phase of the project.

To ensure that metadata are available for all tracefiles in the Testbed, they must be supplied as part of the uploading mechanism. As discussed previously, information such as system- and application-level metadata does not exist *a priori* in the tracefiles, but must be provided by the programmer or benchmarker. The originator of the tracefiles is also the source of descriptive information about user-defined events in the execution-level metadata. To facilitate the input of that information, we developed a tracefile metadata format and a corresponding parser. Most of the metadata elements are likely applicable to a whole series of tracefiles, so the format and uploading tool were designed to facilitate metadata reuse.

3 Tracefile Segments

While tracefiles are typically quite large, the portion of a tracefile which is of interest for a particular purpose may be only a small fragment of the file. For instance, a researcher wishing to compare the performance of FFT implementations may want to work with a fragment that brackets the routine(s) in which the FFT is implemented. Similarly, a tool developer may be interested in testing tool functionality in the presence of broadcast operations; the remainder of the trace may be largely irrelevant. If the source code is appropriately instrumented at the time of tracefile creation, the sections of interest will be easily identifiable, but locating them in a large corpus of tracefile data may still be an onerous task. In order to simplify identification of tracefile fragments which are of interest, it is convenient to maintain a description of the internal structure of tracefiles. Some of this structure may be automatically generated from information in the tracefile, but the remainder must be supplied as metadata, typically by the programmer who contributes the file to the repository.

3.1 Dividing Tracefiles into Segments

Since a tracefile is essentially a list of timestamped events (with some descriptive header information), it is easy to identify a subset of a tracefile corresponding to the events occurring during a particular time interval. The obvious choices for defining the interval are the begin and end timestamps of a user-defined event (such as the FFT routine mentioned above). We discuss user-defined events because system-defined events are typically atomic; that is, they do not have start and end markers. However, such a view may be an oversimplification that does not capture the behavior of interest during the time interval. Since the tracefile is a straightforward list of per-processor events, it is considerably more difficult to define events which pertain to the entire parallel machine. The idealized view of a data-parallel application would have all processors participating in all events (i.e., executing the same segment of code) approximately simultaneously; however, there is no guarantee in an actual application that any event will include all processors, simultaneously or not.

Consequently, a user who wishes to extract a subset of a tracefile to capture system performance during a particular event is faced with a difficulty. Although the user may know that particular events on one processor correspond to events on other processors, it is not clear from the tracefile how these correspondences can be automatically inferred. We have used a heuristic approach to identifying machine-wide events. A *machine-wide event* includes all of the same-type per-processor events whose starting markers in the tracefile are separated by fewer than K*N events, where N is the number of processors in the machine, and K is a definable constant (currently set to 4). The per-processor events which comprise a machine-wide event may or may not overlap in time, but discussion with users of parallel performance evaluation systems indicate that they expect this criterion to effectively capture the corresponding events.

The machine-wide event, defined as a starting timestamp (and, for user-defined events, an ending timestamp) in a particular tracefile, is the basic unit of tracefile data which our system maintains; we allow users to attach descriptions, keywords, and source code references to these events.

3.2 Using Tracefile Segments

To many HPC users, the principal reason for creating and maintaining tracefiles is to be able to use them as input to performance-analysis software. To support this requirement, the Tracefile Testbed provides single-keystroke operations for downloading tracefiles to the user's local machine via http or ftp.

The issue of tracefile segments introduces problems with respect to tool compatibility. Trace-based performance tools require "legal" tracefiles as input; while there is no single *de facto* standard for tracefile format, we assume that a tracefile which is usable by popular performance analysis packages will also be suitable for HPC users who write their own analysis tools. A fragment naively extracted from a tracefile will not, in general, be of a legal format. In particular, it will lack header information and will probably contain unmatched markers of entry to and exit from instrumented program regions. To make segments useful, the Tracefile Testbed modifies the fragment in order to generate a legal tracefile which describes as closely as possible the behavior of the application in the region which the user has selected.

4 User Interface

Users faced with powerful software and complex interfaces quickly discover the features they need to solve their immediate problems and ignore the rest. We made a conscious decision to supply only those features which will be useful to a large community of users, allowing us to develop a concise and intuitive user interface. Our choice for user interface platform was the ubiquitous web browser, which offers near-universal portability.   The interface was developed using an existing web-to-database middleware package, QML (Query Markup Language [5]. QML allowed us to quickly develop prototype implementations of interfaces exploring various search strategies and to support a "drilling-down" style of search. It also supplies a builtin mechanism for downloading tracefiles or segments for further analysis.

Figure 3 shows an example of a tracefile query page. Selectable lists (machine types, application types, keywords, etc.) are generated dynamically from the metadata database contents; the user interface requires no updating to accommodate new additions to the database. The user may drill down through the layers supported in the data mode –system, application, and execution– refining his search at each layer. At any stage in the search, the options are limited to those still available based on the constraints which the user has supplied in earlier layers.

4.1 Performance Summaries

In many cases, the information a user wants from one or more tracefiles may be easily summarized without recourse to other performance analysis software.   This is particularly the case when an application programmer wishes to compare some measure of "overall" performance across several different files.   To simplify such tasks, the Tracefile Testbed provides a - statistical performance summary functions which may be performed on selected tracefiles or segments.

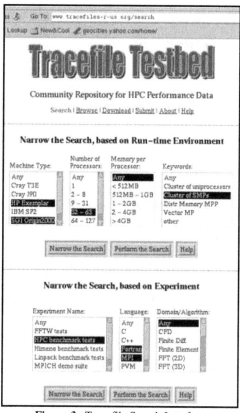

Figure 3. Tracefile Search Interface

Over the next year, we are scheduled to add features providing graphical summaries of application behavior. These will allow the user to compare tracefiles at a synopsis level before deciding to examine them in more detail.

5 Summary

Responding directly to a requirement that has been expressed in a variety of community forums, the Tracefile Testbed uses web-to-database technology to provide HPC programmers and tool developers with access to a repository of tracefiles. A database of metadata describing the systems, applications, and execution-level information of each tracefile supports a variety of search approaches. Performance summaries assist users to assess the relevance of files and segments before they are examined in detail. Individual files and/or segments may be downloaded to the user's local system for further analysis and comparison. Application programmers should find this community repository useful both in predicting the behavior of existing programs and in the development and optimization of new applications. Developers of performance analysis and prediction tools will find the Tracefile Testbed to be a convenient source of tracefiles for testing the functionality and display capabilities of their tool.

Acknowledgments

Scott Harrah and Tom Lieuallen of Oregon State University, and Luisa Massari of the Università di Pavia contributed to the implementation of the Tracefile Testbed. The project was supported by the NAVO (Naval Oceanographic Office) MSRC, through the PET program of the HPC Modernization Office.

Bibliography

1. R. Eigenmann and S. Hassanzadeh. Benchmarking with Real Industrial Applications: The SPEC High-Performance Group. *IEEE Computational Science and Engineering,* Spring Issue, 1996.
2. T. Fahringer and A. Pozgaj. P3T+: A Performance Estimator for Distributed and Parallel Programs. *Journal of Scientific Programming,* 7(1), 2000.
3. B.P. Miller et al. The Paradyn Parallel Measurement Performance Tool. *IEEE Computer,* 28 (11):37-46, 1995.
4. K.L. Karavanic and B.P. Miller. Improving Online Performance Diagnosis by the Use of Historical Performance Data. In *Proc. SC'99,* 1999.
5. C. M. Pancake, M. Newsome and J. Hanus. 'Split Personalities' for Scientific Databases: Targeting Database Middleware and Interfaces to Specific Audiences. *Future Generation Computing Systems,* 6: 135-152, 1999.
6. S.E. Perl, W.E. Weihl, and B. Noble. Continuous Monitoring and Performance Specification. Technical Report 153, Digital Systems Research Center, June 1998.
7. D.A. Reed et al. Performance Analysis of Parallel Systems: Approaches and Open Problems. In *Joint Symposium on Parallel Processing,* pages 239-256, 1998.
8. S. Shende and A. Malony and J. Cuny and K. Lindlan and P. Beckman and S. Karmesin, Portable Profiling and Tracing for Parallel Scientific Applications using C++. In *Proc. SPDT'98: ACM SIGMETRICS Symposium on Parallel and Distributed Tools,* pages 134-145, 1998.
9. J. Yan, S. Sarukhai, and P. Mehra, "Performance Measurement, Visualization and Modeling of Parallel and Distributed Programs Using the AIMS Toolkit," *Software – Practice and Experience,* 25 (4): 429--461, 1995.
10. O. Zaki, E. Lusk, W. Gropp, and D. Swider. Toward Scalable Performance Visualization with Jumpshot. *The International Journal of High Performance Computing Applications,* 13(2):277-288, 1999.

Review of Performance Analysis Tools for MPI Parallel Programs

Shirley Moore, David Cronk, Kevin London, and Jack Dongarra

Computer Science Department, University of Tennessee Knoxville, TN 37996-3450, USA
{shirley,cronk,london,dongarra}@cs.utk.edu

Abstract. In order to produce MPI applications that perform well on today's parallel architectures, programmers need effective tools for collecting and analyzing performance data. A variety of such tools, both commercial and research, are becoming available. This paper reviews and evaluations the available cross-platform MPI performance analysis tools.

1 Introduction

The reasons for poor performance of parallel message-passing codes can be varied and complex, and users need to be able to understand and correct performance problems. Performance tools can help by monitoring a program's execution and producing performance data that can be analyzed to locate and understand areas of poor performance. We are investigating DEEP/MPI, Jumpshot, the Pablo Performance Analysis Tools, Paradyn, TAU, and Vampir. These are research and commercial tools that are available for monitoring and/or analyzing the performance of MPI message-passing parallel programs. This paper describes an on-going review of those tools that we are conducting. We are restricting our investigation to tools that are either publicly or commercially available and that are being maintained and supported by the developer or vendor. We are also focusing on tools that work across different platforms. To give our review continuity and focus, we are following a similar procedure for testing each tool and using a common set of evaluation criteria. We first build and install the software using the instructions provided. After the software has been installed successfully, we work through any tutorial or examples that are provided so that we can become familiar with the tool. Finally, we attempt to use the tool to analyze a number of test programs.

Our set of evaluation criteria includes robustness, usability, scalability, portability, and versatility. In addition to these general criteria, we include the following more specific criteria: support for hybrid environments, support for distributed heterogeneous environments, and support for analysis of MPI-2 I/O. By hybrid environments, we mean a combination of shared and distributed memory, both with respect to hardware and parallel programming models. An emerging parallel programming paradigm is to use message passing between nodes and

Y. Cotronis and J. Dongarra (Eds.): Euro PVM/MPI 2001, LNCS 2131, pp. 241–248, 2001.

shared memory (e.g., OpenMP) within a node. Most performance analysis tools were originally developed to support only one of these models, but tools that support both models simultaneously are highly desirable. Another emerging model of parallel computation is that of running large computations across a computational grid, which consists of a distributed collection of possibly heterogeneous machines [6,5]. MPI I/O, which is part of the MPI-2 standard, simplifies parallel I/O programming and potentially improves performance. Tools are needed that not only trace MPI I/O calls but also assist the user specifically with analysis of MPI I/O performance.

2   Tools Reviewed

2.1   DEEP/MPI

URL	http://www.psrv.com/deep_mpi_top.html
Supported languages	Fortran 77/90/95, C, mixed Fortran and C
Supported platforms	Linux x86, SGI IRIX, Sun Solaris Sparc, IBM RS/6000 AIX, Windows NT x86

DEEP and DEEP/MPI are commercial parallel program analysis tools from Veridian/Pacific-Sierra Research. DEEP (Development Environment for Parallel Programs) provides an integrated graphical interface for performance analysis of shared and distributed memory parallel programs. DEEP/MPI is a special version of DEEP that supports MPI program analysis. To use DEEP/MPI, one must first compile the MPI program with the DEEP profiling driver **mpiprof**. This step collects compile-time information and instruments the code. Then after executing the program in the usual manner the user can view performance information using the DEEP/MPI interface. The DEEP/MPI interface includes a call tree viewer for program structure browsing and tools for examining profiling data at various levels. DEEP/MPI displays whole program data such as the wallclock time used by procedures. After identifying procedures of interest, the user can bring up additional information for those procedures, such as loop performance tables. The DEEP Performance Advisor suggests which procedures or loops the user should examine first. MPI performance data views allow users to identify MPI calls that may constitute a bottleneck. Clicking on a loop in a loop performance table or on an MPI call site takes the user to the relevant source code. CPU balance and message balance displays show the distribution of work and number of messages, respectively, among the processes. DEEP supports the PAPI interface to hardware counters and can do profiling based on any of the PAPI metrics [1,2]. DEEP supports analysis of shared memory parallelism. Thus DEEP/MPI can be used to analyze performance of mixed MPI and shared-memory parallel programs.

2.2   MPE Logging and Jumpshot

URL	http://www-unix.mcs.anl.gov/mpi/mpich/
Version	MPICH 1.2.1, Jumpshot-3

The MPE (Multi-Processing Environment) library is distributed with the freely available MPICH implementation and provides a number of useful facilities, including debugging, logging, graphics, and some common utility routines. MPE was developed for use with MPICH but can be used with other MPI implementations. MPE provides several ways to generate logfiles which can then be viewed with graphical tools also distributed with MPE. The easiest way to generate logfiles is to link with an MPE library that uses the MPI profiling interface. The user can also insert calls to the MPE logging routines into his or her code. MPE provides two different log formats, CLOG and SLOG. CLOG files consist of a simple list of timestamped events and are understood by the Jumpshot-2 viewer. SLOG stands for Scalable LOGfile format and is based on doubly timestamped states. SLOG files are understood by the Jumpshot-3 viewer. The states in an SLOG file are partitioned into frames of data, each of which is small enough to be processed efficiently by the viewer. SLOG and Jumpshot-3 are said to be capable of handling logfiles in the gigabyte range. After a user loads a logfile into Jumpshot-3, a statistical preview is displayed to help the user select a frame for more detailed viewing. Selecting a frame and clicking on the Display button causes Jumpshot-3 to display a timeline view with rectangles that represent the various MPI and user defined states for each process and arrows that represent messages exchanged between those states. The user may select and deselect states so that only those states that are of interested are displayed. In the case of a multithreaded environment such as an SMP node, the logfile may contain thread information and Jumpshot-3 will illustrate how the threads are dispatched and used in the MPI program. The timeline view can be zoomed and scrolled. Clicking on a rectangle or an arrow brings up a box that displays more detailed information about the state or message, respectively.

2.3   Pablo Performance Analysis Tools

URL	http://www-pablo.cs.uiuc.edu/
Version	Trace Library 5.1.3, SvPablo 4.1
Languages	Fortran 77/90, C, HPF (SvPablo)

The Pablo Trace Library includes a base library for recording timestamped event records and extensions for recording performance information about MPI calls, MPI I/O calls, and I/O requests. All performance records generated by the Trace Library are in the Pablo SDDF (Self Defining Data Format) format. SDDF is a data description language that specifies both data record structures and data record instances. The MPI extension is an MPI profiling library [8]. The MPI I/O extension provides additional wrapper routines for recording information about MPI I/O calls [9]. The user may choose between a detailed tracing mode that writes records for each MPI I/O event or a summary tracing mode that summarizes the performance data for each type of MPI I/O call. Utilities are provided that analyze the MPI I/O trace files and produce reports.

SvPablo is a graphical user interface for instrumenting source code and browsing runtime performance data. Applications can be instrumented either interactively or automatically. To interactively instrument an application code, SvPablo parses each source file and flags constructs (outer loops and function or subroutine calls) that can be instrumented. The user then selects the events to be instrumented and SvPablo generates a new version of the source code containing instrumentations calls. Alternatively, the SvPablo stand-alone parser can be used to instrument all subroutines and outer loops, including MPI calls. After the instrumented application has been run, SvPablo correlates the runtime performance data with application source code and performs statistical analyses. The SvPablo data capture library supports the use of the MIPS R10000 hardware performance counters to allow visualization of hardware events on SGI architectures. Plans are to use the PAPI portable hardware counter interface [1,2] in future versions of SvPablo.

2.4 Paradyn

URL	http://www.cs.wisc.edu/paradyn/
Version	3.2
Languages	Fortran, C, C++, Java
Supported platforms	Solaris (SPARC and x86), IRIX (MIPS), Linux (x86), Windows NT (x86), AIX (RS6000), Tru64 Unix (Alpha), heterogeneous combinations

Paradyn is a tool for measuring the performance of parallel and distributed programs. Paradyn dynamically inserts instrumentation into a running application and analyzes and displays performance in real-time. Paradyn decides what performance data to collect while the program is running. The user does not have to modify application source code or use a special compiler because Paradyn directly instruments the binary image of the running program using the DyninstAPI dynamic instrumentation library [4]. MPI programs can only be run under the POE environment on the IBM SP2, under IRIX on the SGI Origin, and under MPICH 1.2 on Linux and Solaris platforms. Additional MPI support is under development. The user specifies what performance data Paradyn is to collect in two parts: the type of performance data and the parts of the program for which to collect this data. The types of performance data that can be collected include metrics such as CPU times, wallclock times, and relevant quantities for I/O, communication, and synchronization operations. The performance data can be displayed using various visualizations that include barcharts, histograms, and tables. As an alternative to manually specifying what performance data to collect and analyze, the user can invoke the Paradyn Performance Consultant. The Performance Consultant attempts to automatically identify the types of performance problems a program is having, where in the program these problems occur, and the time during which they occur.

2.5   TAU

URL	http://www.acl.lanl.gov/tau/
Version	2.9.11 (beta)
Supported languages	Fortran, C, C++, Java
Supported platforms	SGI IRIX 6.x, Linux/x86, Sun Solaris, IBM AIX, HP HP-UX, Compaq Alpha Tru64 UNIX, Compaq Alpha Linux, Cray T3E, Microsoft Windows

TAU (Tuning and Analysis Utilities) is a portable profiling and tracing toolkit for performance analysis of parallel programs written in Java, C++, C, and Fortran. TAU's profile visualization tool, Racy, provides graphical displays of performance analysis results. In addition, TAU can generate event traces that can be displayed with the Vampir trace visualization tool. TAU is configured by running the configure script with appropriate options that select the profiling and tracing components to be used to build the TAU library. The default option specifies that summary profile files are to be generated at the end of execution. Profiling generates aggregate statistics which can then be analyzed using Racy. Alternatively, the tracing option can be specified to generate event trace logs that record when and where an event occurred in terms of source code location and the process that executed it. Traces can be merged and converted to Vampitrace format so that they can be visualized using Vampir. The profiling and tracing options can be used together.

In order to generate per-routine profiling data, TAU instrumentation must be added to the source code. This can be done automatically for C++ programs using the TAU Program Database Toolkit, manually using the TAU instrumentation API, or using the **tau_run** runtime instrumentor which is based on the DyninstAPI dynamic instrumentation package. An automatic instrumentor for Fortran 90 is under development. In addition to time-based profiling, TAU can use PAPI [1,2] to do profiling based on hardware performance counters. Generation of MPI trace data may either by using the TAU instrumentation API or by linking to the TAU MPI wrapper library which uses the MPI profiling interface. Hybrid execution models can be traced in TAU by enabling support for both MPI and the thread package used (e.g., Pthreads or OpenMP).

2.6   Vampir

URL	http://www.pallas.de/pages/vampir.htm
Version	Vampitrace 2.0, Vampir 2.5
Supported languages	Fortran 77/90, C, C++
Supported platforms	All major HPC platforms

Vampir is a commercially available MPI analysis tool from Pallas GmbH. Vampitrace, also from Pallas, is an MPI profiling library that produces trace files that can be analyzed with Vampir. The Vampirtrace library also has an API for stopping and starting tracing and for inserting user-defined events into the trace file. Instrumentation is done by linking your application with the Vampirtrace library after optionally adding Vampirtrace calls to your source code. Vampirtrace records all MPI calls, including MPI I/O calls. A runtime filtering mechanism can be used to limit the amount of trace data and focus on relevant events. For systems without a globally consistent clock, Vampirtrace automatically corrects clock offset and skew.

Vampir provides several graphical displays for visualizing application runtime behavior. The timeline and parallelism display shows per-process application activities and message passing along a time axis. Source-code click-back is available on platforms with the required compiler support. Other displays include statistical analysis of program execution, statistical analysis of communication operations, and a dynamic calling tree display. Most displays are available in global and per-process variants. The timeline view can be zoomed and scrolled. Statistics can be restricted to arbitrary parts of the timeline display.

Although the current version of Vampir can display up to 512 processes, that number of processes in a timeline display can be overwhelming to the user and simultaneous display of thousands of processes would clearly be impractical. A new version of Vampir that is under development uses a hierarchical display to address display scalability [3]. The hierarchical display allows the user to navigate through the trace data at different levels of abstraction. The display is targeted at clustered SMP nodes and has three layers (cluster, node, process) that can each hold a maximum of 200 objects, with the process layer displaying trace data for individual threads. To handle hybrid parallel programming models (e.g., MPI + OpenMP), a new tool is under development that combines Vampir with the Intel/KAI GuideView tool [7].

3   Evaluation Summary

We now address the evaluation criteria that were described in section 1 with respect to each of the tools reviewed in section 2. Although we have encountered problems with some of the tools not working properly in certain situations, it is too early to report on robustness until we have tried to work out these problems with the tool developers.

With the exception of Jumpshot, all the tools have fairly complete user guides and other supporting material such as tutorials and examples. The documentation provided with Jumpshot is scant and poorly written. Jumpshot itself has a fairly intuitive interface as well as online help. However, some of the features remain unclear such as the difference between connected and disconnected states and between process and thread views. A nice usability feature of MPE logging and jumpshot is their use of autoconf so that they install easily and correctly on most systems. Paradyn has the drawback that it is difficult or impossible to use in the batch queuing environments in use on most large production parallel machines. For example, we have been unable to get Paradyn to run an MPI program in such an environment. This is not surprising since Paradyn is an interactive tool, but with sufficient effort other tools such as interactive debuggers have been made to work in such environments. Manual instrumentation is required to produce subroutine-level performance data with the MPE and Pablo trace libraries and with TAU for Fortran 90, and this can be tedious to the point of being impractical for large programs. The source-code click-back capability provided by DEEP/MPI, SvPablo, and Vampir is very helpful for relating performance data to program constructs.

Scalable log file formats such as MPE's SLOG format and the new Vampirtrace format [3] are essential for reducing the time required to load and display very large trace files. The summary option provided by the Pablo MPI I/O trace library allows generation of summary trace files whose size is independent of the runtime of the program. Although Paradyn achieves scalability in one sense by using dynamic instrumentation and by adjusting instrumentation granularity, some of the user interface displays such as the Where Axis and the Performance Consultant graph become unwieldy for large programs. The new hierarchical Vampir display is a promising approach for visualizing performance on large SMP clusters in a scalable manner.

Of all the tools reviewed, Vampir is the only one that has been tested extensively on all major platforms and with most MPI implementations. The MPE and Pablo trace libraries are designed to work with any MPI implementation, and explicit directions are given in the MPE documentation for use with specific vendor MPI implementations. However, on untested platforms there will undoubtedly be glitches. Because of platform dependencies in the dynamic instrumentation technology used by Paradyn, especially for threaded programs, implementation of some Paradyn features tends to lag behind on all except their main development platforms.

TAU provides a utility for converting TAU trace files to Vampir and SDDF formats. SDDF is intended to promote interoperability by providing a common performance data meta-format, but most other tools have not adopted SDDF. Except for TAU and Vampir, tool interoperability has not been achieved. However, several of the tools make use of the PAPI interface to hardware performance counters and make use of hardware performance data in their performance displays.

DEEP/MPI, MPE/Jumpshot, and TAU all support mixed MPI+OpenMP programming, as will the next version of Vampir. The MPE and Pablo trace libraries, when used with MPICH, can generate trace files for heterogeneous MPI programs. Paradyn supports heterogeneous MPI programs with MPICH. None of these tools has been tested or used with Globus or other grid computing systems. Although profiling of MPI I/O operations is possible with several of the tools, the only tools that explicitly address MPI I/O performance analysis are the Pablo I/O analysis tools.

Considerable effort has been expended on research and development of MPI performance analysis tools. A number of tools are now available that allow MPI programmers to easily collect and analyze performance information about their MPI programs. We hope that this review will encourage further development and refinement of these tools as well as increase user awareness and knowledge of their capabilities and limitations. Detailed reviews of these and other tools, along with testing results and screen shots, will be available at the National High-performance Software Exchange (NHSE) web site at http://www.nhse.org/. The current review is an update to a similar review conducted four years ago that has been widely referenced.

References

1. S. Browne, J. J. Dongarra, N. Garner, G. Ho and P. Mucci, *A Portable Programming Interface for Performance Evaluation on Modern Processors*, International Journal of High Performance Computing Applications, 14 (2000), pp. 189-204.
2. S. Browne, J. J. Dongarra, N. Garner, K. London and P. Mucci, *A Scalable Cross-Platform Infrastructure for Application Performance Optimization Using Hardware Counters*, SC'2000, Dallas, Texas, November 2000.
3. H. Brunst, M. Winkler, W. E. Nagel and H.-C. Hoppe, *Performance Optimization for Large Scale Computing: The Scalable VAMPIR Approach,*, International Conferene on Computational Science (ICCS2001) Workshop on Tools and Environments for Parallel and Distributed Programming, San Francisco, CA, 2001.
4. B. Buck and J. K. Hollingsworth, *An API for Runtime Code Patching*, Journal of High Performance Computing Applications, 14 (2000), pp. 317-329.
5. I. Foster and N. Karonis, *A Grid-Enabled MPI: Message Passing in Heterogeneous Distributed Computing Systems*, SC '98, November 1998.
6. I. Foster and C. Kesselman, *The Grid: Blueprint for a Future Computing Infrastructure*, Morgan Kaufmann, 1998.
7. I. Kuck and Associates, *The GuideView performance analysis tool*, http://www.kai.com.
8. H. Simitci, *Pablo MPI Instrumentation User's Guide*, Department of Computer Science, University of Illinois, 1996.
9. Y. Zhang, *A User's Guide to Pablo MPI I/O Instrumentation*, Department of Computer Science, University of Illinois, 1999.

PVM Computation of the Transitive Closure: The Dependency Graph Approach

Aris Pagourtzis[1,3], Igor Potapov[1], and Wojciech Rytter[1,2]

[1] Department of Computer Science, University of Liverpool,
Chadwick Building, Peach Street, Liverpool L69 7ZF, UK
Work partially supported by GR/N09855 EPSRC grant.
{aris, igor, rytter}@csc.liv.ac.uk
[2] Institute of Informatics, Warsaw University, Poland
[3] Computer Science Division, Dept. of ECE,
National Technical University of Athens, Greece

Abstract. We investigate experimentally a dependency graph approach to the distributed parallel computation of the *generic transitive closure* problem. A parallel coarse-grained algorithm is derived from a fine-grained algorithm. Its advantage is that approximately half of the work is organised as totally independent computation sequences of several processes. We consider conceptual description of dependencies between operations as partial order graphs of events. Such graphs can be split into disjoint subgraphs corresponding to different phases of the computation. This approach helps to design a parallel algorithm in a way which guarantees large independence of actions. We also show that a transformation of the fine-grained algorithm into the coarse-grained is rather nontrivial, and that the straightforward approach does not work.

1 Introduction

The generic transitive closure problem (*TC*, for short) is a useful algorithmic scheme having important applications. Our main issue is how to increase the efficiency by increasing the independence between the main parallel operations and how it affects efficiency of the PVM implementation. The issue of gaining independence in the parallel TC computation has been considered in [2], where it has been shown that approximately $\frac{1}{3}$ of basic operations can be done in an independent way. In this paper, we consider conceptual description of dependencies between operations as a partial order graph of events. Such a graph can be split into disjoint subgraphs corresponding to different phases of the computation. Using this approach, we show that we can achieve large independence for approximately $\frac{1}{2}$ of basic operations.

The TC problem can be defined using an abstract notion of a *semi-ring* with operations $\oplus$ and $\otimes$, see [1] for formal definitions. There are given two abstract operations $\oplus$ and $\otimes$, we sum the values of the simple paths from each vertex i to each other vertex j using the generalised summation operation $\oplus$, the value of each path is the $\otimes$-multiplication of the values of its edges. The *simple path* means that no internal vertex appears twice.

Y. Cotronis and J. Dongarra (Eds.): Euro PVM/MPI 2001, LNCS 2131, pp. 249–256, 2001.

Our main problem is the **Generic Transitive Closure Problem TC**:
   Instance: a matrix A with elements from a semi-ring $\mathcal{S}$;
   Output: the matrix A^*, $A^*(i, j)$ is the sum of all *simple* paths from i to j.

Basic semirings correspond to the *boolean closure* of the matrix (where $\oplus$ and $\otimes$ are the boolean OR and boolean AND operations respectively) and *all-pairs shortest paths* computation in graphs (where $\oplus$ and $\otimes$ are the MIN and $+$ operations). Another nontrivial application is the computation of *minimum spanning trees* [7]: if we consider $\oplus$ and $\otimes$ as MIN and MAX then the minimum spanning tree consists of all edges (i, j) such that $A(i, j) = A^*(i, j)$, assuming initial weights are distinct.

2   The Fine-Grained Algorithm

An interesting systolic algorithm for the TC problem is presented in the book of Leighton, see [6], we refer to this algorithm as Leighton's algorithm. We adapt the idea of Leighton's systolic algorithm to create a parallel algorithm for the TC problem.

   We discuss the main ideas behind Leighton's algorithm on the (basic) example of transitive closure of a directed graph, or equivalently, the closure of a boolean matrix. The main concept in Leighton's algorithm is the *restricted-path*: a path from a node i to some other node such that all intermediate nodes have indices smaller than i. Let us define three predicates:

   $i \to j$: there is a single edge from i to j
   $i \rightsquigarrow j$: there is a restricted path from i to j
   $i \overset{*}{\to} j$: there is a path from i to j.

The transitive closure of a directed graph can be obtained by computing the predicate $i \overset{*}{\to} j$, for each i, j. Leighton's algorithm is based on the following properties of the above defined predicates:

Property1. $i \rightsquigarrow j \Leftrightarrow (\exists\, 1 \le i_1 < \ldots i_k < i)\; i \to i_1 \rightsquigarrow \ldots \rightsquigarrow i_k \rightsquigarrow j$ or $i \to j$;
Property2. $i \overset{*}{\to} j \Leftrightarrow (\exists\, i < i_1 < \ldots i_k \le n)\; i \rightsquigarrow i_1 \rightsquigarrow \ldots \rightsquigarrow i_k \rightsquigarrow j$ or $i \rightsquigarrow j$.

In the algorithm all predicates are stored in the same table $val(i, j)$, whose values are monotonically growing. We abstract the main operation in Leighton's algorithm and investigate the dependency graph representing which operations must be executed before other operations (unrelated operations can be done independently in parallel). Let us introduce two basic operations:

   procedure AtomicOperation(i, k, j)
      $val(i, j) \;:=\; val(i, j) \oplus val(i, k) \otimes val(k, j)$;

   procedure MainOperation(i, k)
      for all j **do** in parallel AtomicOperation(i, k, j);

We denote by $\mathcal{M}(n)$ the set of all **main operations** (i.e. all MainOperation(i, k), where $1 \le i, k \le n$, $i \ne k$).

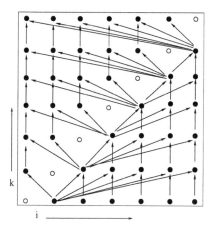

Fig. 1. The dependency graph $\mathcal{G}(7)$. The black node at (i, k) is the operation MainOperation(i, k). The main diagonal is not included.

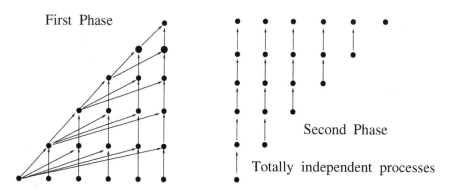

Fig. 2. $\mathcal{G}(n)$ is split into two parts corresponding to the two phases of the algorithm. We obtain a large degree of independence in the second phase.

Definition 1. *A dependency graph is a graph* $\mathcal{G} = \{\mathcal{V}, \mathcal{E}\}$ *where the set of nodes* $\mathcal{V}$ *is a set of operations and each edge* $(u, v) \in \mathcal{E}$ *means that operation* u *should be executed before operation* v.

We will consider the dependency graph $\mathcal{G}(n)$, whose nodes are elements of $\mathcal{M}(n)$, i.e. each node (i, k) corresponds to MainOperation(i, k). The structure of the graph is shown in Figure 1. For each element above the diagonal, except for the top row, there is an edge to its next-upper neighbour. For each element strictly below the diagonal there are edges to all elements in the next-upper row. The nodes below the diagonal correspond to the set of operations related to Property1, and the nodes above the diagonal correspond to Property2.

Observation. The graph $\mathcal{G}(n)$ is one of possible dependency graphs for the TC problem. For example, a different graph corresponds to Warshall's algorithm.

Each dependency graph specifies a class of parallel algorithms (based on this graph).

Definition 2. *A dependency graph $\mathcal{G}$ is* **correct** *for a problem $\mathcal{P}$ if after performing the operations in any order consistent with $\mathcal{G}$ the resulting output will be the correct solution of $\mathcal{P}$. The graph is* **inclusion minimal** *for $\mathcal{P}$ if removal of any of its edges results in incorrectness with respect to $\mathcal{P}$.*

We omit rather technical proofs of the following properties of $\mathcal{G}(n)$.

Theorem 1. *The dependency graph $\mathcal{G}(n)$ is correct and inclusion minimal for the Transitive Closure problem.*

We transform $\mathcal{G}(n)$ into a parallel algorithm by splitting its node-set into two subsets: the first contains the nodes above the diagonal and the second contains the nodes below it. We obtain two dependency subgraphs, as illustrated in Figure 2. The corresponding parallel algorithm has two phases. The operations corresponding to nodes strictly below the diagonal in $\mathcal{G}(n)$ are done in the first phase, and the remaining operations are done in the second phase. The nodes on the diagonal are idle. The algorithm Fine-Grained performs main operations in order consistent with the dependency graph $\mathcal{G}(n)$. This guarantees its correctness, due to Theorem 1. The main point however is that the second phase has a large degree of independence, see Figure 2, which is the very useful property in PVM implementation.

Fine-Grained Algorithm

Pass I {Compute restricted paths $i \rightsquigarrow j$, use Property1 }
for k := 1 **to** n **do**
    for all $i > k$ **in parallel do** MainOperation(i, k);

Pass II {Compute all paths $i \overset{*}{\rightarrow} j$, use Property2 }
for all $i < n$ **in parallel do**
    Process i: **for** $k := i + 1$ **to** n **do** MainOperation(i, k)

---

3   The Coarse-Grained Algorithm

Now each process/processor handles a submatrix of size $s \times s$. In this way we need N^2 processes where $N = \lceil n/s \rceil$. Consider the N subsets of $\{1, \dots, n\}$ each containing s elements: $V_I = \{(I-1) \cdot s + 1, \dots, I \cdot s\}, 1 \leq I \leq N$. By $\tilde{X}_{I,J}$ we denote the sub-matrix of A that contains all elements $A(i, j)$ such that $i \in V_I$ and $j \in V_J$. We will use the convention that uppercase letters represent indices of sub-matrixes and subsets, while lowercase is used for indices of elements. Firstly, we attempt to make a direct transformation of the fine-grained algorithm to a coarse-grained one, by simply replacing elements by sub-matrices. Let us note here that the generalised operations of addition and multiplication for matrices

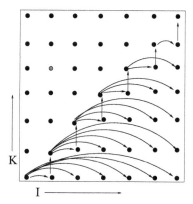

Fig. 3. Additional (compared with $\mathcal{G}(7)$) edges in the partial-order graph of events for coarse-grain computation. The nodes correspond to main block operations BlockMain. The elements on diagonal are no longer idle. First phase has more dependencies, but the second phase has the same degree of independence as in the fine-grained case.

are defined in the same way as the corresponding arithmetic operations. We define the basic block operations.

procedure BlockAtomic(I, K, J):
   $\tilde{X}_{I,J} := \tilde{X}_{I,J} \oplus \tilde{X}_{I,K} \otimes \tilde{X}_{K,J}$

procedure BlockMain(I, K):
   if $I = K$ then $\tilde{X}_{K,K} := (\tilde{X}_{K,K})^*$;
   for all J do in parallel BlockAtomic(I, K, J)

The fine-grained algorithm can be almost automatically transformed into a coarse-grained one by replacing the operation AtomicOperation and MainOperation by BlockAtomic and BlockMain, respectively. Unfortunately, such naive algorithm which results *directly* from the fine-grained one does not work correctly!

Coarse-Grained Algorithm

Pass I
for K := 1 to N do
   BlockMain(K, K);
   for all $I > K$ **in parallel do** BlockMain(I, K)

Pass II
for all $I \leq N$ **in parallel do**
   Process I: for all $I > K$ **in parallel do** BlockMain(I, K)

We derive another algorithm from the dependency graph $\mathcal{G}'(n)$ related to block main operations. This graph is derived from the graph $\mathcal{G}(n)$ by adding extra edges, see Figure 3. In particular, it suffices to add in the naive coarse-grained algorithm an extra statement which computes the transitive closure of a suitable submatrix to get a correct algorithm for the whole matrix. The correctness of the next algorithm is nontrivial, we omit a technical proof.

4   PVM Implementation

The coarse-grained algorithm was implemented using the PVM MASTER/SLAVE model. In the following we give an abstract description of this implementation; these ideas can be applied to any message passing distributed system.

The basic idea is to have N^2 SLAVE programs, one for each block $\tilde{X}_{I,J}$. Each SLAVE handles the data of its corresponding block according to the requests sent by the MASTER.

In the beginning the MASTER program sends to each SLAVE the data of the corresponding block. Then, MASTER repeatedly calls a routine **mult**$(\cdot, \cdot, \cdot)$ which parallelizes the procedure of multiplying two rectangles of blocks by decomposing the whole multiplication to several multiplications of two blocks and by sending appropriate messages to the corresponding slaves. Data remain in place (slave) and only appropriate messages are sent; this way moving huge amounts of data between the MASTER and the SLAVEs is avoided.

A routine **tc**$(\cdot)$ for computing the transitive closure of a single block is also used; again data remain in place and only a message is sent. Synchronization is achieved by using a routine **expect**() which waits for all active operations to finish. When it is necessary to implement independent parallel processes which cannot be together parallelised by a single **mult**$(\cdot, \cdot, \cdot)$ command (e.g. because they consist of sequential steps) we may follow two strategies:

- *2-Level implementation*: employ a SUBMASTER program; MASTER creates as many SUBMASTERs as the parallel processes are (in our case $N-1$). At the end, MASTER gets all the data from SLAVEs to assemble the resultant matrix.
- *1-Level implementation*: MASTER controls itself synchronization by adding appropriate indices to the messages. Each parallel process has its own index, so each time MASTER receives a message it knows to which process it corresponds. Then MASTER invokes the next step of this process.

5   Experimental Results and Discussion

We compare the practical behaviour of the standard sequential Warshall algorithm, a parallel Block-Warshall algorithm [2] and the two versions of our coarse-grained algorithms (1-Level and 2-Level).

All experiments were performed on a cluster of Hewlett Packard 720 Unix workstations networked via a switched 10Mb Ethernet using PVM version 3.4.6. PVM was used in preference to MPI as previous experience had shown MPI to exhibit unusual anomalies rendering it unusable in our workstation environment.

Table 1. The (sequential) Warshall algorithm.

Matrix Size	240 x 240	480 x 480	960 x 960	1920 x 1920
Time (sec)	31	273	2249	18519

Table 2. Parallel Algorithms: Blocks-Warshall, 1-Level and 2-Level.

	240 x 240			480 x 480			960 x 960			1920 x 1920		
No.	BlW	L1	L2	BlW	L1	L2	BlW	L1	L2	BlW	L1	L2
CPUs	alg.	alg.	alg.	alg.	alg.	alg.	alg.	alg.	alg.	alg.	alg.	alg.
4	6.0	3.0	3.0	20.1	16.3	16.3	160	132	127	2720	1936	2010
9	4.6	3.0	3.3	11.3	9.6	10.0	93	72	72	1633	1189	1275
16	3.2	3.0	4.6	9.2	7.0	9.0	62	50	50	526	351	384
25	2.8	4.6	5.0	8.7	8.0	10.6	52	40	44	507	329	363
36	3.1	6.3	7.0	11.6	9.3	11.6	42	37	36	443	283	296

Table 1 shows the average execution time (in seconds) for the sequential Warshall algorithm to complete on a single CPU for matrices ranging in size from 240×240 to 1920×1920 elements. The results in Table 1 show a cubic increase of the execution time as the matrix dimension increases.

Table 2 shows the average execution time in seconds for the Block Warshall, the 1-Level and the 2-Level algorithms, respectively, for the same sized matrices as above and for varying CPU cluster size. Both our new algorithms perform better than the Block Warshall algorithm. In particular, the 1-level algorithm is the best of three, probably because it avoids the extra cost of SUBMASTER programs.

We can follow a standard approach and describe execution time of our algorithms in terms of computation, communication and synchronization costs. The analysis below explains the non-monotonic behaviour of execution time with respect to the number of processors. We will express the execution time as the number of *cost units* which represent the cost of a basic arithmetic operation or a local memory access. Let W be the parallel computation cost, H be the communication cost and S be the synchronization cost. The total execution cost is then given as:

$$T = W + H \cdot g + S \cdot l \tag{1}$$

where g and l are network parameters. Note that the total execution time is proportional to T. The computation, communication and synchronisation costs are the same for all three algorithms (Blocks Warshall, 1-Level and 2-Level) in terms of O-notation. In particular,

$$W = O(n^3/p) \quad H = O(n^2\sqrt{p}) \quad S = O(p) \tag{2}$$

The above estimation corresponds to the model where message passing is actually sequential, in contrast to other models (e.g. BSP [10]). The model corresponds to a bus-based network, like the one that is used for our experiments.

Although expression (2) is the same for all three algorithms the execution time for each algorithm is different because of different coefficients in (2). Despite the small increase of the computation cost coefficient for the two new algorithms (comparing with Blocks Warshall) the communication and synchronisation coefficients are significantly smaller. This is due to the separation of half the processes into totally independent sequences during the second pass of the new algorithms. The slight difference between 1-Level and 2-Level algorithms can be explained by the additional synchronisation cost of 2-Level algorithm, due to the use of SUBMASTER processes.

On the other hand, these algorithms display a similar behaviour: the time cost function given by 1 has a minimum value for $p > 0$. Usually (for realistic coefficients) this minimum occurs for $p > 1$. This happens in our system too, as shown in table 2. By using the experimental results one might be able to specify the coefficients for a particular parallel computation system. This could give us the possibility to predict the number of processors (depending on the size of the matrix) for which the execution time is minimum.

In this paper we introduced dependency graphs as a tool to design parallel algorithms in more structural way. The usefulness of the dependency graphs follows from the fact that efficient parallelism is closely related to large independence.

References

1. A.Aho, J.Hopcroft, J.Ullman, The design and analysis of computer algorithms, Addison-Wesley (1974)
2. K. Chan. A.Gibbons, M. Marcello, W. Rytter, On The PVM Computations of Transitive Closure and Algebraic Path Problem, EuroPVM'98, Springer Verlag, Lecture Notes in Computing 1998
3. A.Gibbons, W.Rytter, Efficient parallel algorithms, Cambridge University Press (1988)
4. H.T.Kung and Jaspal Subhlok, A new approach for automatic parallelization of blocked linear algebra computations, 122–129, Supercomputing '91. Proceedings of the 1991 Conference on Supercomputing, 1992
5. Hans-Werner Lang, Transitive Closure on the Instrucion Systolic Array, 295–304, Proc. Int. Conf. on Systolic Arrays, San Diego, K. Bromley, S.Y. Kung, E. Swartzlander, 1991
6. F.T. Leighton, Introduction to Parallel Algorithms and Architectures : Arrays, Trees, Hypercubes, Morgan Kaufmann Publishers 1991
7. B. Maggs, S. Plotkin, Minimum cost spanning trees as a path finding problem, IPL 26 (1987) 191-293
8. Gunter Rote, A systolic array algorithm for the algebraic path problem, Computing, 34, 3, 191–219, 1985
9. Alexandre Tiskin, All-pairs shortest paths computation in the BSP model, Proc. ICALP'01, 2001
10. Leslie G. Valiant, A bridging model for parallel computation, Communications of the ACM, 33, 8, 103–111, 1990

Parallizing 1-Dimensional Estuarine Model

Jun Luo[1], Sanguthevar Rajasekaran[1], and Chenxia Qiu[2]

[1] Computer & Information Science & Engineering Department, University of Florida
[2] Coastal & Oceanographic Engineering Department, University of Florida,
Gainesville, Florida, 32611, USA

Abstract. Wave simulation is an important problem in engineering. Wave simulation models play a significant role in environment protection. Thus far wave simulations have been mostly done in serial. In order to meet the demand for increased spatial and temporal resolution and uncertainty analysis in environmental models for ecological assessment and water resources management, it is essential to develop high-performance hydrodynamics and water quality models using parallel techniques. In this paper, we propose algorithms for parallelizing 1-dimensional estuarine models. We have implemented these algorithms using PVM. Our experiments show that these algorithms are efficient and the performance of the parallel estuarine model is excellent.

1 Introduction

Wave simulation is very important in the study of how waves behave in the real world. It plays a significant role in coastal control and environment protection. As wave simulation deals with a large amount of information and massive calculation, a long term esturine simulation program would run for hours or even days in a row. In order to meet the demand for increased spatial and temporal resolution and uncertainty analysis in environmental models for ecological assessment and water resources management, it is essential to develop high-performance hydrodynamics and water quality models using parallel techniques.

This paper focuses on parallelizing 1-dimensional estuarine models. The 1-dimensional is abbreviated 1-d in the rest of the paper. The 1-d model is described as follows.

Definition 1. *The tide propagates in a rectangular channel with constant water depth. At one open end of the channel, the water level varies with tide. The other end of the channel is closed, and the velocity of it is set 0.*

The task of the 1-d model is to find out how the tide propagates in the esutary with time. As Figure 1 shows, the 1-d estuarine model space is divided into subsections, called grids. For each grid, the variables to be calculated are the water surface elevation h and horizontal velocity n. h is defined at the center of the grid and n is defined at the middle of the vertical boundaries of the grid's. Intrinsically, both h and n have two attributes: space position and time point. The superscript, such as n in Figure 1, stands for the nth simulation step. The

Y. Cotronis and J. Dongarra (Eds.): Euro PVM/MPI 2001, LNCS 2131, pp. 257–264, 2001.

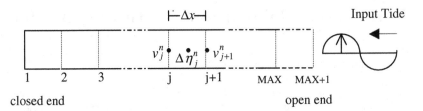

Fig. 1. 1-d estuarine model illustration

subscript, such as j in Figure 1, represents the jth grid in the estuary. The grids are numbered from left to right starting with 1.

Generally speaking, there are two schemes in implementing the 1-d estuarine model: explicit scheme and implicit scheme. The difference between these two schemes are discussed in section 3. From a computational point of view, in the explicit scheme, the values of h and n can be simply calculated by solving a stand-alone equation. In contrast, in the implicit scheme, the values of h and n are calculated solving a tri-diagonal system of linear equations.

The difference between the two schemes results in the quite different parallel algorithms. For parallelizing the explicit scheme, we introduce a novel paradigm called *LessTalk*. This paradigm reduces the number of inter-processor communication steps in an elegant way. We hope that this technique can be applied in the design of efficient parallel algorithms in other domains as well. The idea is to perform some redundant computations in return for reducing the total number of communication steps. In the parallelization of the implicit scheme, solving a tri-diagonal system of linear equations is a key step. We employ the algorithm given in [1].

This paper is organized as follows. In section 2 we overview the 1-d estuarine model. The parallel algorithms are given in section 3. We discuss the experimental results in section 4. Finally, concluding remarks are made in section 5.

2   1-d Estuarine Model Overview

The 1-d wave model consists of two partial differential equations (PDE), and the solution of these equations gives us a time-changing wave form. The equations are shown below.

$$\frac{\partial \eta}{\partial t} + \frac{\partial U}{\partial x} = 0 \tag{1}$$

$$\frac{\partial U}{\partial t} + \frac{\partial \frac{U^2}{H}}{\partial x} = -gH\frac{\partial \eta}{\partial x} + \tau_{wx} - g\rho\frac{U|U|}{C^2 H^2} + A_H\frac{\partial^2 U}{\partial x^2} \tag{2}$$

where t is the time. x is the grid size. U is calculated as the integration of the horizontal velocity, v, over the total depth H at some point x in the estuary. g, r, C and A_H are given as constants, where g is the gravity acceleration, r is the water density, C is related with the estuary bottom friction, and A_H denotes the

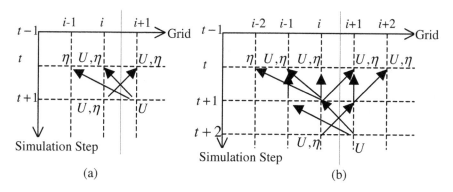

Fig. 2. Data Dependency of 1-d Explicit Numerical Scheme

diffusion coefficient. H is the total depth of the water from the water surface to the water bottom. τ_{wx} stands for the wind stress along the longitudinal axis.

Equations (1) and (2) are also called the continuity equation and the momentum equation, respectively, which describe the properties of the mass conservation and the momentum conservation in the estuary. Due to limitations on space, we omit the discussion of 1-d explicit and implicit numerical schemes here. A complete description of these numerical schemes could be found in [2].

3   Parallel Algorithms

3.1   1-d Explicit Scheme

A straightforward parallel algorithm for 1-d explicit scheme works as follows. Partition the grids into N parts as evenly as possible. Assign each part to a processor and each processor performs the same calculations on disjoint grids. However, there is a subtle question: can each processor work alone? In other words, at each simulation step, can each grid be computed without any information from its neighboring grids? To answer the question, it is necessary to analyze the data dependency between the consecutive simulation steps in the numerical scheme. The analysis result is summarized in Figure 2.

In Figure 2, $t-2, \cdots, t+1$ represent simulation steps. $i-1, \cdots, i+2$ represent grids. The calculation of the variable values at the end of any arrow depends on the variable values at the head of the arrow. Figure 2 only shows the dependency chains that come across the different partitions. In Figure 2(a), if grids i and $i+1$ belong to different partitions, say p_1 and p_2, then at simulation step $t+1$, p_1 and p_2 must communicate to exchange the variable values of η_{i-1}^t, η_i^t, η_{i+1}^t, U_i^t and U_{i+1}^t.

The total communication times needed by the straightforward parallel algorithm equals to the number of simulation steps. As the I/O operation is expensive, the performance of the parallel program can be improved by reducing the total I/O operations. We define the communication factor as the length of the simulation step intervals during which no inter-processor communication takes place. The parallel algorithm using the concept of the communication factor is called *LessTalk*. The intuition behind *LessTalk* is to reduce the number of communication times at the expense of some redundant local computations. The number of communication times is reduced as follows. The neighboring partitions exchange more variable values in any given communication step. As a result, the processors need not communicate with each other frequently.

Consider the case where the communication factor is 2. At the step t, p_1 and p_2 communicate with each other. The variable values that are needed to be exchanged could be found out by looking at Figure 2(b), which is expanded from Figure 2(a). It shows that if p_1 and p_2 exchange the values of η_{i-2}^t, η_{i-1}^t, η_i^t, η_{i+1}^t, η_{i+2}^t, U_{i-1}^t, U_i^t, U_{i+1}^t, and U_{i+2}^t then p_1 and p_2 need not communicate until step $t + 2$. It could also be observed that the number of variables that have to be exchanged in any simulation step is a function of the communication factor. From this observation, *LessTalk* algorithm is described below.

Algorithm1: *LessTalk* Parallel Algorithm

Step 1) Calculate the workload for each processor as follows. Assign 1st MAX/N grids to $processor_1$, 2nd MAX/N grids to $processor_2$, $\cdots$, $(n-1)^{th}$ MAX/N grids to $processor_{n-1}$ and the remaining grids to $processor_N$;

Step 2) Determine the value of the communication factor, denoted as K;

Step 3) For the processor p whose workload is denoted as p_w , do the following.

calculate $\partial t/\partial x$ and $(\partial t/\partial x)^2$;
forall grid i which falls under p_w do
    $\eta_i := 0$; $U_i := 0$; *count* $:= K$;
for simulation step $n := 1$ to M do
{
    If *count* $= K$ then
    {
        If $p = N$ then
            update η_{MAX}^n at the open end boundary;
        If $p \neq N$ then
            send $\eta_{p_w}^{n-1}$, $\eta_{p_w-1}^{n-1}$, $\cdots$, $\eta_{p_w-K}^{n-1}$, $U_{p_w}^{n-1}$, $\cdots$, $U_{p_w-K+1}^{n-1}$ to $processor_{p+1}$;
        If $p \neq 1$ then
            send $\eta_{(p-1)_w+1}^{n-1}$, $\cdots$, $\eta_{(p-1)_w+K}^{n-1}$, $U_{(p-1)_w+1}^{n-1}$, $\cdots$, $U_{(p-1)_w-K}^{n-1}$
            to $processor_{p-1}$;
        If $p \neq N$ then
            wait until $\eta_{p_w+1}^{n-1}$, $\cdots$, $\eta_{p_w+K}^{n-1}$, $U_{p_w+1}^{n-1}$, $\cdots$, $U_{p_w-K}^{n-1}$
            arrive from $processor_{p+1}$;
        else if $p \neq 1$ then
            wait until $\eta_{(p-1)_w}^{n-1}$, $\eta_{(p-1)_w-1}^{n-1}$, $\cdots$, $\eta_{(p-1)_w-K}^{n-1}$, $U_{(p-1)_w}^{n-1}$, $\cdots$,
            $U_{(p-1)_w-K+1}^{n-1}$ arrive from $processor_{p-1}$;
    }
    forall grid i which falls under p_w do
    {

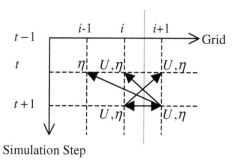

Fig. 3. Data Dependency of 1-d Implicit Numerical Scheme

update H_i^n by using η_{i-1}^{n-1} and η_i^{n-1};
calculate η_i^n;
calculate $(advection)_i^n$, $(propagation)_i^n$, $(diffusion)_i^n$
and $(bottomfriction)_i^n$;
calculate U_i^n and then calculate v_i^n;
    }
}

Let C stand for the time overhead per communication, B represent the size of the data to be exchanged, E stand for the time for transmitting one unit data, and T denote the number of simulation steps. If the communication factor is set to 1, then the total run time for the communications is $(B \times E + C) \times T$. If the communication factor is set to K, then the run time would be $(B \times K \times E + C) \times T/K = (B \times E + C/K) \times T$. So, the reduced run time is $C \times T \times (1 - 1/K)$. The data exchanged in 1-d model is not large. Thus the above analysis does not consider the time overhead for data contentions in the network and overhead C is thought of being constant.

3.2   1-d Implicit Scheme

The data dependency analysis of 1-d implicit scheme is shown in Figure 3, which is similar to that of 1-d explicit scheme. Also, Figure 3 only displays the dependencies that come across the different partitions. In 1-d implicit scheme, the values of U and h of grid i at simulation step $t+1$ also depend on the values of U and h of grid $i+1$ at simulation step $t+1$, which is the most significant difference from the explicit scheme. So, the processors have to communicate every simulation step.

Furthermore, in order to calculate U and h of every grid at each simulation step, a tri-diagonal system of linear equations have to be solved. Assume the current simulation step is n, the tri-diagonal system will have the following form.

$$A^n \times \eta^n = D^n \qquad (3)$$

where A^n is a $MAX \times MAX$ matrix and $A_{i,j} \neq 0$ only if $\mid i - j \mid \leq 1$; η^n and D^n are $MAX \times 1$ column vectors, respectively.

In the rest of the section, we focus on solving the above tri-diagonal system. The parallel algorithm is described below, which is on the basis of [1].

Algorithm2: Solving Tri-diagonal System of Linear Equations

Step 1) The tri-diagonal system is divided into N parts along its rows. The first processor receives the 1st MAX/N equations. The second processor receives the 2nd MAX/N equations. The nth processor except the last one receives the nth MAX/N equations. The last processor receives the remaining equations;

Step 2) Each processor uses elimination strategy to eliminate as many unknowns as possible from the equations it is assigned. At the end of the step 2, each processor is left with 2 equations;

Step 3) All the processors except the first one send their equations got at the end of the step 2 to the first processor. The first processor then organizes the $2N$ equations into a new tri-diagonal system of linear equations. The first processor solves this new tri-diagonal system by using conjugate gradient method or recursively do the same procedure as this description. According to which equations it receives from which processor, the first processor sends the values of the unknowns contained in the equations back to that processor;

Step 4) After receiving the results from the first processor, each processor uses back-substitution strategy, which is the reverse process of the second phase, to use the known values to find out the values of all the unknowns.

4   Performance Evaluation

We use the master/slave computation model in implementing the parallel algorithms discussed in the paper. We use PVM (Parallel Virtual Machine) [3] as the message passing method between the processes. The experiments were run on the distributed-memory multicomputer system which consists of 11-nodes. Each node has two processors whose CPUs are PII 450 MHz. Each node has 256MB main memory and 19GB hard disk. All the nodes in the system are connected through 100Mbps fast Ethernet. The OS of the system is Red Hat Linux release 6.1.

4.1   1-d Explicit Scheme

In the first set of experiments, we tested the speedup of the parallel programs. We kept the length of the estuary and the number of grids constant. We changed the number of processors and the value of the communication factor. The total number of grids in the estuary is set to 2,000. Figure 4 (a) shows the run time of the parallel programs and the sequential program. Figure 4 (b) shows the corresponding relative speedup.

In Figure 4, the performance of the PVM program initially increased and then gradually decreased. This phenomenon can be explained as follows. Starting with the value of 1, when the communication factor increased, the total run time of

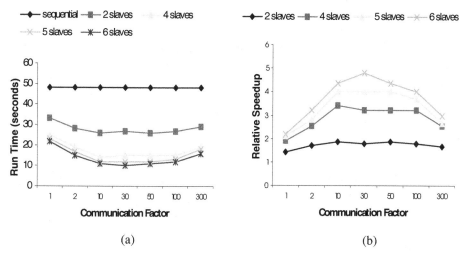

Fig. 4. Experiment 1

the parallel program decreased, which means the benefit brought from reducing the total communication time is greater than the increased local computation time. When the communication factor reached a certain value, the total run time of the program stopped decreasing even though the communication factor still increases, which means that the benefit ties up to the burden.

In the second set of experiments, we set the communication factor to 20 and kept the communication factor constant. We grew the length of the estuary by increasing the number of grids and kept the size of each grid constant. The run time results are shown in Figure 5(a) and Figure 5(b) shows the corresponding relative speedup. From Figure 5, it can be seen that the more data a node processes, the less significant becomes the communication time.

4.2   1-d Implicit Scheme

We tested the speedup performance for the 1-d implicit parallel programs. We grew the length of the estuary by increasing the number of grids and kept the size of each grid constant. The run time results are similar to that of 1-d explicit scheme. Due to limitations on space, we omit the figures. The details can be found [2].

5   Conclusions

In this paper, we have proposed two parallel algorithms for the simulation of 1-d estuarine models. Our experimental results show that these two parallel algorithms perform well. The *LessTalk* paradigm introduced in this paper has the potential of finding independent applications.

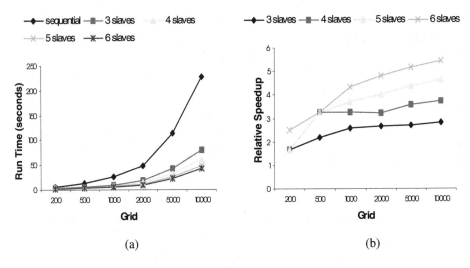

Fig. 5. Experiment 2

References

1. Rajasekaran S., Luo J., and Sheng Y.P. A simple parallel algorithm for solving banded systems. In: Proceedings of the International Conference on Parallel and Distributed Computing and Systems (PDCS '99), Boston, MA, November 1999.
2. http://www.cise.ufl.edu/~jluo/coastal/1dnumericalscheme.html
3. Al Geist. PVM : Parallel Virtual Machine : A Users' Guide and Tutorial for Networked Parallel Computing,. Cambridge, Mass. : MIT Press, 1994.

A Parallel ADI and Steepest Descent Methods

I.V. Schevtschenko

Rostov State University
Laboratory of Computational Experiments on Super Computers
Zorge street 5, Rostov-on-Don, 344090, Russia
ishevtch@uic.rnd.runnet.ru

Abstract. The aim of the paper is an improvement of a parallel alternating-direction implicit, or ADI, method for solving the two-dimensional
equation of ground water flow and its implementation on a distributed-
memory MIMD-computer under the MPI message-passing system. Besides, the paper represents evaluation of the parallel algorithm in terms
of relative efficiency and speedup. The obtained results show that for
reasonably large discretization grids the parallel ADI method is effective
enough on a large number of processors.

Key-Words: finite difference method, parallel alternating-direction implicit method, parallel steepest descent method, conjugate gradient
method, the problem of ground water flow.

1 Introduction

In solving partial differential equations (PDEs) with the help of the ADI method
we obtain two systems of linear algebraic equations (SLAEs) with special matrices. One of those matrices obtained in the $(n+\frac{1}{2})$ time-layer is a band tridiagonal
matrix while the second one obtained in the $(n + 1) - st$ time-layer is a block
tridiagonal matrix. In order to solve the SLAEs in the $(n + \frac{1}{2})$ time-layer in parallel we have to transform the corresponding matrix by means of permutation of
rows to a block tridiagonal matrix. This transformation involves transposition of
the right hand side (RHS) of the equation we solve when the RHS is represented
as a matrix. It is clear that such a transposition entails extra communication
which may be reduced if a parallel SD method is exploited.

An outline of the paper is as follows. Section 2 introduces to general formulation and numerical approximation of the original mathematical model, represents
some results on accuracy and stability of the used difference scheme and sub-
stantiates applicability of the steepest descent (SD) and conjugate gradient (CG)
methods to solving systems of linear algebraic equations (SLAEs) generated by
Peaceman-Rachford difference scheme [8]. Section 3 is represented by a parallel
implementation of the ADI method. In the same place we evaluate the parallel
algorithm in terms of relative efficiency and speedup and compare it with its
improved version. Finally, in section 4, we give our conclusions.

Y. Cotronis and J. Dongarra (Eds.): Euro PVM/MPI 2001, LNCS 2131, pp. 265–271, 2001.

2   General Formulation and Numerical Approximation

The underlying equation for the current paper is the linear model describing gravitational flow of ground water in an element of a water-bearing stratum Ω which, in according to [2], can be represented in the form

$$a^* \frac{\partial h}{\partial t} = \sum_{i=1}^{2} \frac{\partial^2 h}{\partial x_i^2} + \phi, \tag{1}$$

Here $x = x_1$, $y = x_2$; $\phi = \frac{1}{kh_{av}}\left(v_{(x_1)} + v_{(x_2)}\right)$, $h = h(x, y, t)$ represents a ground water table, k denotes the filtrational coefficient, $v_{(x)}$ and $v_{(y)}$ are filtration velocities from below and from above of the water-bearing stratum respectively. Parameter $a^* > 0$ depends on h_{av} and physical characteristics of the water-bearing stratum.

The initial condition and Dirichlet boundary value problem for equation (1) are

$$h(x, y, t = 0) = h_0(x, y), \ h|_{\partial\Omega} = f(x, y).$$

Here $h_0(x, y)$ is a given function, $f(x, y)$ is a function prescribed on the boundary of the considered field.

Approximation of equation (1) bases on Peaceman-Rachford difference scheme, where along with the prescribed grid functions $h(x, y, t)$ and $h(x, y, t+\tau)$ an intermediate function $h(x, y, t + \frac{\tau}{2})$ is introduced. Thus, passing (from the n time-layer to the $(n + 1) - st$ time-layer) is performed in two stages with steps 0.5τ, where τ is a time step.

Let us introduce a grid of a size $M \times N$ in a simply connected area $\Omega = [0, a] \times [0, b]$ with nodes $x_i = i\Delta x$, $y_j = j\Delta y$, where $i = 1, 2, \ldots M$, $j = 1, 2, \ldots N$, $\Delta x = \frac{a}{M}$, $\Delta y = \frac{b}{N}$. By denoting $h = h^n$, $\bar{h} = h^{n+\frac{1}{2}}$, $\hat{h} = h^{n+1}$ let us write out the difference approximation for equation (1)

$$\begin{cases} a^* \frac{\bar{h}_{ij} - h_{ij}}{0.5\tau} = \frac{\bar{h}_{i-1,j} - 2\bar{h}_{ij} + \bar{h}_{i+1,j}}{\Delta x^2} + \frac{h_{i,j-1} - 2h_{ij} + h_{i,j+1}}{\Delta y^2} + \phi, \\[2mm] a^* \frac{\hat{h}_{ij} - \bar{h}_{ij}}{0.5\tau} = \frac{\bar{h}_{i-1,j} - 2\bar{h}_{ij} + \bar{h}_{i+1,j}}{\Delta x^2} + \frac{\hat{h}_{i,j-1} - 2\hat{h}_{ij} + \hat{h}_{i,j+1}}{\Delta y^2} + \phi. \end{cases} \tag{2}$$

The difference approximation of the initial condition and Dirichlet boundary value problem can be represented as

$$h|_{t=0} = h_{(0)ij}, \quad h|_{\partial\Omega} = f_{ij}. \tag{3}$$

By addressing stability investigation of difference scheme (2) we formulate the the following lemma

Lemma 1 *Peaceman-Rachford difference scheme for equation (1) with Dirichlet's boundary conditions at $a^* > 0$ is unconditionally stable.*

Concerning the difference scheme (2) it can be noted that it has second approximation order [9] both in time and in space.

By using natural regulating of unknown values in the computed field let us reduce difference problem (2),(3) to the necessity of solving SLAEs $A_k u_k = f_k, k = 1, 2$ with special matrices.

The obtained SLAEs have been scaled and solved with CG method [9] afterwards. Here the scaling means that the elements of the coefficient matrices and the Right Hand Sides (RHSs) have the following form

$$\hat{a}_{ij} = \frac{a_{ij}}{\sqrt{a_{ii}a_{jj}}}, \ \hat{f}_i = \frac{f_i}{a_{ii}}, \ i, j = 1, 2, \ldots, MN.$$

The selection of the CG method is based on its acceptable calculation time [5] in comparison with the simple iteration method, the Seidel method, the minimal residual method and the steepest descent method.

To proceed, it is necessary to note that from previous lemma we can infer the appropriateness of using the SD method and the CG method since

$$A_k = (A_k)^T, \ A_k > 0, \ k = 1, 2.$$

3   Algorithm Parallel Scheme

Before passage to the description of the parallel algorithms we would like to say a few words about the library with the help of which the parallel algorithms have been implemented and the computational platform on which they have been run.

The implementation of the parallel algorithms was carried out using C and MPI. Relative to the paradigm of message passing it can be noted that it is used widely on certain classes of parallel machines, especially those with distributed memory. One of the representatives of this conception is MPI (Message Passing Interface) [3], [4], [6], [7]. The MPI standard is intended for supporting parallel applications working in terms of the message passing system and it allows to use its functions in C/C++ and Fortran 77/90.

All the computational experiments took place on an nCube 2S, a MIMD-computer, the interconnection network of which is a hypercube. The number of processing elements (PEs) is 2^d, $d \leq 13$. These PEs are unified into the hypercube communication network with maximum length of a communication line equaled d. A d-dimensional hypercube connects each of 2^d PEs to d other PEs. Such communication scheme allows to transmit messages fast enough irrespective of computational process since each PE has a communication coprocessor aside from a computational processor. At our disposal we have the described system with 64 PEs.

The parallel algorithm for solving equation (2) is based on inherent parallelism which is suggested by Peaceman-Rachford difference scheme allowing find the numerical solution of the SLAEs in each time-layer independently, i.e. irrespective of communication process. The main communication loading lies on connection between two time-layers. Thus, one simulation step of the parallel algorithm to be executed requires two interchanges of data at passage to the $n + \frac{1}{2}$ and to the $(n + 1) - st$ time-layers. Besides, the ADI method gives us an opportunity to exploit any method to solve the SLAEs obtained in the $(n + \frac{1}{2})$ and the $(n + 1) - st$ time-layers. In this paper we use the CG method and the SD method.

Taking into account aforesaid one step of the parallel algorithm (let us denote it by **A**) for solving system (2) with SLAEs $A_k X_k = B_k$, $k = 1, 2$ can be represented in the following manner

1. Compute the value $h_{av} = \frac{1}{MN} \sum\limits_{i=1}^{M} \sum\limits_{j=1}^{N} h_{ij}$.

2. Compute B_1 in the $(n + \frac{1}{2})$ time-layer.

3. Make the permutation of vectors $X_1^{(0)}$, B_1, where $X_1^{(0)}$ is an initial guess of the CG method in the $(n + \frac{1}{2})$ time-layer.

4. Solve equation $A_1 X_1 = B_1$ in the $(n + \frac{1}{2})$ time-layer with the CG method.

5. Compute B_2 in the $(n+1) - st$ time-layer partially, i.e. without the last item of the second equation (2).

6. Make the permutation of vectors $X_2^{(0)}$, B_2, where $X_2^{(0)}$ is an initial guess of the CG method in the $(n + 1) - st$ time-layer.

7. Compute the missing item so as the computation of B_2 in the $(n + 1) - st$ time-layer has been completed.

8. Solve equation $A_2 X_2 = B_2$ in the $(n+1) - st$ time-layer with the CG method.

First step of the algorithm is well-understood (to gather all the values h_{ij} from all the PEs, sum and broadcast them to all the PEs afterwards) while the second claims more attention. Let vectors $X_1^{(0)}$, B_1 be matrices (which are distributed uniformly in the rowwise manner) consist of elements of the corresponding vectors, then to solve equation $A_1 X_1 = B_1$ in the $(n + \frac{1}{2})$ time-layer with the CG method in parallel we need to transpose the matrix corresponding to vector $B_1 = \{b_1, b_2, \ldots, b_{MN}\}$

$$
\begin{pmatrix}
b_1 & b_2 & \cdots & b_N \\
b_{N+1} & b_{N+2} & \cdots & b_{2N} \\
\cdots & \cdots & \cdots\cdots & \\
b_{(M-1)N+1} & b_{(M-1)N+2} & \cdots & b_{MN}
\end{pmatrix}
\rightarrow
\begin{pmatrix}
b_1 & b_{N+1} & \cdots & b_{(M-1)N+1} \\
b_2 & b_{N+2} & \cdots & b_{(M-1)N+1} \\
\cdots & \cdots & \cdots\cdots & \\
b_N & b_{2N} & \cdots & b_{MN}
\end{pmatrix}
$$

and the matrix corresponding to vector $X_1^{(0)}$. Of course, such a transposition requires transmission of some sub-matrices ($\frac{M}{p} \times \frac{N}{p}$ size, p is the number of available PEs) to the corresponding PEs. By denoting the number of send/receive operations, the amount of transmitted data and computational complexity of the algorithm by $C_{s/r}^A$, C_t^A and C_c^A we get that one step of algorithm **A** to be executed requires

$$
C_{s/r}^A = 2p + 4p(p - 1), \quad C_t^A = 2p + 4M\left(N - \frac{N}{p}\right),
$$

$$
C_c^A = (66NM - 20N)I_{CG^A} + M\left(\frac{33N}{p} + 12\right) + N(10M - 4) + 31,
$$

where I_{CG^A} is the iteration number of the CG method.

Now, we present test experiments all of which are given for one step of the algorithm and only for one iteration of the CG method. The horizontal axis, in all the pictures, is the function $p(x) = 2^x$, $x = 0, 1, \dots, 6$ which denotes the number of PEs.

By following [10] let us consider relative efficiency and speedup

$$S_p = \frac{T_1}{T_p}, \quad E_p = \frac{S_p}{p},$$

where T_p is a time to run a concurrent algorithm on a computer with p PEs $(p > 1)$, T_1 is a time to run a sequential algorithm on one PE of the same computer.

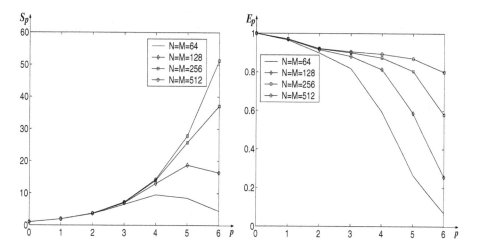

Fig. 1. Relative speedup (to the left) and efficiency (to the right) of the parallel algorithm at various grid sizes.

The Fig.1 shows that relative speedup and efficiency are high enough even for a grid of $N = M = 512$ size.

As mentioned above the lack of the previous ADI method is a number of transmitting data at passing from one time-layer to the other time-layer. To reduce communication time we solve the SLAEs in the $(n + \frac{1}{2})$ time-layer by a parallel SD method.

The pseudo code of the SD method [1] can be written down in the form

Initial guess is $x^{(0)}$

for $k = 0, 1, 2, \ldots$

$\quad r^{(k)} = b - Ax^{(k)}$;

$\quad \alpha_k = \dfrac{(r^{(k)}, r^{(k)})}{(Ar^{(k)}, r^{(k)})}$;

$\quad x^{(k+1)} = x^{(k)} + \alpha_k r^{(k)}$;

$\quad$ convergence test $\|r^{(k)}\| \le \varepsilon$.

$\quad$ continue if necessary.

end

Fig. 2. The steepest descent method.

As we can see from Fig.2 to parallelize the SD method we have to parallelize the two inner products and the two matrix-vector multiplications. Parallel schemes of an inner product and a matrix-vector multiplication are well known and we do not concentrate our attention on them. Thus, one step of the improved algorithm (let us denote it by **B**) can be represented as

1. Compute the value $h_{av} = \frac{1}{MN} \sum\limits_{i=1}^{M} \sum\limits_{j=1}^{N} h_{ij}$.
2. Compute B_1 in the $(n + \frac{1}{2})$ time-layer.
3. Solve equation $A_1 X_1 = B_1$ in the $(n + \frac{1}{2})$ time-layer with the parallel SD method ($X_1^{(0)}$ is an initial guess of the SD method in solving equation $A_1 X_1 = B_1$).
4. Transmit missing rows of the matrix describing the current level of ground water (taking into account rowwise distribution of this matrix) to the corresponding PEs and compute B_2 in the $(n + 1) - st$ time-layer.
5. Solve equation $A_2 X_2 = B_2$ in the $(n+1) - st$ time-layer with the CG method ($X_2^{(0)}$ is an initial guess of the CG method in solving equation $A_2 X_2 = B_2$).

In accordance with the previous section let us write $C_{s/r}^B$, C_t^B and C_c^B. As a result, for one step of algorithm **B** we have

$$C_{s/r}^B = 2p + 2(p - 1) + I(4p + 4(p - 1)),$$
$$C_t^B = 2p + 2(p - 1)N + I_{SD^B}(4p + 4(p - 1)N),$$
$$C_c^B = N(33M - 10)I_{CG^B} + \left(15\frac{NM}{p} + 1\right) I_{SD^B} + 29\frac{NM}{p} + M(12 + 5N) - 2N.$$

Here I_{CG^B} is the iteration number of the CG method, and I_{SD^B} is the iteration number of the parallel SD method.

With respect to computational complexity, algorithm **B** is two time faster than algorithm **A** since

$$\lim_{M \to \infty} \frac{C_c^A}{C_c^B} = 2, \ N = I_{CG^A} = I_{SD^B} = I_{CG^B} = M, \ p = 1$$

but we cannot say the same about overall computation time of the algorithms, because their communication schemes are different. In this work we do not conduct theoretical analysis of computational complexity and communication time of the algorithms sinsce this theme deserves particular attention, but we give numerical comparison of the algorithms for $p = 64$ and grids of $N = M = 512$ and $N = M = 1024$ sizes.

The numerical experiments show that algorithm **B** is faster than algorithm **A** under following conditions:

1. The number of transmitted data is large enough, and I is an arbitrary number.
2. The number of transmitted data and the iteration number I are small enough.

4   Concluding Remarks

In the present paper an improved ADI method has been implemented. The improved method bases on substitution the CG method (in the $(n + \frac{1}{2})$ time-layer) for the parallel SD method. Such an improvement is conditional, i.e. it is justified under the two previous conditions. In that way, further work will be aimed at the deduction of a special condition (depending on $N, M, I_{SD^B}, I_{CG^B}, p$) which will allow us to determine the applicability of algorithm **B** more exactly.

References

1. Barrett, R., Berry, M., Chan, T.F., Demmel, J., Donato, J.M., Dongarra, J., Eijkhout, V., Pozo, R., Romine, C., Henk Van der Vorst: Templates for the Solution of Linear Systems: Building Blocks for Iterative Methods,
   http://www.netlib.org/templates/Templates.html
2. Bear, J., Zaslavsky, D., Irmay, S.: Physical principles of water percolation and seepage. UNESCO, (1968)
3. Foster, I.: Designing and Building Parallel Programs: Concepts and Tools for Parallel Software Engineering, Addison-Wesley Pub. Co., (1995)
4. Gropp, W., Lusk, E., Skjellum, A., Rajeev, T.: Using MPI: Portable Parallel Programming With the Message-Passing Interface, Mit press, (1999)
5. Krukier L.A., Schevtschenko I.V.: Modeling Gravitational Flow of Ground Water. Proceedings of the Eighth All-Russian Conference on Modern Problems of Mathematical Modeling, Durso, Russia, September 6-12, RSU Press, (1999), 125-130
6. *MPI: A Message-Passing Interface Standard, Message Passing Interface Forum,* (1994)
7. Pacheco, P.: Parallel Programming With MPI, Morgan Kaufamnn Publishers, (1996)
8. Peaceman, D., and Rachford, J. H.H.: The numerical solution of parabolic and elliptic differential equations. J. Soc. Indust. Appl. Math., No.3 (1955), 28-41
9. Samarskii, A.A., and Goolin, A.V.: Numerical Methods, Main Editorial Bord for Physical and Mathematical Literature, (1989)
10. Wallach, Y.: Alternating Sequential/Parallel Processing, Springer-Verlag, (1982)

Distributed Numerical Markov Chain Analysis

Markus Fischer* and Peter Kemper*

Dept. of Computer Science IV, University of Dortmund,
D-44221 Dortmund, Germany
{fischer,kemper}@ls4.cs.uni-dortmund.de

Abstract. Very large systems of linear equations arise from numerous fields of application, e.g. analysis of continuous time Markov chains yields homogeneous, singular systems with millions of unknowns. Despite the availability of high computational power sophisticated solution methods like distributed iterative methods combined with space-efficient matrix representations are necessary to make the solution of such systems feasible. In this paper we combine block-structured matrices represented by Kronecker operators [3,4] with synchronous and asynchronous two-stage iterative methods [11] using the *PVM* message-passing tool. We describe, how these methods profit from the proposed matrix representation, how these methods perform in wide-spread local area networks and what difficulties arise from this setting.

1   Introduction

Continuous time Markov chains (CTMCs) [14] represent the behavior of a system by describing all the different states the system may occupy and by indicating how state transitions take place in time. CTMCs support performance analysis in a model-based design of complex systems. Typical application areas are computer, communication and transportation systems.

Analysis can aim at the transient or the long term, stationary behavior of the CTMC. We focus on the latter and compute the so-called steady state distribution π which is characterized as the solution of a homogeneous system of linear equations $\pi \cdot \mathbf{Q} = 0$ subject to $\|\pi\|_1 = 1$. Matrix $\mathbf{Q}$ is singular and usually non-symmetric. In practice extremely large state spaces resp. matrix dimensions appear with millions of states. Therefore from an algorithmic point of view CTMC analysis leads to large systems and sophisticated methods are necessary to make their solution feasible. Since $\mathbf{Q}$ is typically sparse, iterative solution methods which use successive approximations $\pi(t)$ to obtain more accurate solutions and preserve sparsity of $\mathbf{Q}$ during computation are advantageous compared to conventional direct methods like Gauss elimination. Iterative methods are obtained by appropriate matrix splittings: one or sometimes several splitting(s) of type $\mathbf{Q} = \mathbf{M} - \mathbf{N}$ with non-singular matrix $\mathbf{M}$ are used to transform $\pi \cdot \mathbf{Q} = 0$ into fixed point equations $\pi = \pi \cdot \mathbf{T}$ with iteration operator $\mathbf{T} = \mathbf{N} \cdot \mathbf{M}^{-1}$. Hence,

---

* This research is supported by DFG, collaborative research center 559 'Modeling of Large Logistic Networks'

each iteration step $\pi(t + 1) = \pi(t) \cdot \mathbf{T}$ requires an efficient multiplication of vector π with a representation of matrix $\mathbf{T}$.

The paper is structured as follows: section 2 describes space efficient representations of appropriately structured CTCMs and corresponding iterative solution methods. Section 3 considers implementation issues for a distributed computation based on PVM. Finally, in section 4 we give results of experiments we exercised on a workstation cluster. Section 5 gives conclusions.

2   Distributed Asynchronous Two-Stage Iterations

In this section we describe a space efficient, well structured representation of $\mathbf{Q}$ and a corresponding iterative solution methods that can cope with very large equation systems. Distributed methods gain efficiency from a suitable problem decomposition. Hence we focus on a matrix $\mathbf{Q}$ in a block-structured form containing $N \times N$ sub-matrices

$$
\mathbf{Q} = \begin{pmatrix} \mathbf{Q}[0,0] & \cdots & \mathbf{Q}[0, N-1] \\ \vdots & \ddots & \vdots \\ \mathbf{Q}[N-1, 0] & \cdots & \mathbf{Q}[N-1, N-1] \end{pmatrix}.
\tag{1}
$$

The iteration vector $\pi(t)$ as well as other intermediate vectors are partitioned in a way consistent with the structure of $\mathbf{Q}$, i.e. into sub-iterates $\pi[0](t), \ldots, \pi[N-1](t)$. Nearly completely decomposable CTMCs [14], in which the nonzero elements of off-diagonal blocks are small compared with those of the diagonal blocks are of special interest. In that case, a decompositional approach leads to subproblems that can be solved nearly independently. Experimental studies [10] investigate block partitioning techniques which hopefully achieve low degrees of coupling and help to understand the effect of the degree of coupling on the convergence characteristics of iterative methods.

An alternative and very promising approach is to achieve a block structure of $\mathbf{Q}$ by a so-called hierarchical Kronecker representation [3,4,6]. The key idea is to describe block-matrices of $\mathbf{Q}$ in a compositional manner by a set of smaller matrices combined by Kronecker products and sums [14]. This reduces space requirements of $\mathbf{Q}$ significantly, such that $\pi(t)$ becomes the dominant factor for the space complexity. Kronecker representations trade space for time, so computation times for a vector-matrix multiplication slightly increase [8]. A Kronecker representation of $\mathbf{Q}$ can be directly generated from several modeling formalisms without explicit generation of matrix $\mathbf{Q}$. Methods for an automatic generation are e.g. known for queueing networks, GSPNs and we refer the interested reader to [3,5,6]. The assignment of N results directly from the matrix generation and represents the smallest possible value for the granularity which can be reached. Let a Kronecker representation of $\mathbf{Q}$ with N sub-matrices be given in the following. Since N usually exceeds the number of available processes, an initial load distribution must be computed, see section 3. Although our current implementation does not make use of it, runtime load re-balancing strategies are an issue as well. Runtime load balancing profits from the fact, that the space efficiency of a Kronecker representation allows to keep the complete information of $\mathbf{Q}$ at each process, so no portions of $\mathbf{Q}$ need to be communicated after initialization. With or without load balancing, only sub-iterates need to be exchanged among processes during iterations. Thus communication profits from zero-block-matrices in equation 1. Hierarchical Kronecker representations derived from Petri net models [6] use structural information about weak interacting sub-models and result in numerous zero-blocks. This is very advantageous with respect to the communication overhead which is necessary to perform distributed iterations, because nonzero-blocks indicate potential communication dependencies between processes. In experiments we often observed that roughly 70 % of N^2 blocks are in fact zero-blocks.

Many iteration schemes exist and block variants [1,14] of well known classical iterative methods are some of the most competitive approaches for the distributed iterative solution of CTMCs. Block methods follow a decompositional approach, the system is subdivided into subsystems by the outer matrix-splitting, each of which is analyzed in isolation and the global approximation is constructed from the partial solutions. Distributed computations make block Jacobi (BJAC) outer matrix splittings

$$\boldsymbol{\pi}[x](t+1) \cdot \mathbf{Q}[x,x] = - \sum_{y<N, y\neq x} \boldsymbol{\pi}[y](t) \cdot \mathbf{Q}[y,x] \stackrel{def}{=} -\mathbf{b}[x](t) \qquad (2)$$

particularly attractive. In many situations subsystems, as in equation 2, are still too large for direct solutions and in turn matrix splittings are applied to main diagonal blocks $\mathbf{Q}[x,x]$. In this case one is in the presence of so-called two-stage (or inner-outer) iterative methods [13].

The application of block two-stage methods is not only restricted to purely synchronized outer iterations but also to outer asynchronous iterations [2,11]. Asynchronous iterations arise in distributed iterations if one wants to minimize or even to abolish idle times at synchronization points. Therefore asynchronous methods have the potential to outperform their synchronous counterparts. Synchronization usually appears when vectors are exchanged. Therefore in asynchronous iterations currently computed vectors are sent to other processes as soon as possible and conversely, processes use whatever vectors are available locally when needed.

Convergence theory of outer asynchronous two-stage methods is of theoretical interest [13], but results are weak from a practical point of view. Therefore there is a need to gain understanding of how these methods actually perform in parallel settings.

3   Implementation Issues

Our target hardware platform is a local area network (LAN) which consists of a set of heterogeneous workstations which are interconnected by a 100-Mbps

Ethernet in a conventional client/server architecture. Access to the LAN is non-exclusive, with the effect of unpredictable variations in work load. Such a scenario is rather unfavorable to achieve impressive speedups in distributed computing, but realistic and we aim for a sufficiently robust implementation to perform in this setting.

The application of the PVM tool is very attractive for our intention to solve very large systems of linear equations using the proposed distributed iterative approach, because PVM supports parallel programming of style 'single-program multiple-data' (SPMD) and provides an explicit message-passing model for data exchange. In fact for our workload decomposition, here given by sub-iterates (multiple-data) we need a straightforward way to implement local iterations (single-programs) which receive and send vectors using an Ethernet-LAN setting (message-passing). A distributed implementation of an iteration scheme using PVM becomes challenging due to the following observations.

1. The hierarchical Kronecker representation of matrix $\mathbf{Q}$ resp. $\mathbf{T}$ allows to describe blocks of $\mathbf{Q}$ resp. $\mathbf{T}$ in a very space-efficient manner, but requires more sophisticated vector-matrix multiplications.

2. In equation systems with millions of unknowns, a single iteration step causes a high communication effort by the number of messages as well as by the amount of data. The example in Section 4 illustrates this aspect.

3. The proposed method is based on two-stage splittings, i.e. each outer block-iteration step encapsulates an inner iteration process. Block-iteration may also perform in an asynchronous fashion. From an algorithmic point of view such computational models 1) offer numerous degrees of freedom which have a strong impact on the numerical convergence, e.g. the number of inner iterations, 2) induce different and increased communication rates, because some processes compute faster and communicate more frequently than others and 3) give more flexibility to decide when computed vectors are sent to others.

We distinguish two different kinds of processes, conceptually termed master and worker processes. The latter realizes the iteration, while the master process performs some preprocessing tasks and monitors the progress of the iteration. Its tasks are merely the following:

1. **Configuration of Parallel Virtual Machine:** Worker are spawned with respect to a user-given pool of workstations. For simplicity assume that we assign only one process to a workstation, let $\mathcal{P} = \{1, \ldots, P\}$ be the set of available processes.

2. **Domain Decomposition:** Matrix $\mathbf{Q}$ is generated in a block-structured fashion using Kronecker operators to represent block-matrices. The partitioning of iteration vector $\pi(t)$ into sub-iterates follows this structure.

3. **Mapping decomposed domains to processes:** Since the number N usually exceeds the number of processes P the assignment of sub-iterates to processes is not trivial and requires to compute a balanced load distribution. Such problems are usually represented by a directed graph in which nodes describe

a computational load (here derived from the number of non-zeros matrix entries) and edges indicate data exchange between processes, whereas edges are weighted by a metric measuring communication costs (here the size of the vector to be sent). Graph partitioning techniques try to identify disjunct subgraphs representing nearly equal loads with low communication overhead. Beside self-implemented heuristics we use standard techniques offered by the Chaco 2.0 software package [12] in order to find an initial load distribution. Let $\mathcal{R}(p)$ for some process $p \in \mathcal{P}$ be the subset of block indices assigned to process p, i.e., process p computes approximations of $\boldsymbol{\pi}[x]$ for all $x \in \mathcal{R}(p)$. Finally, the master process informs the worker processes about the computed load distribution $\mathcal{R}(p)$.

4. **Monitoring:** During the iteration, the master monitors its progress by receiving norms of local residual vectors. The iteration is stopped in a safe way when the maximum time has elapsed or when the solution is estimated to be sufficiently accurate.

The following pseudo-code describes the behavior of a single worker process.

Basic algorithm of worker process p

```
1.  while (convergence criterion not satisfied)
2.     for all p' ∈ P\{p}
3.        for all x ∈ R(p')
4.           compute h = Σ_{y∈R(p)} π[y](t) · Q[y,x];
5.           send triple (h,x,t) to process p';
6.        end for
7.     end for
8.     for all x ∈ R(p)
9.        compute b_p[x] = Σ_{y∈R(p),y≠x} π[y](t) · Q[y,x];
10.    end for
11.    if (receiving message buffer not empty)
12.       receive (h,x,·) from process p';
13.       update b_{p'}[x] = h;
14.    fi
15.    for all x ∈ R(p)
16.       update b[x] = Σ_{p'∈P} b_{p'}[x];
17.       π[x](t+1) = solve(π[x](t), Q[x,x], b[x]);
18.    end for
19. end while
```

The worker process is divided into several phases: lines **2-10** prepare the subsystems which result from the BJAC splitting, cf. equation 2. This enforces collaboration among worker processes. Worker processes have restricted access to the global iteration vector, because only dedicated sub-iterates from $\mathcal{R}(p)$ are assigned to a particular worker process p. Therefore every worker process is requested to compute and to send intermediate results to other worker processes, see lines **2-7**. Intermediate results which are required for its own local iteration are directly buffered in $\mathbf{b}_p[x]$, see line **9**. Lines **11-14** realize the receiving phase.

Worker p has to receive vectors in order to iterate $\pi[x]$ with $x \in \mathcal{R}(p)$. In a distributed setting, equation 2 becomes to

$$\pi[x](t+1) \cdot \mathbf{Q}[x,x] = \sum_{p' \in \mathcal{P}} \sum_{y \in \mathcal{R}(p'),y \neq x} \pi[y](t) \cdot \mathbf{Q}[y,x] \overset{def}{=} \sum_{p' \in \mathcal{P}} \mathbf{b}_{p'}[x]. \qquad (3)$$

In case of outer *asynchronous* iterations the worker process cannot expect to receive particular vectors of equation 3 and must therefore buffer vectors recently received from other worker processes in $\mathbf{b}_{p'}[x]$. Of course buffering requires supplementary local memory. Line **15** describes the choice of the next local iterate. In the synchronous case all $x \in \mathcal{R}(p)$ have to be considered. In the asynchronous case we are in principle free to choose x. We may use a loop over all $x \in \mathcal{R}(p)$, or alternatively we can develop an adaptive approach. For instance we may choose the sub-iterate with the largest estimated error or we may choose x whenever the right side $\mathbf{b}[x]$ has been modified. Line **17** describes the computation of a new approximation for $\pi[x]$. We apply in turn an iterative method (two-stage) which is indicated by the call of the function 'solve'. In principle we are free to choose a method, but usually we apply Jacobi with relaxation (JOR) [1].

4   Application Example

The approach is implemented in C, uses the PVM 3.4.3 library and runs on an Ethernet-LAN in a multi-user environment. The implementation is integrated in the APNN-toolbox [7], a selection of several techniques for the analysis of CTMCs. First experiences and parallel performance results are reported in [5] and we highlight, that CTMCs with about 8 million states were analyzed and a speedup of approximately 1.5 was achieved. The implementation is stable, but in asynchronous settings there is still potential for improvement, i.e. the buffering of transferred but not yet received vectors requires an effective overwrite mechanism within PVM which we approximated by a polling mechanism. However, if the polling frequency does not match the frequency iteration vectors are communicated, we observe PVM daemon processes that hold a lot of unreceived vectors, in some cases to an extent that the whole application collapses due to thrashing at one worker's site. Vectors which are not processed immediately upon reception are natural in asynchronous iterations and an overwrite-mechanism ensures that only the most recent vector is kept in the buffer. The implementation of overwrite mechanisms is subject of current work.

We exercise two examples from the area of performance analysis to illustrate applicability of our approach, i.e. a flexible manufacturing system (FMS) [9] and a simple tandem queueing network (TQN). For a particular FMS [TQN] model configuration we obtain a CTMC with $1,639,440$ $[164,076]$ states with $N = 8$ $[N = 51]$ and 22 $[100]$ non-zero blocks and $\mathbf{Q}$ contains $14,099,976$ $[718960]$ non-zero off-diagonals entries.

The results given in Tables 1 [2] are achieved on a LAN of Sun workstations in a client/server architecture. We use up to 10 workstations with different computational power (Ultra Sparc IIi, ..., Ultra 10).

Table 1. FMS: performance results

No	Meth	B	P	in_it	out_it	Time
1	JOR	0	1	1	849	8313
2	BSOR	0	1	1	820	7175
3	PAR	0	4	1	849	5130
4	PAR	0	4	2	475	4676
5	PAR	0	4	4	272	4409
6	PAR	0	4	6	234	4302
7	PAR	0	4	8	212	4290
8	PAR	0	4	10	198	4512
9	PAR	0	4	12	186	5420
10	PAR	∞	4	8	227-330	4920
11	PAR	∞	4	8	228-408	4890

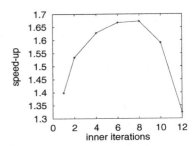

Fig. 1. FMS: speed-up with respect to #inner iterations

Table 2. TQN: performance results

No	Meth	B	P	in_it	out_it	Time
12	PAR	0	1	6	1080	2278
13	PAR	0	2	6	1080	2017
14	PAR	0	5	6	1080	1832
15	PAR	0	8	6	1080	1433
16	PAR	0	10	6	1080	1348

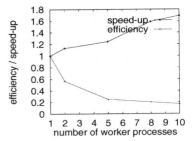

Fig. 2. TQN: speed-up resp. efficiency with respect to #processes

The second column in both tables gives the method: JOR is a sequential JOR implementation, BSOR is a sequential block SOR method and PAR indicates the distributed iteration as described above. Column B gives a limiting constant for the difference of (outer) asynchronous iterations: $B = 0$ denotes synchronous iterations and $B = \infty$ is a totally asynchronous iteration. Column in_it gives the user defined fixed number of inner iterations to solve subsystems and out_it resp. Time contains the observed number of outer iterations resp. wall-clock time in seconds necessary to achieve the desired accuracy. In the asynchronous case, column out_it gives the minimum and maximum number of iterations performed by different worker processes. In table 1, rows 3 to 9 vary the number of inner iterations for a synchronized iteration scheme; the optimal value is 8. For 8 inner iterations, rows 10 and 11 give results for a repeated experiment of a totally asynchronous block iteration. Due to the inherent stochastic character of asynchronous iterations the number of outer iterations varies for fixed parameter settings in repeated experiments. For the FMS example a synchronous computation gives slightly better solution times than an asynchronous iteration. Compared to the sequential implementation, we observe for the FMS model a speedup by 1.96 [1.67] with respect to a similar [the fastest] sequential method (experiments No 1 [2] and 7) we have at hand, see figure 1. Times for sequential methods refer to the fastest machine of the set. The number of messages sent per iteration step varies between 40 and 64 and the overall volume varies between 6.7 MB and 10.4 MB. In table 2, rows 12 to 16, we vary the number of worker processes. For 10 processes we observe a speed-up of 1.69 compared to the sequential implementation of PAR (row 12), see figure 2. The maximum communication complexity was reached in row 16, here we observe 45 messages with an overall volume of 0.84 MB.

5   Conclusions

We presented distributed, synchronous and asychronous two-stage iterative methods for the solution of very large systems of linear equations arising in the analysis of Markov chains. The approach is implemented with the help of PVM and runs on an Ethernet-LAN. Future work is dedicated to implement overwrite-mechanisms for message buffers, to introduce adaptive adjustments for the number of inner iterations and for the communication rate in asynchronous settings.

References

1. Berman, A., Plemmons, R.J.: Nonnegative matrices in the Mathematical Sciences. Academic Press New York, Reprinted by SIAM Philadelphia (1994)
2. Bru, R., Migallón, V., Penadés J., Szyld, D.B.: Parallel, synchronous and asynchronous two-stage multisplitting methods. Electronic Transaction on Numerical Analysis **3** (1995) 24-38
3. Buchholz, P.: Hierarchical structuring of superposed GSPNs. IEEE Transactions on Software Engineering **25:2** (1999) 166-181
4. Buchholz, P.: Structured analysis approaches for large Markov chains. Applied Numerical Mathematics **31:4** (1999) 375-404.
5. Buchholz, P., Fischer, M., Kemper, P.: Distributed steady state analysis using Kronecker Algebra. Proc. of the 3rd Int. Workshop on Numerical Solution of Markov Chains (1999) 76-95
6. Buchholz, P., Kemper, P.: On generating a hierarchy for GSPN analysis. ACM Performance Evaluation Review **26:2** (1998) 5-14
7. Buchholz, P., Kemper, P.: A toolbox for the analysis of discrete event dynamic systems. Proc. of Computer Aided Verification (CAV), LNCS **1633** (1999) 483-487
8. Buchholz, P., Ciardo, G., Donatelli, S., Kemper, P.: Kronecker operations and sparse matrices with applications to the solution of Markov models. INFORMS J. on Computing **13:2** (2000) 203-222
9. Ciardo, G., Trivedi, K.: A decomposition approach for stochastic reward net models. Performance Evaluation **18** (1994) 37-59
10. Dayar, T., Stewart, J.S.: Comparison of partitioning techniques for two-level iterative solvers on large, sparse Markov chains SIAM Journal on Scientific Computing **21:5** (2000) 1691–1705
11. Frommer, A., Szyld, D.B.: Asynchronous two-stage iterative methods. Numerische Mathematik **69** (1994) 141-153
12. Hendrickson, B., Leland, R.: The Chaco user's guide version 2.0. Technical Report NM 87185-1110, Sandia National Laboratories (1995)
13. Migallón, V., Penadés, J., Szyld, D.B.: Block two-stage methods for singular systems and Markov chains. Num. Linear Algebra with Applications **3** (1996) 413–426
14. Stewart, W.J.: Introduction to the Numerical Solution of Markov Chains. Princeton University Press, Princeton, N.J. (1994)

A Parallel Algorithm for Connected Components on Distributed Memory Machines*

Libor Buš and Pavel Tvrdík

Department of Computer Science and Engineering
Czech Technical University, Karlovo nám. 13
121 35 Prague, Czech Republic
{xbus,tvrdik}@fel.cvut.cz

Abstract. Finding connected components (CC) of an undirected graph is a fundamental computational problem. Various CC algorithms exist for PRAM models. An implementation of a PRAM CC algorithm on a coarse-grain MIMD machine with distributed memory brings many problems, since the communication overhead is substantial compared to the local computation. Several implementations of CC algorithms on distributed memory machines have been described in the literature, all in Split-C. We have designed and implemented a CC algorithm in C++ and MPI, by combining the ideas of the previous PRAM and distributed memory algorithms. Our main optimization is based on replacing the conditional hooking by rules for reducing nontrivial cycles during the contraction of components. We have also implemented a method for reducing the number of exchanged messages which is based on buffering messages and on deferred processing of answers.

1 Introduction

Consider an undirected graph $G = (V, E)$, $|V| = n$, and $|E| = m$. The problem of finding the connected components (CC) of G can be solved sequentially in $O(n + m)$ time by the depth-first search.

There exist several PRAM CC algorithms. The simplest ones require $O(\log n)$ parallel steps using $n + m$ CRCW PRAM processors [7]. Although a simulation of a CRCW PRAM algorithm on an EREW PRAM slows down the time by the factor of $\log n$, there are some $o(\log^2 n)$-time algorithms running on a linear number of EREW processors. In particular, an algorithm running in $O(\log n \log \log n)$ time on $n + m$ EREW PRAM processors [2] and an algorithm running in $O(\log^{1.5} n)$ time on $n + m$ CREW PRAM processors [5]. However, these algorithms are quite complicated to be efficiently implemented on a coarse-grain MIMD machine.

PRAM CC algorithms have been directly implemented only in the data-parallel computing model on SIMD machines [3] or on shared memory MIMD

---

* This research has been supported by MSMT Czech Republic under research program #J04/98:2123000

Y. Cotronis and J. Dongarra (Eds.): Euro PVM/MPI 2001, LNCS 2131, pp. 280–287, 2001.

machines [4]. Any naive direct implementation on a MIMD machine with distributed memory is inefficient, because it results in a huge number of remote accesses. A distributed memory algorithm could be efficient if much of the work can be performed locally. For some classes of input graphs, such as graphs with underlying 2- or 3-dimensional meshes, the graph can be decomposed into smaller parts (tiles) with only a small fraction of edges crossing between two different parts. Therefore, most of the components can be found locally: each processor labels its local vertices and ignores remote edges. The two algorithms in [1,6] use exactly this approach. They differ in the global phase in which the processors merge the partial information about the components. The algorithm published in [1] is designed for graphs corresponding to 2-dimensional images. In the global phase, processors merge the vertical and horizontal borders of their local tiles in $\log p$ iterations, where p is the number of processors. The algorithm [6] simulates the Shiloach and Vishkin PRAM algorithm [7] on graphs consisting of remote edges and of roots of local components produced by local DFS's. It was implemented in Split-C on CM-5.

Our algorithm, denoted by MPI-CC in the further text, is an extension and optimalization of the algorithm from [6]. It was implemented in C++ and MPI, and therefore, it is portable to most message-passing coarse-grain parallel machines, including clusters. Its main optimization consists in replacing conditional hooking with rules for reducing nontrivial cycles (called *CR rules* in the further text) during the contraction of components. This approach comes from the CREW PRAM algorithm published in [5]. This replacement has provided significant performance improvement. Next, we have implemented a method how to reduce the number of exchanged messages. It consists in buffering messages and deferred processing of the answers.

2   The Description of the MPI-CC Algorithm

2.1   Outline

A graph G is represented by an array $vertices[1 \ldots n]$. Each vertex u has a pointer to a list of adjacent edges, $edges(u)$, and a pointer to its parent, $parent(u)$. Initially, $parent(u) = u$ for all u. The vertices are split uniformly among local memories of the processors. An edge (u, v) is *local* if u and v belong to the same processor and is *remote* otherwise.

The basic idea of the algorithm is as follows. It consists of 4 phases. Phases 1 and 4 are local, whereas 2 and 3 are global. In Phase 1, each processor marks sequentially local components, groups local vertices into rooted stars, moves local endpoints of remote edges from the leaves of the formed stars to their roots, and deletes local edges from adjacency lists. In Phase 2, all remaining edges (u, v) in the adjacency list of each local vertex u are updated to $(u, parent(v))$. This involves some communication among processors, but as a result, the communication work of Phase 3 is significantly reduced. Phase 3 does not distinguish between local and remote edges. Hooking and contraction go across boundaries of processors. Each processor starts with the list of roots resulting from Phase 1 and produces leaves from some of them as a result of hooking and contracting. More specifically, one iteration of Phase 3 consists of 5 steps.

In the *hooking step*, each vertex u from the list of roots is hooked to the end-vertex v of the first edge from its list $edges(u)$ (i.e., $parent(u)$ is set to v).

In the *contraction step*, the components are collapsed into stars by the pointer jumping based on the CR rules (see the next section). As was shown in previous papers, the expensive testing of stars during the contraction can be avoided if roots and their children are successfully eliminated from the contraction process by moving them into temporary lists of roots and leaves, respectively. The contraction empties the lists of roots and leaves. At the end of this step, the temporary lists are converted into lists of new roots and leaves.

Then in the *internal-edges-removing step*, all internal edges are removed. An edge (u, v) is *internal* if $parent(u) = parent(v)$. Then in the *edge-list-concatenation step*, edge lists of leaves of stars are concatenated with the edge lists of their roots. In the final step, vertices of complete components (i.e, stars with no remaining remote edges) are removed from the lists.

These 5-step iterations are repeated until the lists of roots on each processor are empty.

In Phase 4, the leaves from Phase 1, which were ignored in Phase 3, are updated.

2.2   CR Rules

In the standard CC algorithms, cycles in constructed CCs cannot arise, since the star hooking is performed first conditionally and then unconditionally. In the CRCW PRAM algorithm [5], an alternative approach based on *CR rules* was proposed. Stars are hooked unconditionally, so that they can form cycles. During the contraction step, the CR rules guarantee that all such cycles will be broken.

Each vertex u is assigned a unique identification number $id(u)$ and a flag $bold(u)$. The flag $bold(u)$ is set if and only if $id(u) < id(parent(u))$ (see Fig. 1). At the beginning of the first iteration of Phase 3, $bold(u)$ is reset for each vertex u from the list of roots, because $parent(u) = u$, and it is recomputed in the hooking step. New roots are chosen to be the vertices with the smallest identification numbers within cycles during contraction steps. At the end of the contraction step, the flag $bold(u)$ of each vertex u from the list of leaves must be recomputed.

The original paper [5] defined six CR rules, depicted on Fig. 1. The contraction of vertex u satisfying conditions of Rule 1 and 2 can be terminated. The next three rules serve for jumping over the parent of u. Jumping over the new root is prevented by Rule 6.

The original definition of CR rules in [5], as shown on Fig. 1, however, is not correct. As the implementation on a distributed memory machine revealed and successive inductive proof confirmed, Rule 2 must be modified so that it applies to vertex u even if $bold(u)$ is not set. Otherwise, the contraction step of Phase 3 cannot complete.

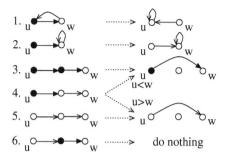

Fig. 1. Pointer jumping based on the CR rules. Filled circles denote vertices with the flag *bold* set. The arrows correspond to the parent pointers.

3   Implementation Issues on a MIMD Machine

3.1   CR Rules

A correct implementation of the CR rules on a MIMD machine is an interesting problem, since the rules were designed for the synchronous PRAM model. In our algorithm, local computation can be interleaved with passing messages with read requests to other processors. To guarantee the consistency of the CR rules when asynchronous MIMD processors allow time racing, an update of the flag *bold* and the pointer jumping over a given vertex must be indivisible. Each pair of these operations must be executed atomicly. This is guaranteed by placement of the code for message requests and responses into well defined entry points within the code of Phase 3.

3.2   Input Graphs and Their Distribution Among Processors

We have tested the behavior of the algorithm on input graphs with either high or low locality of neighborhoods of vertices. We have used 2 mesh-like graphs, incomplete 2D and 3D meshes, and two kinds of general random graphs, graphs with average degree k (ADk) and k-regular graphs (REGk). The incomplete meshes are denoted by 2Dρ and 3Dρ, where ρ is the probability of including edges when the graphs are generated by adding edges to mesh vertices. The definition of all the graphs except REGk were taken from [6].

Among others, the performance of the algorithm for mesh graphs is affected by 2 factors: the unique numbering *id* and the distribution of the vertices among the local memories of processors. If vertices with smallest identification numbers were concentrated on one processor, the probability of choosing roots on that processor would be the highest and the computation would be imbalanced.

To avoid such a distribution, we have experimented with several numbering schemes. For the 2D meshes, the optimal mapping scheme was surprisingly simple. The vertices were distributed among processors using the block row-major mapping, while the identification numbers were assigned in column-major way. Similarly for the 3D meshes.

3.3 Buffering of Messages

In Phase 3, processors must read some attributes of remote vertices and edges. We have optimized the corresponding message passing protocol by using the fact that in each step of Phase 3, the number of needed types of the attributes is one or two (e.g., the attribute *parent* in the internal-edges-removing step or the attributes *parent* and *bold* in the contraction step). Each request message for each processor consists of the type of the attribute and a list of numbers of remote nodes. The response message contains a list of attribute values. Each processor has $2(p-1)$ buffers (p is the number of processors), $p-1$ buffers for messages that are currently assembled for the next communication round and $p-1$ buffers for messages currently exchanged. Deferred processing of response messages requires context saving.

3.4 Synchronization

Individual phases and steps must be separated by nonblocking barriers, since each processor after reaching the end of its phase or step must continue with processing of incoming messages. We had to implement this type of barriers in SW, they were not supported in our MPI libraries.

4   Performance Results

4.1 Computing Platforms

We have evaluated the performance of our MPI-CC algorithm on an IBM SP2 machine and on a Beowulf workstation cluster with Fast Ethernet. The nodes of the IBM SP2 machine had Power II processors at 66, 120, and 160 MHz and the nodes of the cluster had AMD Athlon 600 MHz processors. We have used the IBM implementation of MPI-1 on the IBM SP2 and MPICH on the Beowulf. The code was compiled with the xlC compiler on IBM SP2 and with g++ on the Beowulf, both with the option -O3.

4.2 Methodology

All the measurements were averaged over several runs of the algorithm using distinct random seeds. The time to distribute an input graph from the first processor P_0 to other processors and the time to collect results by P_0 were not included into the measured time of the algorithm. Therefore, the measured times correspond to the run times of the algorithm from Phase 1 to Phase 4.

4.3 Speedup and Efficiency

We have measured the speedup of the MPI-CC algorithm (Figure 2) by comparing the running time of parallel and sequential versions of the algorithm on graphs with $n = 10^6$ vertices. The sequential time includes just depth-first

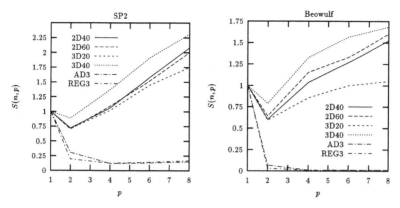

Fig. 2. Speedups for input graphs with $n = 10^6$ vertices.

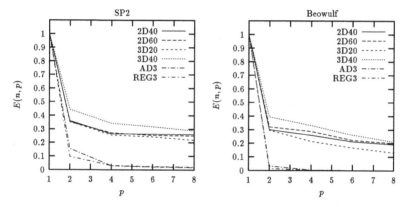

Fig. 3. Efficiencies for input graphs with $n = 10^6$ vertices.

search of the input graph by one processor. The MPI-CC algorithm has a linear speedup for meshes as was already mentioned. It follows from the fact that the most of the components are constructed locally in the Phase 1. The speedup for input graphs AD3 and REG3 falls down to zero indicating overwhelming communication overhead due to the nonlocal structure of these graphs.

The efficiency, defined as the speedup per one processor, is shown on Figure 3 for the same data set. We see that for meshes the algorithm achieves efficiency around 30% for a range of processors.

4.4   Scalability

Figure 4 shows the scalability of the algorithm for meshes. We measured the scalability by fixing the efficiency $E(n,p) = 0.2$ and by searching the appropriate size of the graphs for given number of at most $p \leq 8$ IBM SP2 processors. The measured data indicate that the isoefficiency function is linear in the size of the mesh side. That is, the efficiency is constant if p scales linearly with $\sqrt{n}$ ($\sqrt[3]{n}$) for 2Dρ (3Dρ, respectively) graphs.

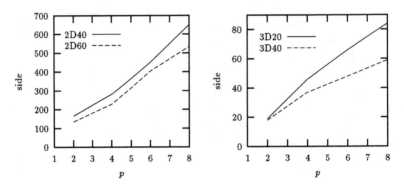

Fig. 4. Scalability for mesh input graphs on IBM SP2.

Table 1. Influence of message buffering on the parallel time in case of AD3 graphs with $n = 10^5$ vertices.

Buffer size	$T(n, p)$ [s]					
	SP2			Beowulf		
[messages]	$p = 2$	$p = 4$	$p = 8$	$p = 2$	$p = 4$	$p = 8$
1	28.91	36.47	47.69	60.66	140.77	161.44
10	3.31	4.38	5.54	8.19	22.97	23.62
100	0.81	1.48	1.63	3.02	19.27	16.51
1000	0.66	1.29	1.23	2.95	20.95	17.77
10000	0.67	1.49	1.48	2.99	21.81	17.99

4.5   Impact of Message Buffering on the Performance

Because only a small fraction of edges of a ADk graph is processed locally and the communication overhead in Phase 3 is critical, we have chosen ADk graphs as testbeds for evaluation of the impact of message buffering on the running time, even though parallel construction of CCs makes little practical sense for these graphs (assuming that all data fit into main memories). Table 1 shows how the size of the buffers influences the running time of the algorithm for ADk graphs with $n = 10^5$ vertices and $k = 3$. The first line corresponds to the implementation without buffering. The optimal buffer size is clearly 100.

4.6   CR Rules vs. Conditional Hooking

Comparison of our MPI-CC algorithm with the "classical" CC algorithm with conditional hooking is shown in Table 2. We used input graphs with $n = 10^6$ vertices and the buffer of the size 100 and the machine was an 8-processor SP2. The application of the CR rules speeds up the CC algorithm by 60% in case of 3D20 graphs and to 30% in case of AD3 and REG3 graphs. The speedup depends on the relative number of vertices entering Phase 3.

Table 2. Parallel time of our MPI-CC algorithm compared with the "classical" CC algorithm with conditional hooking on SP2.

Input graph	CR rules		Conditional hooking		T_1/T_2
	$T_1(n,p)$ [s]	# iterations	$T_2(n,p)$ [s]	# iterations	
2D40	0.38	2	0.44	2	0.86
2D60	0.50	2	0.52	2	0.94
3D20	0.40	2	0.68	2	0.60
3D40	0.65	2	0.72	2	0.90
AD3	14.46	4	38.33	5	0.38
REG3	15.06	4	50.81	5	0.30

5   Conclusion

Our course-grain implementation of the fine-grain CC problem has confirmed similar older results, which, however, were achieved on totally different computer architectures. Our parallel implementation on distributed memory machine with very fast local nodes still exhibits expected performance and scalability for graphs with local neighborhood, and, as might be expected, fails for graphs with global neighborhood. But even for these graphs, parallel implementation of the CC algorithm is the only way how to process graphs that are too large to fit into the main memory of 1 processor. Swapping data to disks is much slower than network communication. Currently, we are investigating this dimension of the problem.

References

1. D. Bader and J. JáJá. Parallel algorithms for image histogramming and connected components with an experimental study. *Journal of Parallel and Distributed Computing*, 35(2):173–190, 1996.
2. K. W. Chong and T. W. Lam. Finding connected components in $O(\log n \log \log n)$ time on the EREW PRAM. In *SODA: ACM-SIAM Symposium on Discrete Algorithms*, 1993.
3. J. Greiner. A comparison of data-parallel algorithms for connected components. In *Proceedings of SPAA*, pages 16–25, June 94.
4. T. S. Hsu, V. Ramachandran, and N. Dean. Parallel implementation of algorithms for finding connected components. In *Proceedings of the 3rd Annual DIMACS Challenge*, 1995.
5. D. B. Johnson and P. Metaxas. Connected components in $O(\log^{3/2} n)$ parallel time for the CREW PRAM. In *32nd FOCS*, pages 688–697, 1991. PCS-TR91-160.
6. A. Krishnamurthy, S. Lumetta, D. Culler, and K. Yelick. Connected components on distributed memory machines. In *Proceedings of the 3rd Annual DIMACS Challenge*, 1995.
7. Y. Shiloach and U. Vishkin. An $O(\log n)$ parallel connectivity algorithm. *Journal of Algorithms*, 3(1):57–67, March 1982.

Biharmonic Many Body Calculations for Fast Evaluation of Radial Basis Function Interpolants in Cluster Environments

George Roussos and B.J.C. Baxter

Imperial College, 180 Queen's Gate, London SW7 2BZ, UK

Abstract. This paper discusses the scalability properties of a novel algorithm for the rapid evaluation of radial basis function interpolants. The algorithm is associated with the problem of force calculation in many-body calculations. Contrary to previously developed fast summation schemes including treecodes and fast multipole methods, this algorithm has simple communication patterns which are achieved by exploiting the localisation and smoothness properties of radial basis functions. Thus, the algorithm is scalable even in low bandwidth environments like clusters of workstations and even for relatively small problem sizes.

1 Introduction

Recent results show that the radial basis function method produces high quality solutions to the multivariate scattered data interpolation problem, especially in higher dimensions. On the other hand, the method is associated with higher computational cost when compared against alternative methods, such as finite elements or multivariate spline interpolation. Indeed, solving directly the interpolation equations for N data points requires $\mathcal{O}(N^3)$ floating-point operations, while direct evaluation of the resulting interpolant at M locations requires $\mathcal{O}(MN)$ floating-point operations.

Several attempts aiming at the reduction of the computational complexity of both the calculation and the evaluation tasks have produced a number of novel algorithms. For example, rapid evaluation of radial basis function interpolants has been achieved using a variety of methods including subtabulation on a regular grid, treecodes, moment based methods and the fast multipole method. In [9] a method based on the Fast Gauss Transform and suitable quadrature rules has been developed. This method exploits fundamental smoothness and localisation properties of the radial basis function method to achieve high performance.

In this paper, we explore the performance of the method developed in [9] in the context of a cluster of workstations. Indeed, we discover that due to the regularity of its computational structure the method performs very well even in low bandwidth environments and small problem sizes.

Y. Cotronis and J. Dongarra (Eds.): Euro PVM/MPI 2001, LNCS 2131, pp. 288–295, 2001.

2   Fast Evaluation of Radial Basis Function Interpolants

Radial Basis Function interpolation has the form

$$s(y) = \sum_{i=1}^{N} \lambda_i \varphi(\|y - x_i\|), \tag{1}$$

where λ_i are real coefficients, x_i points in $\mathbb{R}^d$ called *centres*, $\| \cdot \|$ the Euclidean norm and φ the *basis function*. The function $\varphi : \mathbb{R}^+ \to \mathbb{R}$ is unary and radially symmetric with respect to the norm, in the sense that it has the symmetries of the unit ball in $\mathbb{R}^d$. The coefficients λ_i are chosen so that the interpolation conditions are satisfied, that is

$$s(x_i) = f(x_i) = f_i, \quad i = 1, 2, \ldots, N, \tag{2}$$

where f_i is the value of the interpolated function f at location $x_i \in \mathbb{R}^d$.

Examples of useful choices of φ include the *Gauss kernel*, the *Euclidean distance*, the biharmonic *Hardy Multiquadric*

$$\varphi(r) = \sqrt{r^2 + c^2}, \tag{3}$$

and the *Inverse Multiquadric* which is exactly the inverse of (3).

Solving the interpolation problem (2) is equivalent to solving the system of linear equations $A\lambda = f$, where A is the *interpolation matrix* $A_{ij} = \varphi(\|x_i - x_j\|)$, λ is the vector of coefficients $(\lambda_1 \ \lambda_2 \ \ldots \ \lambda_N)^T$ and f is the vector of function values $(f_1 \ f_2 \ \ldots \ f_N)^T$. It is clear that the solution of the interpolation equations directly requires $\mathcal{O}(N^3)$ floating point operations, while the form of the interpolant (1) implies that evaluating s at M points $y_1, y_2, \ldots, y_M$, directly requires $\mathcal{O}(MN)$ operations.

Beatson and Newsam [2] first noted the intimate relation between the many body problem and the evaluation of the radial basis function interpolant. Furthermore, they exploited this observation to develop the mathematical framework required for the introduction of fast evaluation methods for a particular radial basis function in two dimensions. Indeed, in [2] the Laurent and Taylor expansions required by the Fast Multipole Method (FMM) in two dimensions were constructed. These results were used by Powell [8] to introduce a fast algorithm for the evaluation of radial basis function interpolants which is a higher order treecode. This method uses a decomposition of the set of interpolation centers similar to Appel [1] which results in observed computational complexity of $\mathcal{O}(N \log N)$ for fairly uniform distributions of centers. More recently, Beatson has implemented a full FMM based on the results of [2] with reported performance similar to that discussed in [4]. A variant of this method, whereby the coefficients of the multipole expansion are not calculated directly but approximated is discussed by Suter [10].

In particular, the relation between the Hardy Multiquadric and many body computations has received a physical justification. Indeed, Hardy [6] relates the solution of an interpolation problem to simulation of the Earth's geomagnetic field by a biharmonic potential. The biharmonic approach has the advantage over the use of a harmonic potential that the Earth is considered as a solid, rather than a hollow body.

Finally, it is worth pointing out that the force calculation step of the many body computation with the Plummer potential is exactly the evaluation of an Inverse Multiquadric interpolant, where $y_i = x_i$ are the locations of the point masses λ_i.

3   Algorithm Description

One of the features that makes radial basis function interpolation a useful technique is the fact that a unique interpolant is guaranteed under weak conditions on the location of the centres. Micchelli [7] specified these conditions by relating the interpolation matrix of several of the radial basis functions with almost positive (or negative) definite functions. A by-product of this proof is a way to represent certain radial basis functions as an integral of a Gaussian by a suitable measure. For example, we can prove that the Hardy Multiquadric can be rewritten in terms of Gaussian kernel sums in the following way

$$s(y) = c\sum_{i=1}^{N}\lambda_i + \frac{1}{\sqrt{2\pi}}\int_0^\infty \frac{e^{-sc^2}}{\sqrt{s}}\cdot\frac{c\sum_{i=1}^{N}\lambda_i - \sum_{i=1}^{N}\lambda_i e^{-s\|y-x_i\|^2}}{s}\,ds, \quad (4)$$

for centres $x_i, i = 1, 2, \ldots, N$ and evaluation point y in $\mathbb{R}^d$.

Formula (4) implies that two ingredients are required for the construction of a fast evaluation algorithm: the first ingredient is a rapid summation scheme for Gaussian kernels and the second a suitable quadrature rule for the approximation of the integral. The first ingredient is provided by the Fast Gauss Transform of Greengard and Strain [5] which will be discussed in following paragraphs. The second ingredient in this case is provided by Gauss-Laguerre quadrature. Of course, each radial basis function has a somewhat different integral representation and thus requires a suitable choice of quadrature rule (in [9] we have identified such rules for the most commonly used functions). It is also worth noting that the resulting algorithm works in any d-dimensional setting. This is in contrast to the FMM where significant differences exist between the two and the three dimensional cases. For briefness of exposition, in this paper we will consider only the three-dimensional case and will not provide the error estimates for the employed approximations.

By shifting the origin and re-scaling, we may assume that all the interpolation centers and all the evaluation points lie within the unit cube $B_0 = [0, 1] \times [0, 1] \times [0, 1]$. This is a a convenient normalization and does not restrict the generality of the method.

The first element of the Fast Gauss Transform is the observation that we may express a Gaussian in $\mathbb{R}^3$ as the Hermite expansion

$$e^{-s\|y-x\|^2} = \sum_{\beta\geq 0}\frac{1}{\beta!}\left(\sqrt{s}(x - C)\right)^\beta h_\beta\left(\sqrt{s}(y - C)\right). \quad (5)$$

with h_β a Hermite function. Also, we may assume that the point y is contained in the box $B = \{y \in [0,1]^3 : \|y - C\|_\infty < r/\sqrt{2s}\}$ of side length $r\sqrt{2/s}$ for some $r < 1$ centered at C.

For centres $x_1, x_2, \ldots, x_N \in \mathbb{R}^3$ inside box B we can precompute the moments

$$A_\beta = \frac{1}{\beta!} \sum_{i=1}^N \lambda_i \left(\sqrt{s}(x_i - C)\right)^\beta, \tag{6}$$

which we can then use to evaluate the Gaussian sum at a point y by

$$\sum_{i=1}^N \lambda_i \exp\left(-s\|y - x_i\|^2\right) = \sum_{\beta \geq 0} A_\beta h_\beta \left(\sqrt{s}(y - C)\right). \tag{7}$$

Thus, it is possible to approximate the Gaussian (7) in terms of the moments (6). The second element of the FGT is the decomposition of the computational space B_0 into subboxes B of side length $r\sqrt{2/s}$ parallel to the axes, for some fixed parameter r. Each centre is assigned to the subbox B that contains it and contributes only to the p^3 moments of subbox B. At the end of the precomputation step, the p^3 moments for each of the subboxes B have been computed. The precomputation requires $\mathcal{O}(p^3 N)$ operations.

For the estimation of the FGT at a particular evaluation point y contained in subbox D, we need to consider the influence of only some of the nearest neighbour boxes of D. Indeed, due to the exponential decay of the Gauss kernel, its effect on subboxes away from its centre may be insignificant within certain accuracy. For example, taking into account only the $(2l + 1)^3$ nearest neighbours to D, introduces error bounded by $Qe^{-2r^2 l^2}$. Hence, for $r = 1/2$ and $l = 6$ relative accuracy of 10^{-7} is obtained. We will call the set of $(2l+1)^3$ nearest neighbours the *interaction list* of box D. Thus, in order to estimate the Gaussian sum on the left side of (7) at point y, we have to accumulate the p^3 moments for each of the boxes B in the interaction list of D. Evaluation at a single point requires $\mathcal{O}((2l + 1)^3 p^3)$ operations. Overall, the computational complexity of the FGT is $\mathcal{O}(p^3 N + p^3(2l + 1)^3 M)$.

We now turn our attention to the second ingredient of our method, that is the calculation of the integral form (4). We can approximate s using a q-term generalised Gauss-Laguerre quadrature rule

$$s(y) = c \sum_{i=1}^N + \frac{1}{\sqrt{2\pi}} \sum_{k=1}^q w_k f(t_k), \tag{8}$$

where $f(t) = (c\sum_{i=1}^N \lambda_i + \sum_{i=1}^N \lambda_i e^{-t\|y - x_i\|^2})/t$. Thus, rather than evaluating directly the sum of Inverse Multiquadrics at overall cost of $\mathcal{O}(MN)$ operations, we may evaluate q sums of Gaussians (one for each quadrature node t_k) via the Fast Gauss Transform in $\mathcal{O}(q(N + M))$ operations. Recall that the decrease in the computational complexity of the latter task is due to the decoupling of the precomputation of the moments of the points x_i and the estimation of the interpolant at points y_j through the already computed moments.

1: choose q, p and r to guarantee the required precision
2: compute the weights w_k and nodes t_k of the quadrature
3: **for** each quadrature node t_k **do**
4:    subdivide B_0 into boxes of side at most $\sqrt{2/t_k}$
5: **end for**
{start first stage: precompute moments}
6: **for** each centre x_j **do**
7:    **for** each quadrature node t_k **do**
8:       find the box C which contains x_j
9:       **for** $\beta < p$ **do**
10:          compute the contribution of x_j to the moments A_β
             of box C using (6) on page 4
11:       **end for**
12:    **end for**
13: **end for**
{start second stage: evaluate moments}
14: **for** each evaluation point y_i **do**
15:    **for** each quadrature node t_k **do**
16:       find the box B that contains y_j
17:       **for** each of the $(2l+1)^3$ nearest neighbours of B **do**
18:          accumulate the series (7) on page 4 truncated after p^3
             terms to obtain an approximation to the Gaussian with parameter t_k
19:       **end for**
20:       accumulate the contribution of the k-th point of the quadrature rule (8)
21:    **end for**
22: **end for**

Algorithm 1: Fast Summation of Hardy Multiquadrics.

The quadrature nodes t_k are the zeros of the generalised Laguerre polynomial $L_q^{(-1/2)}(t)$ and the weights may be computed by a well-known formula. Overall, the fast evaluation of Inverse Multiquadric interpolants may be performed by Algorithm 1. Overall, the fast evaluation algorithm requires $\mathcal{O}(qp^3N + qp^3(2l+1)^3M)$ operations.

The Gaussian quadrature nodes and the corresponding weights may be computed using one of a number of standard methods, for example using Gautschi's ORTHOPOL package [3]. From these, the weights and nodes for quadrature when the Multiquadric or Inverse Multiquadric constant c is not the unit are calculated by $t_k^* = t_k/c^2$ and $w_k^* = w_k/c$.

4   Parallelism

In a clusted based environment Algorithm 1 has to compete against treecodes and Fast Multipole Methods. Due to the regular communication/computation patterns exhibited by Algorithm 1, we believe that it is significantly more scalable even in relatively small problem sizes. Indeed, in this section we discuss the observed performance of the algorithm in practise and discuss the implications of this result.

In treecodes parallelism may be exploited in the tree building, moment accumulation and moment evaluation stages. Cell-level synchronisation is required in the tree building phase, when different processors try to simultaneously modify the same part of the tree. For moment precomputation, a processor calculating the moments of a specific cell needs to wait until the moments for all its children have been computed. This requires the use of cell level mutexes to avoid dependency conflicts. On the other hand, the moment evaluation stage requires only communication between processors: the evaluation of the moments at a certain point requires information of nearby centre locations and coefficients, along with moment information for well separated cells. This information may be replicated and there is no need for write backs.

Orthogonal Recursive Bisection (ORB) [11] is currently the most successful algorithm in achieving good data locality while preserving load balance. The aim of the method is to provide data locality by explicitly partitioning the computational space and assigning the parts to the available workstations. The algorithm subdivides recursively the computational domain in two parts with equal computational cost. In this context, the computational cost of a particular region is defined to be the total number of interactions between each point in this region either with a centre or a cell. The assignment of domains to processors is done using the following rule: Initially all processors are associated with the entire domain. At each ORB step, the processors are split in two groups and assigned to one of the two subspaces. This process builds an ORB tree[1] which is separate from the cell tree used in the sequential algorithm.

On the other hand, the algorithm discussed in this paper provides for a clear approach to achieving both load balancing and data locality. Indeed, the algorithm has two distinct stages: first the precomputation of the moments and second the evaluation of the quadrature at a point. It is sufficient to split the set of centres (for the first stage) and the set of evaluation points (second stage) into subgroups of approximately equal size with arbitrarily selected members. Each group may be stored locally to avoid communication overheads. At the end of the first step a broadcast of each nodes moments is required and the accumulation those computed at the other nodes. This is a relatively small amount of data, independent of the number of centres and evaluation points.

An implementation of the above approach to exploit parallelism using a message passing paradigm is straightforward and particularly effective given a static computational environment. On the other hand, it is often desirable to employ unused processor cycles in idle or under-used machines to complete a large scale computation. This algorithm offers the opportunity to do so, using its potential for adaptive parallelism. The problem decomposition approach we favour is based on Piranha-type parallelism implemented using MPI 2.0 dynamic processes. Our approach requires that for a given problem, client machines offering

---

[1] This is only one of several data structures introduced specifically for the distributed treecode and are not used by the sequential algorithm.

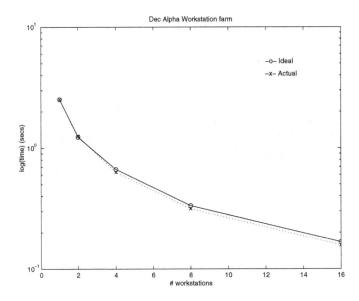

Fig. 1. Performance on the DEC Alpha cluster.

a compute service request a part of the computation and return the result. Thus each client consuming a part of the problem, the property which gives the name to the method.

The testbed for the algorithm implementation consists of two clusters of workstations. The first is assembled from commodity components and employs ten Intel based personal workstations running Linux 2.2, a UNIX-like operating system. The nodes are connected over a standard Ethernet network (10Mbits/ sec), organised in a one dimensional torus topology. The second cluster consists of sixteen high end Digital Unix 4.0 workstations connected over fast Ethernet and organised in a star topology.

We have implemented the static version of the algorithm on the Alpha workstation farm. Synchronisation was implemented with allreduce operations in a SPMD model. Even on relatively small problems ($N = M = 32,000$) the method scales very well. In this case, the I/O is local at the filesystem of each workstation and data are distributed and collected using standard operating system services rather than a distributed filesystem. The actual performance of the method is shown in Figure 1. Note that for a problem of similar size the best scalable implementation of a treecode achieves approximately 65% efficiency [12,11]. On the other hand, the fast evaluation method examined here achieves in excess of 94% efficiency. A treecode has achieved similar speedup only on a shared memory multiprocessor.

The dynamic parallelism variant of the method also offers competitive performance. In test runs between one and twenty workstations participated to the computation. For a problem of size $N = M = 10^8$ it was possible to reduce the computation time by a factor of eight on the low bandwidth cluster.

5   Conclusions

In this paper we have discussed the scalability properties of a rapid evaluation method for Radial Basis Function. In contrast to hierarchical methods, this algorithm exhibits a regular computation and communication structure due to the localisation and smoothness characteristics of the radial basis function method. This regularity results in predictable patterns which can be exploited to provide for a scalable implementation even on low bandwidth clusters of workstations. In particular, due to the intimate relation between the many body problem using the Plummer potential and the evaluation of the Inverse Multiquadric, we anticipate that a hybrid method can be devised which will benefit from the scalability properties of Algorithm 1 for the computation of far filed interactions in many body calculations.

References

1. A.W. APPEL (1985) "An Efficient Program for Many-body Simulation", *SIAM J. Sci. Stat. Comp.*, Vol. 6, No. 1, pp. 85-103.
2. R.K. BEATSON AND G.N. NEWSAM (1992) "Fast Evaluation of Radial Basis Functions: Part I", *Comp. Math. Applic.*, Vol. 24, No. 12, pp. 7-19.
3. W. GAUTSCHI (1994) "Algorithm 726: ORTHOPOL – A Package of Routines for Generating Orthogonal Polynomials and Gauss-type Quadrature Rules", *ACM Trans. Math. Soft.*, Vol. 20, No. 1, pp. 21-62.
4. L. GREENGARD (1987) *The Rapid Evaluation of Potential Fields in Particle Systems*, The MIT Press.
5. L. GREENGARD AND J. STRAIN (1991) *The Fast Gauss Transform*, SIAM J. Sci. Stat. Comput, Vol. 12(1), pp 79 - 94.
6. R.L. HARDY (1997) "The Mathematical Physics of a Biharmonic Approach to Disturbing Potential based on Multiquadric Summation", *IMACS Conference on Radial Basis Functions*, May 27-29, Pacific Grove, CA.
7. C.A. MICCHELLI (1986) "Interpolation of Scattered Data: Distance Matrices and Conditionally Positive Functions", *Constr. Approx.*, Vol. 2, pp. 11-22.
8. M.J.D. POWELL (1993) "Truncated Laurent Expansions for the Fast Evaluation of Thin-plate Splines", *Num. Alg.*, Vol. 5, No. 2, pp. 99-120.
9. GEORGE ROUSSOS (1999) "Computation with Radial Basis Functions", *Ph.D. Thesis, Imperial College of Science, Technology and Medicine*, London, UK.
10. D. SUTER (1993) "Multipole Methods for Visual Reconstruction", *SPIE Geometric Methods in Computer Vision II*, Vol. 2031, pp. 16-26.
11. M.S. WARREN AND J.K. SALMON (1992) "Astrophysical N-body Simulations using Hierarchical Tree Data Structures", *Proceedings of Supercomputing 92*, ACM Press, pp. 570-572.
12. M.S. WARREN AND J.K. SALMON (1993) "A Parallel Hashed Oct-tree N-body Algorithm", *Proceedings of Supercomputing 93*, ACM Press, pp. 12-21.

Heterogeneous Networks of Workstations and the Parallel Matrix Multiplication

Fernando Tinetti[1,2], Antonio Quijano[2], Armando De Giusti[1], Emilio Luque[3]

[1] Facultad de Informática, Universidad Nacional de La Plata, 50 y 115
1900 La Plata, Argentina
fernando@ada.info.unlp.edu.ar, degiusti@lidi.info.unlp.edu.ar

[2] Facultad de Ingeniería, Universidad Nacional de La Plata, 48 y 116
1900 La Plata, Argentina
quijano@ing.unlp.edu.ar

[3] Universidad Autónoma de Barcelona
Facultad de Ciencias
08193 Barcelona, España
e.luque@cc.uab.es

Abstract. This paper outlines the general considerations to be taken into account when a network of heterogeneous workstations is used for scientific parallel computing. Special emphasis is made on already installed networks of workstations, which provide a "zero cost" parallel computer at least in hardware. Existing matrix multiplication algorithms are analyzed, taking them as representative enough for this application area. A new algorithm is proposed in order to take advantage of the specific characteristics of networks of workstations. This parallel multiplication algorithm is presented, and its performance is shown by preliminary experimentation using the PVM library to construct a parallel (virtual) computer from a network of workstations. Also, new features concerning data communication performance are identified from the execution profile of the experiments.

1 Introduction

The growing processing power of standard workstations, along with their low cost and the relatively easy way in which they can be available for parallel processing, have contributed to their increasing use in computation intensive application areas. Usually, computation intensive areas have been referred to as scientific processing; one of them being linear algebra, where a great effort has been made to optimize solution methods for serial as well as for parallel computing [1] [3] [2] [16]. Also, there is a high number of publications showing their effectiveness (from the performance point of view) when used as parallel computers, and this give a sound reason to keep researching active. Installed networks of workstations constitute the lowest cost as well as the most available parallel computers.

Topics such as processing as well as communication hardware heterogeneity are considered solved by the use of parallel processing libraries such as PVM (Parallel Virtual Machine) [8] and implementations of MPI (Message Passing Interface) [11]. However, experimentation about performance under these circumstances seems to be

Y. Cotronis and J. Dongarra (Eds.): Euro PVM/MPI 2001, LNCS 2131, pp. 296–303, 2001.

necessary [14]. Also, installed networks of workstations (NOWs) are specially attractive due to its extremely low cost for parallel processing as well as their availability. Performance of such networks of workstations can be fully analyzed by means of a simple application: matrix multiplication. A parallel algorithm is proposed for matrix multiplication derived from two main sources: a) previous proposed algorithms for this task in traditional parallel computers, and b) the bus based broadcast interconnection network of workstations. This parallel algorithm is analyzed experimentally in terms of workstations workload and data communication, two main factors in overall parallel computing performance.

From the whole area of linear algebra applications, the most challenging (in terms of performance) operations to be solved are the so called Level 3 BLAS (Basic Linear Algebra Subprograms). In Level 3 BLAS, all of the processing can be expressed (and solved) in terms of matrix-matrix operations. Even more specifically, the most studied operation has been matrix multiplication, which is in fact a benchmark in this application area.

Many common and well-known characteristics of traditional parallel computers are not found on installed networks of workstations used for parallel (distributed) computing. Networks of workstations were not designed as parallel computers and most of the differences are originated from their intended use: a set of stand-alone computers which are able to communicate temporarily with each other. Furthermore, installed networks carry out their own evolution at least about adding and replacing of workstations.

Section 2 summarizes most of the matrix multiplication algorithms proposed in the literature as well as a short description of networks of workstations from the parallel processing point of view. The parallel matrix multiplication algorithm designed for NOW is described in Section 3 in terms of data layout/distribution and workstation local processing and communication. The experimental work in a real heterogeneous NOW with the proposed algorithm is described and presented in Section 4. Section 5 presents the immediate conclusions based on the experimental data as well as further work to improve/explore algorithmic and performance issues.

2 Parallel Matrix Multiplication Algorithms and NOWs

Many parallel algorithms have been designed, implemented, and tested on different parallel computers for matrix multiplication [17]. For simplicity, the algorithms are usually described in terms of $C = A \times B$, where the three matrices A, B, and C are dense and square of order n.

The so called "direct implementation" [17], in which every processor computes a portion of the resulting matrix C having (or receiving) the necessary portions of matrices A and B is used mainly for introductory and teaching purposes, and it is hardly ever used in practice, mainly due to its high requirement of memory.

Divide-and-conquer and recursive algorithms are considered specially suited for multiprocessor parallel computers. In fact, matrix multiplication is inherently good for shared memory multiprocessors because there is no data dependence. Matrices A and B are accessed only for reading to calculate every element of matrix C, and no element of matrix C has any relation (from the processing point of view) with any other element of the same matrix C. Unfortunately, networks of workstations used for parallel computing are not shared memory architectures, and implementations of divide-and-conquer and recursive algorithms are far from optimal in networks of workstations. The main reason for the loss of performance is found in the need of a shared (uniform) memory view of a distributed and loosely coupled memory architecture.

One of the most *innovative* algorithms for sequential matrix multiplication is due to Strassen [13] and its parallelization is straightforward on shared memory parallel computers. Again, the architecture of a network of workstations is not well suited (form the performance point of view) for this algorithm and its "immediate" ways of parallelization. Also, the Strassen method is defined in terms of different (for matrix multiplication) arithmetic operations such as subtraction, and then special care has to be taken for computer numeric rounding errors.

It is possible that most of the reported parallel algorithms used in practice are based on parallel multicomputers where the processors are arranged (interconnected) in a two dimensional mesh or torus [10]. These algorithms may be roughly classified as "broadcast and shift algorithms" initially presented in [9] and "align and shift algorithms", initially presented in [4]. Both kinds of algorithms are described in terms of a PxP square processor grid where each processor holds a large consecutive block of data.

Many algorithms have been proposed as rearrangements and/or modifications from those two initial ones ([4] [9]). The underlying concept of blocking factor became intensively and successfully used from two different but related standpoints: load balance and processor local performance. It can be proved that distributing relatively small blocks of matrices the job to be done on each processor is almost the same. Furthermore, the blocking factor is essential in achieving the best performance on each processor given the current memory hierarchies, where cache memories have to be taken into account with special care.

It is very interesting how the broadcast based algorithms have been successively and successfully adapted to the point-to-point interconnection of two dimensional torus. A very long list of publications is available about this subject (e.g. [5] [15] [6]), and all of them aim to obtain the best performance for the whole algorithm by implementing broadcast over the point-to-point communication links of distributed memory (mesh interconnected) parallel computers.

The DNS (Dekel, Nassimi, Sahni) [7] [12] algorithm is proposed specifically for distributed memory computers with their processors interconnected as a three-dimensional hypercube. Even if the relationship between matrix size and number of processors defined by this algorithm is not taken into account as a drawback, it is clear its orientation towards point-to-point communication in a three dimensional array.

2.1 Characteristics of NOWs as Parallel Computers

Form the point of view of parallelization, the main characteristics distinguishing NOWs from traditional parallel computers can be stated as:
- Loosely coupled parallel architecture.
- Computing heterogeneity, involving different processors (including architecture and speed), memory hierarchies, memory sizes, etc.

- Low communication performance, specifically at the startup time taken for every data communication.
- Communication homogeneity, mostly based on Ethernet, thus having a logical bus topology where broadcast is implemented by hardware and every point-to-point communication prevents other data transmission.

The parallel (matrix multiplication) algorithms should be analyzed on this parallel computers. At least, already designed parallel algorithms are the first alternatives to be considered for implementation.

Shared memory based parallel algorithms should be discarded because they are not suitable for implementation on a loosely coupled parallel architecture such as a NOW. Mesh based class of parallel algorithms are the following to be analyzed. Initially, the installed networks of workstations do not seem to be related with meshes, because of their interconnection network usually based on a single bus (Ethernet - Fast Ethernet). This leads to eliminate at least shift based parallel algorithms. The basic idea of broadcast message should be used because it has a direct relationship with the bus interconnection. Unfortunately, as it has been explained, the initial "broadcast and shift algorithms" have evolved to "shift and shift algorithms". The point-to-point communications should be minimal in parallel programs executed on networks of workstations because they imply a bottleneck on a single and shared communication medium as the LAN bus.

It is important to remark that most (if not all) of the libraries designed for parallel computing on networks of workstations have the capability of every possible logical arrangement (and interconnection) of processes with relatively small effort. However, given that performance is the main issue in parallel computing, the overhead of mapping those topologies on a bus becomes unacceptable most of the times.

3 Parallel Matrix Multiplication on NOWs

The parallel algorithm proposed for installed networks of workstations has two main characteristics:
1. Easy workload distribution for heterogeneous processing power.
2. Based only on broadcasting data.

The description of the proposed parallel algorithm to compute $C = A \times B$ will be made taking into account:
- A, B, and C are $n \times n$ matrices,
- P workstations, $ws_1, ..., ws_P$,
- pw_i is the normalized relative processing power of workstation ws_i, $\forall\ i = 1...P$, where normalized implies $pw_1 + ... + pw_P = 1$.

3.1 Data Layout

Data partition and assignment to workstations is defined in terms of row blocks for matrices A and C and column blocks for matrix B:
- ws_i contains $rA_i = n \times pw_i$ rows of matrix A, and
- ws_i contains $cB_i = n/P$ columns of matrix B.

Thus, the number of rows of matrix A assigned to each workstation (rA_i) is proportional to the workstation relative processing power. This data distribution is not uniform when the workstations have different processing power. Due to rounding errors for the operation $rA_i = n \times pw_i$ $(0 < pw_i < 1 \; \forall \; i = 1, ..., P)$ it is possible that $dr = rA_1 + ... + rA_P < n$. The remaining rows can be uniformly distributed among workstations $ws_1, ..., ws_{(n-dr)}$ one row for each workstation. Given that the usual case is $P << n$, this reassignment of rows can be considered non relevant from the point of view of proportional (according to workstations relative processing power) data distribution.

Matrix B is equally distributed by columns among workstations, as the usual case in (homogeneous parallel computers) bibliography. Matrix B data distribution is made by columns and uniformly. Each workstation contains the same amount of data, cB_i columns, because

- computing workload is already achieved by matrix A data distribution, and
- matrix B elements will be involved in broadcast messages of uniform size.

3.2 Computing

Workstation ws_i computes a portion of matrix C, $C^{(i)}$, proportional to its relative processing power:

$$C^{(i)}_{rA_i \times n} = A^{(i)}_{rA_i \times n} \times B \tag{1}$$

where only a submatrix of B is held locally (cB_i columns). Let $A^{(i)}$ ($A_{rA_i \times n}$), $B^{(i)}$ ($B_{n \times cB_i}$), and $C^{(i)}$ ($C_{rA_i \times n}$) the local portions of matrices A, B and C assigned to ws_i respectively, and let n multiple of P, the algorithm in pseudocode for ws_i is shown in Fig. 1.

```
1B(i) = B(i);        /* Save local B(i) */
C(i)(i)=A(i)×B(i);   /* First partial matrix-local data */
For j = 1 to P {
  if (j == i)
     Broadcast 1B(i);/* Bcast local data (in turn) */
  else {
     Receive B(i);   /* Recv data broadcasted from wsj */
     C(i)(j)=A(i)×B(i);  /* Compute partial matrix */
  }
}
```

Figure 1. Parallel Matrix Multiplication Algorithm.

where $C^{(i)}_{(j)}$ denotes cB_i columns of the local portion of matrix C, $C^{(i)}$, columns $cB_i \times (j-1)+1$ to $j \times cB_i$ inclusive. The two main characteristics of the algorithm are:

1. It follows the SPMD (Single Program - Multiple Data) parallel computing model. Every workstation ws_i ($i = 1, ..., P$) executes the same code. On each iteration,

every workstation assigns cB_i columns of its local portion of matrix C, $C^{(i)}$.

2. It follows the message passing parallel programming model which is the best suited for networks of workstations (loosely coupled parallel machines).

As PVM has the facility for overlapped communications with processing, the broadcast routine should be called (conditionally) before $C^{(i)}$ partial computing. Note that local computing is independent on each workstation. Local optimization, such as the selection of a blocking factor for matrix multiplication, can be made independently on each workstation.

4 Experimentation

Workstations used for experimentation are described in Table 1, they are interconnected by an Ethernet 10 Mb/s Ethernet LAN, and the PVM library [8] was used as the message passing as well as parallel computing library tool.

Table 1. Characteristics of the Workstations used in the Experiments.

Name	CPU / Mem	Mflop/s
purmamarca	Pentium II 400 MHz / 64 MB	316
cetadfomec1	Celeron 300 MHz / 32 MB	243
cetadfomec2	Celeron 300 MHz / 32 MB	243
sofia	PPC604e 200 MHz / 64 MB	225
Josrap	AMD K6-2 450 MHz / 62 MB	99

The (balanced) workload was verified by means of a synthetic "embarrassingly parallel" version of the matrix multiplication. In this version, the broadcast is eliminated and local assignment of $B^{(i)}$ is made instead of receiving a message. The library PVM was still used to "spawn" the processes remotely as well as to synchronize at the end to record time completion of the whole parallel application. The performance of the synthetic version as well as the proposed algorithm was evaluated in terms of local measured times for different squared matrix sizes ranging from $n = 500$ to $n = 3500$. A representative execution example of the synthetic version for squared matrices of size $n = 2000$ is shown in Table 2. Local measured times (in seconds) are discriminated by iteration as if communication were made.

The difference between maximum and minimum local computing times (14.48 and 14.31 respectively) is about 1% which could be highly acceptable given the relative processing power differences. The proposed algorithm (Figure 1) was used, it was verified also for squared matrix sizes between $n = 500$ y $n = 3500$, and Table 3 shows a representative execution example for matrix size of $n = 2000$.

The main conclusions taken from comparison of Table 2 and Table 3, are

1. The overall computing performance is reduced by approximately 13% when there is data communication. Communication processes have a minimum memory (cache memory, in particular) requirement by which there is contention with computing processes, and it is reflected on the computing performance reduction.

2. The processing workload is also affected by communication processes but the balance loss can be considered acceptable. The difference between the maximum and minimum local computing times (16.42 and 15.6 respectively) is about 5%.

Table 2. Workload Verification for the Synthetic Parallel Algorithm.

Name	Assigned Rows	Local Comp. Time	Per It.
purmamarca	562	14,36	2,87
cetadfomec1	431	14,31	2,86
cetadfomec2	431	14,35	2,87
sofia	400	14,48	2,90
Josrap	176	14,39	2,88

Table 3: Algorithm Computing Performance.

Name	Assigned Rows	Local Comp. Time	Per It.
purmamarca	562	16,22	3,24
cetadfomec1	431	16,37	3,27
cetadfomec2	431	16,18	3,24
sofia	400	16,42	3,28
Josrap	176	15,60	3,12

The analysis on computing workload becomes non relevant when the communication times are taken into consideration (Table 2 and Table 3 just show local computing times). The amount of time taken by the communication routines is between 76 and 108 seconds for this problem size ($n = 2000$). Then, most of the total execution time is spent on communication. The first question on this context is about communication performance: are these communication times (between 76 and 108 seconds) the expected ones? In the best case, a whole matrix of 2000x2000 single precision floating point numbers takes 76 seconds for communication, this implies having a network of about 210 Kb/s, which in turn means a communication overhead of about 4/5 (80%) over a 10 Mb/s bus based communication hardware.

5 Conclusions and Further Work

It was described a new SPMD algorithm for parallel matrix multiplication on heterogeneous NOWs based on previous algorithms and adapted to the specific characteristics of this parallel architecture. Computing workload has been shown to be balanced on experimental work over a real heterogeneous NOW. Even when the algorithm communication pattern was designed for the specific interconnection network found on NOWs, the *real* communication performance achieved is poor and it is directly translated to the total running time.

Having seen the high communication overhead on the experiments, it becomes necessary to analyze the communication performance achieved from the PVM library for this application. One of the first alternatives to be investigated is the way in which the broadcast message routine is implemented. Other message passing parallel computing libraries such as the available MPI implementations could be considered for experimental measurement of communication performance, since the algorithm is easily adapted to use MPI.

Other parallel algorithms designed for traditional homogeneous parallel computers with specific interconnection hardware (e.g. meshes) could be adapted and/or redesigned to cope the specific characteristics of heterogeneous NOWs used for parallel processing. Linear algebra problems and numerical scientific algorithms in general are directly related to the matrix multiplication example shown, and are the first candidates to be approached for analysis, adaptation and/or redesign.

References

1. Anderson E., Z. Bai, C. Bischof, J. Demmel, J. Dongarra, J. DuCroz, A. Greenbaum, S. Hammarling, A. McKenney, D. Sorensen, LAPACK: A Portable Linear Algebra Library for High-Performance Computers, Proceedings of Supercomputing '90, pages 1-10, IEEE Press, 1990.
2. Bilmes J., K. Asanovic, C. Chin, J. Demmel, Optimizing matrix multiply using phipac: a portable, high-performance, ansi c coding methodology, Proc. Int. Conf. on Supercomputing, Vienna, Austria, July 1997, ACM SIGARC.
3. Blackford L., J. Choi, A. Cleary, E. D'Azevedo, J. Demmel, I. Dhillon, J. Dongarra, S. Hammarling, G. Henry, A. Petitet, K. Stanley, D. Walker, R. Whaley, ScaLAPACK Users' Guide, SIAM, Philadelphia, 1997.
4. Cannon L. E., A Cellular Computer to Implement the Kalman Filter Algorithm, Ph.D. Thesis, Montana State University, Bozman, Montana, 1969.
5. Choi J., J. Dongarra, D. Walker, PUMMA: Parallel Universal Matrix Multiplication Algorithm on Distributed Memory Concurrent Computers, in Concurrency: Practice and Experience, 6:543-570, 1994.
6. Choi J., "A New Parallel Matrix Multiplication Algorithm on Distributed-Memory Concurrent Computers", Proc. of the High-Perf. Comp. on the Information Superhighway, IEEE, HPC-Asia '97.
7. Dekel E., D. Nassimi, S. Sahni, Parallel matrix and graph algorithms, SIAM Journal on Computing, 10:657-673, 1981.
8. Dongarra J., A. Geist, R. Manchek, V. Sunderam, Integrated pvm framework supports heterogeneous network computing, Computers in Physics, (7)2, pp. 166-175, April 1993.
9. Fox G., M. Johnson, G. Lyzenga, S. Otto, J. Salmon, and D. Walker, Solving Problems on Concurrent Processors, Vol. I, Prentice Hall, Englewood Cliffs, New Jersey, 1988.
10. Golub G. H., C. F. Van Loan, Matrix Computation, Second Edition, The John Hopkins University Press, Baltimore, Maryland, 1989.
11. Message Passing Interface Forum, MPI: A Message Passing Interface standard, International Journal of Supercomputer Applications, Volume 8 (3/4), 1994.
12. Ranka S., S. Sahni, Hypercube Algorithms for Image Processing and Pattern Recognition, Springer-Verlag, New York, 1990.
13. Strassen V., Gaussian Elimination Is Not Optimal, Numerische Mathematik, Vol. 13, 1969.
14. Tinetti F., A. Quijano, A. De Giusti, Heterogeneous Networks of Workstations and SPMD Scientific Computing, 1999 ICPP, IEEE Press, Aizu-Wakamatsu, Japan, pp. 338-342, Sep. 1999.
15. van de Geijn R., J. Watts, SUMMA Scalable Universal Matrix Multiplication Algorithm, LAPACK Working Note 99, Technical Report CS-95-286, University of Tenesse, 1995.
16. Whaley R., J. Dongarra, Automatically Tuned Linear Algebra Software, Proceedings of the SC98 Conference, Orlando, FL, IEEE Publications, November, 1998.
17. Wilkinson B., Allen M., Parallel Programming: Techniques and Applications Using Networking Workstations, Prentice-Hall, Inc., 1999.

Collecting Remote Data in Irregular Problems with Hierarchical Representation of the Domain

Fabrizio Baiardi, Paolo Mori, and Laura Ricci

Dipartimento di Informatica, Università di Pisa
Corso Italia 40, 56125 Pisa, Italia
{baiardi,mori,ricci}@di.unipi.it

Abstract. Irregular problems require the computation of some properties of a set of elements irregularly distributed in a domain. These problems satisfy a locality property because the properties of an element e depend upon those of its neighbors according to a problem dependent stencil. This paper proposes two strategies, fault prevention and informed fault prevention, to collect properties of elements mapped onto remote processing nodes that minimize the corresponding overhead. We describe an MPI implementation of informed fault prevention and the experimental results in the case of the adaptive multigrid method.

1 Introduction

Several phenomena, such as the motion of the stars in a galaxy or the illumination of objects in an image, are modeled by time dependent partial differential equations systems, that can be efficiently solved through numerical adaptive iterative algorithms. To apply these algorithms, at least one dimension of the problem, such as the time interval or the space to be analyzed, is discretized. Hence, the phenomenon is modeled as a set of elements in a domain of interest where some properties are computed for each element. The computation satisfies a locality property because the properties of an element depends upon those of a subset of all the other ones, according to a problem dependent neighborhood relation. The main feature of any irregular problem is the *non homogeneous and dynamic* distribution of the elements in the domain. This implies that some subsets of the domain require a larger computational effort than the other ones.

The adoption of parallel architectures for the resolution of these problems is mandatory because of the large number of elements, but this poses the problems of the mapping of the elements onto the processing nodes, p-nodes, and of the communication handling due to the irregularity and the dinamicity of the distribution. Since no information about the element distribution is statically available, a sophisticated run time support is required that is even more important on distributed memory architectures with a sparse interconnection network.

Any parallel implementation of these problems on this class of architectures should define, at least, three strategies that, respectively, *i)* define a load balancing mapping of the elements onto the p-nodes *ii)* update this mapping to

Y. Cotronis and J. Dongarra (Eds.): Euro PVM/MPI 2001, LNCS 2131, pp. 304–311, 2001.

recover an unbalancing due to the changes in the distribution and *iii)* collect remote data. These three strategies are built around a data structure that describe the mapping of the elements. In the following we assume that this structure describes a hierarchical decomposition of the domain. While the first two strategies are strictly correlated, the third one is independent from the other ones. In this paper, we focus on this strategy because it has a large influence on the overall efficiency. The strategies that define and update the mapping have been discussed in [1] and [2]. Alternative approaches to irregular problems are LPARX [4], Chaos and Multiblock Parti, [5] and that presented in [6]. However, the study is focused on the data mapping techniques.

The next section briefly describes the data structure that represents the mapping of the elements. Sect. 3 describes three strategies to collect remote data: fault handling, fault prevention and informed fault prevention. Sect. 4 discusses the MPI implementation of the informed fault prevention strategy. Experimental results are presented in the last section.

2   A Hierarchical Representation of the Element Mapping

The *Hierarchical Tree, H-Tree*, is the data structure that represents both the distribution of the elements in the domain and their mapping onto the p-nodes, assuming that the domain is an *n*-dimensional space. A tree has been adopted because it naturally represents the hierarchical relations and it is intrinsically adaptive. The H-Tree defines a multi level hierarchical representation that, at each level, partitions the domain into a set of equal subdomains, or spaces. The root of the H-Tree represents the whole domain, while each other node N, *hnode*, represents either a space produced by the decomposition, *space(N)*, or an element, and it records the corresponding information, see Fig. 1. If *space(N)* has been partitioned, the resulting subspaces are represented by the sons of N. Each element is paired with a space including it. If an element e is paired with *space(N)*, then the hnode L representing e is a son of N and is a leaf of the H-Tree. As the number of elements and their distribution change during the computation, the decomposition of the domain and the H-Tree are updated to represent the current distribution.

We do not discuss the mapping strategies that can be supported by the H-Tree. Here, we only assume that they do not violate the locality property and we denote by $Do(P_k)$ the subdomain including all and only the spaces mapped onto P_k. The mapping of the spaces defines np subsets of the H-Tree, one for each p-node. The subset assigned to P_h is the private H-Tree of P_h that includes an hnode N iff $space(N) \in Do(P_h)$. Notice that, in general, the hnodes assigned to a p-node could define distinct subtrees of the H-Tree but, for the sake of simplicity, we assume that they belong to just one subtree. A further subset of the H-Tree, the replicated H-Tree, is defined by the mapping strategy. This tree includes all the hnodes in the path from the root of the H-Tree to the root of any private H-Tree and is replicated in all the p-nodes. Each leaf L of the replicated H-Tree points to the p-node storing the private H-Tree rooted in L .

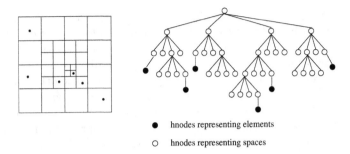

○   hnodes representing elements

○   hnodes representing spaces

Fig. 1. Domain Decomposition and H-Tree

3   Remote Data Collection Techniques

At each iteration of the algorithm, P_h updates the properties of each element e_i in $Do(P_h)$ by applying one or more operators to the current properties of e_i and to those of its neighbors. Hence, P_h needs the properties of the neighbors of e_i, as defined by the neighborhood stencil of each operator. The abstraction level of these properties depends upon several problem dependent features, such as the operator that is applied or the value of the properties of the element. For this reason, in the following, we say that a p-node needs the properties of some spaces to compute the property of an element. Even if the mapping strategy respects the locality property, some of the neighbors of e_i have been allocated onto other p-nodes. Hence, the definition of a strategy to collect remote spaces, i.e. spaces allocated onto other p-nodes, is required.

However, a fundamental problem is that a p-node cannot know in advance where the spaces it needs have been mapped. Moreover, because of the dynamic mapping, even if, at each iteration, a p-node needs the same spaces, they may be mapped onto distinct p-nodes at distinct iterations. If locality is preserved, most of the neighbors of each element e are allocated onto the same p-node of e. This implies that: *i)* most of the spaces to be collected are allocated on the same p-node, and *ii)* the intersection between the set of the neighbors of two neighbors elements is large. Hence, a remote space should be reused to update the properties of several elements mapped onto the p-node (reuse property).

3.1   Fault Handling

The simplest strategy to collect remote spaces is *fault handling*. Each p-node, while computing the properties of its elements, interacts with the other p-nodes as soon as it needs some spaces. To compute the properties of each element e_i in $Do(P_h)$, P_h visits its private H-Tree. If P_h determines that it needs a space A, assigned to another p-node, it suspends the computation of the properties of e_i, and it requests the properties of A to the owner, determined by accessing the replicated H-Tree. The computation can be resumed only when P_h has received all the missing properties. These properties should be cached in the local memory of P_h because they may be used to compute the properties of other elements in $Do(P_h)$ by the current operator. The time inbetween the request of the remote spaces and the reply may be large. If, during this time, P_h is idle, a low efficiency will results. Since the computation of the properties of e_i is independent of that of each other element e_j, the waiting time can be overlapped with the computation of the properties of another element e_j, according to an excess parallelism strategy. If the computatation of e_j is suspended too, and all the spaces required for e_i have been received, then the computation of e_i can be resumed, otherwise the computation of another element is started.

This strategy is very general but, as a counterpart, it requires two communications to collect the properties of a remote space, one for the request and one to transmit the properties. Moreover, it is not trivial to optimize the communications. Since the size of the data sent in either a request or a reply is small, several messages for the same process could be merged into a single message, but this delays some messages. Hence, message merging increases the time between a request and the corresponding reply. In turns, this implies that a larger number of elements has to be computed in parallel by P_h and this may result in a large overhead. Furthermore, to fully exploit the reuse property, only one request for each space should be sent. This requires that each request for a space should be checked against those that have already been sent to the same p-node.

3.2   Fault Prevention

The *fault prevention* strategy generalizes the data exchange in the data parallel programming model where the compiler determines which spaces are required by each p-node and the run time support executes the communications before starting a new iteration. Hence, a p-node does not explicitly request the spaces it needs because the compiler can exactly determine the spaces to be exchanged.

In the fault prevention strategy, for each operator, the proper spaces are exchanged among the p-nodes before executing the operator. In this way, when P_h computes the properties of its elements, all the spaces it requires are available in its local memory. Taking into account the dynamic element mapping, P_k computes, before applying an operator, which of its spaces are required by P_h, $h \neq k$, by applying to each space A in $Do(P_k)$ the inverse of the operator neighborhood stencil. If A is a neighbor of any element in $Do(P_h)$, then P_k sends the properties of A to P_h. This fully exploits the reuse property because each space is sent just once.

Fault prevention assumes that all the p-nodes share some informations about the domain decomposition that is recorded in the replicated H-Tree. The main disavantage of this strategy is that the replicated H-Tree records a partial information only, hence, P_k approximates the set of spaces that P_h requires. However, P_h can execute its computation only if the approximation is *safe*, i.e. P_h receives all the spaces it needs. Safeness requires that, if the replicated H-Tree does not include enough information to determine whether A belongs to the neighborhood stencil of an element in $Do(P_h)$, P_k includes A in the data to be sent. If the approximation is not accurate, most of the spaces sent to P_h are useless.

3.3   Informed Fault Prevention

If the information in the replicated H-Tree does not allow a p-node to compute an accurate approximation, we propose an improved version of fault prevention, *informed fault prevention*. According to this strategy, the p-nodes exchange some information about their private H-Trees in a separate phase, *replicated H-Tree extension* phase, before the fault prevention one. During the fault prevention phase, each process exploits the information received in the replicated H-Tree extention phase to determine the spaces it has to send. Due to the locality property, the information sent by P_h to P_k in the replicated H-Tree extension phase usually describes the distribution of the elements on the boundary of $Do(P_h)$ that intersect that of $Do(P_k)$. In particular, the information sent by P_h to P_k is related to all the elements e_i in $Do(P_h)$ that could have at least one neighbor in $Do(P_k)$. To reduce the overhead of this phase, only the subset of properties of e_i useful to determine its neighborhod stencil are trasmitted. Moreover, this information is sent from P_h to P_k in the first replicated H-Tree extension phase after the creation of e_i and an invalidation message is sent when e_i is destroyed, remapped or when the subset of properties have been updated. When P_k receives the creation message, it stores the properties in its local memory and it will use this information until it receives an invalidation message.

With respect to fault prevention, the computation of the spaces to be sent is simplified, because each process P_k, for each element e_i received in the information phase, determines which of its spaces belongs to the neighborhood stencil of e_i and send them to the owner of e_i. Hence, while the fault prevention has to visit all the private H-Trees, the informed fault prevention considers the elements received in the replicated H-Tree extention phase only, that are considerably less than those in the private H-Trees.

An advantage of both fault and informed fault prevention is that they concentrate the communications in two small sections of the algorithm, the replicated H-Tree extension phase and the fault prevention phase. This implies that the trasmission of a set of a data can be delayed because they are not used immediatly. Hence, a group of data to be sent to the same p-node can be merged into one message. This strategy can be fundamental in cluster of workstations, where little communication support is provided and the time to set up a communication cannot be neglected. A further advantage of fault prevention strategies is that they preserve the sequential code. As a matter of fact, after the fault prevention phase, all the spaces required by the computation are available on each p-node and the sequential code can be executed.

An example of application of the methodology to adaptive multigrid method is showed in fig. 2. In adaptive multigrid method, [3], the final result is basically computed through the application of five operators: restriction, smoothing, prolongation, norm and refinement. The restriction and the prolongation operators are separately applied on each level of the hierarchical representation of the domain, while the smoothing operator is applied on the whole domain. The refinement operator is the only one that modifies the domain by adding new

```
data mapping and replicate and private h-trees creation

while (not global error < threshould)  {
   replicated h-tree extention(union_all_stencils, all_levels)

   for level from max_level downto min_level  {
     fault prevention(restriction_stencil, level)
     restriction(level)  }

   fault prevention(smoothing_stencil, all_levels)
   smoothing(all_levels)

   for level from min_level to max_level  {
     fault prevention(prolongation_stencil, level)
     prolongation(level)  }

   fault prevention(norm_stencil, all_levels)
   norm(all_levels)
   refinement(all_levels)

   data mapping and replicate and private h-trees update  }
```

Fig. 2. Example of Application of the Methodology

elements in those subdomains where the approximation error is too large. Hence the replicated H-Tree extention phase can be executed at the begining of each iteration only, and the values collected are valid until the end of the iteration, when the refinement operator is applied. The fault prevention phase, instead, has to be executed before each operator, because each operator updates the values.

4   MPI Implementation of Informed Fault Prevention

In the following, we focus on informed fault prevention, because this strategy is the most complex one, and consider how it can be implemented through the MPI primitives to manage the data exchange among the p-nodes.

Both in the replicated H-Tree extention phase and in the fault prevention one, each process determines the spaces to be sent while receiving the spaces from the other p-nodes. In this way, the latency of communications is overlapped with some useful computation. P_h issues an MPI_Irecv from each other p-node and starts the computation of the spaces to be trasmitted. The handles returned by each MPI_Irecv are recorded in the request array, in the position corresponding to the rank of the sender process. As soon as P_h determines that a space is to be sent, it issues an MPI_Isend, it records the corresponding handle in a buffer and it polls, through an MPI_Testany invoked on the request array, whether a new message has been received. If a message has been received from P_k, the corresponding hnode is inserted in the replicated H-Tree and a new MPI_Irecv from P_k is posted. The handle returned by MPI_Irecv is recorded in the kth position of the request array. The polling solution avoids any overhead of interrupt based solutions. Periodically, each p-node issues an MPI_testall on the buffer containing the send handles and frees the positions paired with the partners with which the exchange have been completed.

A process synchronization is executed before applying of each operator, to ensure that all the processes have terminated the data exchange. Since the number of messages to be exchanged is not known in advance, a barrier cannot be adopted to synchronize processes because the MPI_barrier is a blocking primitive and, once issued, only messages related to the barrier can be received. Hence, each process, after sending all the data, broadcasts a termination message to each other process through np point to point MPI_Isend primitives and waits for the corresponding termination messages from the other processes. In this way, termination messages are interleaved with the data ones.

If the properties of a space to be exchanged among the p-nodes have different data types, MPI derived data types may be adopted. Moreover, since the set of properties exchanged in the replicated H-Tree extention phase are different from those exchanged in the fault prevention phase, distinct derived data types have to defined. To reduce the communication overhead, message merging has been adopted. Hence, each process P_h defines a buffers of b positions for each of its communication partners and it stores in the kth buffer any data to be sent to P_k to immediatly continue the computation of the data to be sent. As soon as the buffer paired with P_k is full, the content of the buffer is sent to P_k through an MPI_Isend . The value of b is choosen according to the features of the adopted architecture.

5   Experimental Results

In order to evaluate the performances and the effectiveness of the methodology and, in particular, of the informed fault prevention strategy, we have implemented a parallel version of an adaptive multigrid method, using the MPI primitives embedded in the C language.

The parallel architecture we consider is a cluster of workstations. Each workstation is a PC with an Intel Pentium II CPU (266 MHz) and 128 Mbyte of local memory. The interconnection network is a 100Mbit Fast Ethernet switch and the operative system running on each p-node is LINUX 2.2.13.

The adaptive multigrid method we consider solves an highly irregular partial differential equation derived from the Poisson problem. In fact, at some iterations of the computation, a very large number of elements is created in few subdomains. Hence, this is a good test for a parallel implementation.

Figure 3 shows the efficiency of our implementation for a variable number of p-nodes where the informed fault prevention strategy is adopted. The size of the problem, i.e. the initial number of elements in the domain, is the same for all the executions. The optimal size of the buffer used to implement the message merging strategy is 50, because in a cluster of workstations the cost of a communication is dominated by the time to setup the communication.

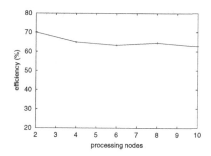

Fig. 3. Efficiency for Adaptive Multigrid Method

Figure 3 shows that our implementation arises an efficiency larger than 60%, even on 10 p-nodes. Moreover, from the point of view of scalability, we observe that the value of the resulting efficiency is almost independent of the number of p-nodes. The main reason of this efficiency value is that, even if the replicated H-Tree extention phase and the fault prevention phase take less than 20% of the total execution time, more than 10% of the time is lost due to an unbalanced load distribution in the refinement operator. An improovement of the methodolgy where, if necessary, two updates of the domain partitioning are executed in the same iteration, is under development.

References

1. Baiardi, F., Chiti, S., Mori, P., Ricci, L.: Integrating Load Balancing and Locality in the Parallelization of Irregular Problems. In: Future Generation Computer Systems, Vol. 17. Elsevier Science (2001) 969–975
2. Baiardi, F., Chiti, S., Mori, P., Ricci, L.: A Hierarchical Approach to Irregular Problems. In: Proc. of Europar 2000: LNCS, Vol. 1900. (2000) 218–222
3. Briggs, W.: A multigrid tutorial. SIAM (1987)
4. Fink, J.S., Baden, S.B., Kohn, S.R.: Efficient run-time support for irregular block-structured applications. In: Journal of Parallel and Distributed Computing, Vol. 50(1) (1998) 61-82
5. Mukherjee, S.S., Sharma, S.D., Hill, M.D., Larus, J.R., Rogers, A., Saltz, J.: Efficient support for irregular applications on distributed-memory machines. In ACM SIG-PLAN Notices, Vol. 30(80) (1995) 68–79.
6. Sohn, A., Biswas, R., Simon, H.D.: A Dynamic Load Balancing Framework for Unstructured Adaptive Computations on Distributed Memory Multiprocessors. In: Proc. 8th Annual ACM Symposium on Parallel Algorithms and Architectures, (SIGARCH, ACM, 1996) 189–192.

Parallel Image Matching on PC Cluster[*]

Henryk Krawczyk[1] and Jamil Saif[1]

Technical University of Gdańsk,
Faculty of Electronics Telecommunications and Informatics,
Dept. of Computer Systems Architecture
G.Narutowicza 11/12 str.
80-952 Gdańsk
Poland
hkrawk@pg.gda.pl, saif@ask.eti.pg.gda.pl

Abstract. Hierarchical image matching using PVM on a local network workstation cluster is considered. Three different parallel matching strategies for endoscopic coloured images are analysed, and their suitability for medical diagnosis are discussed.

1 Introduction

To support medical diagnosis various information processing systems are largely used [1,11]. The recent improvement and proliferation of high performance PCs and high speed network interfaces create favourable opportunities of their development. Many telemedicine systems are used for storing and retrieving medical information or exchange of it for educational reasons [2]. Other such systems are dedicated to support specific medical examinations. The majority of medical data directly connected with description of patient diseases are both white/black or coloured (grey) images [9]. Analysis [13] of these images is very important for disease recognition. However, this is a very complex problem, and many works try to solve some essential subproblems such as image filtering [4], feature extraction (e.g. for image retrieval [6]), image comparison [15] or reconstruction (i.e. deriving 3D shape of an object from its simple 2D grey scale images [10]). Besides, data mining approach is used [12] to discover basic association rules essential for medical diagnosis. The complexity of the problems is very high (this is in many cases an NP - problem). Therefore a new approach is observed, which tries to parallelise existing algorithms in order to achieve high efficiency of medical computer-based systems.

One example of such tendency is Endoscopy Recommender System (ERS) [7], its general architecture is presented in Fig. 1. During patient examination ERS main functions are archivisation of gastroenterological examinations and evaluation of consistency (degree of similarity) of target images registered during a patient examinations with template endoscopic images stored in a data base or data bases. A physician observes a registered endoscopic film on the computer screen and decides to store the most representative images as a document of

---

[*] The paper sponsored by National Grant KBN No. 8T11C00117

Y. Cotronis and J. Dongarra (Eds.): Euro PVM/MPI 2001, LNCS 2131, pp. 312–318, 2001.

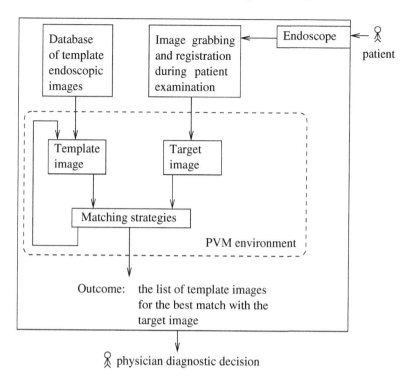

Fig. 1. ERS Architecture

the patient examination. The suitable procedures allows to prepare the recommendation list of template images which are the most consistent with the set of target images. Such procedures are called matching procedures and they can support less experienced physicians (or students) in medical diagnosis. Besides, the accepted consistency of two images means that corresponding disease descriptions of the template image can be rewritten from that template to the currently prepared examination document. In this way physicians can be also less occupied with heavy administration works.

In the paper we present different parallel matching approaches for coloured endoscopic images. In Section 2 we describe the image matching problem and in Section 3 we present suitable algorithms. In Section 4 we give how to configure the PVM platform for efficient execution of these algorithms and we demonstrate some experimented results. Finally, in Section 5 we summarise our proposition and suggest some future investigations.

2   Image Matching Problem Definition

Let us consider two images $I^{(1)}$ and $I^{(2)}$. $I^{(1)}$ is a template image and $I^{(2)}$ is a target one. Let $M^{(1)}$ and $M^{(2)}$ be suitable matrices representing $I^{(1)}$ and $I^{(2)}$ respectively, i.e.:

$$M^{(1)} = \{m_{ij}^1 | i \in H^{(1)} = \{1, 2, \ldots, h_1\}, j \in W^{(1)} = \{1, 2, \ldots, w_1\}\}$$
$$M^{(2)} = \{m_{ij}^2 | i \in H^{(2)} = \{1, 2, \ldots, h_2\}, j \in W^{(2)} = \{1, 2, \ldots, w_2\}\} \quad (1)$$

m_{ij}^k is a scalar or a vector according to the assumed image transformation, image size of $I^{(k)}$ is equal to $h_k \times w_k$; $k = 1, 2$. In practice $k > 2$ if we have many templates. For simplicity, we assume that image sizes are equal if and only if $h_1 = h_2 = h$ and $w_1 = w_2 = w$.

If image size of $I^{(1)}$ equals image size of $I^{(2)}$ then images $I^{(1)}$ and $I^{(2)}$ can be compared on the basis of their matrices $M^{(1)}$ and $M^{(2)}$ and a degree of similarity $d(I^{(1)}, I^{(2)}) = d(M^{(1)}, M^{(2)})$ can be expressed in the following way:

$$d(I^{(1)}, I^{(2)}) = \sum_{i=1}^{h} \sum_{j=1}^{w} \left\| m_{ij}^{(1)} - m_{ij}^{(2)} \right\| \quad (2)$$

where $\| m_{ij}^{(1)} - m_{ij}^{(2)} \|$ is defined in different way in accordance to considered image features.

In practice we use rather relative measure D, where D is defined as:

$$D(I^{(1)}, I^{(2)}) = \left(1 - \frac{d(I^{(1)}, I^{(2)})}{\sum_{i=1}^{h} \sum_{j=1}^{w} m_{ij}^{(1)}} \right) 100\% \quad (3)$$

Let Δ, $\Delta \neq 0$, be the admissible error of image similarity, then the result of image matching can be defined as follows:

$$I^{(1)} \equiv I^{(2)} \text{(images are the same) if and only if } D(I^{(1)}, I^{(2)}) = 0$$
$$I^{(1)} \approx I^{(2)} \text{(images are similar) if and only if } D(I^{(1)}, I^{(2)}) \leq \Delta \quad (4)$$
$$I^{(1)} \neq I^{(2)} \text{(images are different) if and only if } D(I^{(1)}, I^{(2)}) > \Delta$$

Formulas (2),(3),(4) describe formally the *matching problem* of two images. Moreover, all approaches to image matching concern with the following key issues:

1. The way of image description and transformation of I into M. In general, $M = \alpha(I), M \in \{M^{(k)} | k = 1, 2, \ldots, K\}, I \in \{I^{(k)} | k = 1, 2, \ldots, K\}$.
2. Size correction of compared images.
3. Suitable criteria for best matching (see (2) and (3)); i.e. D and Δ.
4. The effective algorithms for the best matching.

In accordance to the assumed image transformation α, image matching problem can be divided into the following three categories:

– colour-pixel scale values, where various histograms of images are compared,
– characteristic image objects, such as points, lines and regions are analysed,
– topological and geometrical features, where shapes, wavelet or texture are taken into considerations.

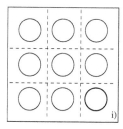 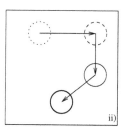 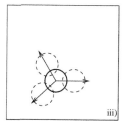

Fig. 2. Matching Strategies: i) solid ii) gradient iii) tuning

For each of the above categories the matching criteria (2) can be properly modified and the lowest distance is regarded as being the best match. Then on the base of some experiments a value of Δ is assessed. In case of endoscopic images we try to find such a value of Δ, which minimises the number of template images consistent with a given target image. It is obvious, that the minimum should equal 1. However, in practice this value strongly depends on the assumed image transformation.

3   Image Matching Strategies

Taking into account specific characteristics of endoscopic images we prepare new matching strategies. We assume that each template image has the most important region of interest (called ROI) marked by the thin circle line. It allows to concentrate on specific details of the image and in consequence can facilitate disease recognition. Moreover, the radius of the circle cannot be the same for all images. Let consider a simple target image and K template images with ROIs. Then the matching problem is to recognise if the target image includes one or more of the regions similar to the ROIs marked in each template image. To solve this, the transformation of rectangle image into circle image is needed and then the following *matching strategies* (see Fig. 2) can be proposed:

solid — where we use a matching algorithm for the a priori located regions in the target image,
gradient — where we start searching from an arbitrary point of the target image, and move in a direction determined by gradient of maximal growth of the degree of similarity,
tuning — where we concentrate on one small point of the target image and deeply examine it by moving the examined template region in many different directions, even changing the sizes, dilations and rotation of template regions,
hybrid — where we use in different way all three strategies defined above.

In this way the complicated image matching task is divided into a number of subtasks, examining different regions in the target image. The number of such subtasks denoted by x can be estimated from the following inequality:

$$S_T \geq x \cdot S_{tr} = x \frac{\pi r^2}{2} \qquad (5)$$

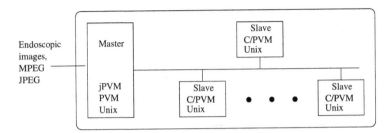

Fig. 3. The considered network environment

where: S_T is size of the target image, and S_{tr} is size of the template region (ROI); all expressed as a number of pixels. The value r represents the radius of the given ROI.

Each of these proposed above matching strategies can be applied to all categories of matching problems mentioned in Section 2.

However, we start our research only for the first matching problem specified as the colour-pixel scale values approach. Based on PVM platform we want to prepare an environment specially oriented on these kinds of matching algorithms. After this, we will concentrate on other categories of the image matching.

4   Architecture & Experiments

To create the dedicated environment for solving the above defined image matching problems we make further assumptions. Let us consider a single target image registered during a patient examination, and its similarity with T template images. Template images are distinguished into two categories:

- T - V pattern images representing typical gastroenterologic diseases, and
- V images corresponding to endoscopic images of alimentary organs for healthy patients.

Besides, all template images of the first category have pointed out the regions of interests, i.e. ROIs. We try to find the degrees of similarity between the target image and all template images. In the case of template images with ROIs we want to determine if the target image contains one of the ROIs, for the given relative similarity (e.g. $D \geq 80\%$). To solve the above problems we use the PC cluster (see Fig. 3) with the total number of processors higher than (xT).

The images in MPEG or JPEG format [8] come from ERS system to the parallel network environment. Because ERS system is implemented in Java we use jPVM [14] as the highest layer of the master. Other layers of the master and slaves are implemented in C and PVM [5].

For template images with ROIs we consider three possible levels of parallelism:

target parallelism where the target image is replicated T times and all pairs of images, i.e. the replicated target image and one of template images (briefly: RTa - TeI), are analysed in parallel.

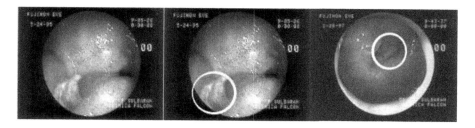

Fig. 4. The example of endoscopic images with similar ROIs

ROI parallelism where for each pair of images RTa - TeI, the region of interest belonging to the template image is replicated x times and allocated on the replicated target images, according to the solid matching strategy described in Section 3.

MA parallelism where we use a parallel matching algorithm (MA) for each allocated ROI on each replicated target image. According to formula (1) and (2), such a parallel algorithm bases on decomposition of matrices M into sub-matrices with suitable transformation, which is a function of image contrast, brightness, delation and rotation.

It can be observed that in case of template and target images without ROIs only third type of parallelism can be realised. We define the suitable endoscopic image matching problem and implement it in the assumed computing environment for all the above types of parallelism. We also try to find the suitable configuration to achieve high performance and scalability. Our first experiments show that the most interesting architecture consists of two layers working independently. The first layer is message passing layer (jPVM, PVM), and it is responsible for T-target parallelism. The second layer bases on PCs with double processors (in general thread processing) and is the most responsible for the next two levels of parallelism.

Fig. 4 illustrates the functionality of the prepared environments. Two first images represent template images without and with ROI. The third image is a target image corresponding to the same organ. The result of calculations is the ROI pointed at the target image which is the most similar to the analysed template ROI. We consider only the simplest image transformation α, i.e. colour-pixel scale values. For the case shown in Fig. 4 we obtained correct location of the target ROI on the template image. In general the open problem is not scalability [3] but effectiveness of the matching algorithm when quality of target images is low and we cannot find any similarities with templates images. Then the other image transformations should be considered.

5   Final Remarks

We consider matching problems from endoscopic images point of view, and define new approach to find similarity between a target image and each of templates image. Three levels of parallelism are considered, and the suitable two layer architecture is suggested. It leads to mixed solution where PVM layer is the high level of computation and the low level is multi-thread processing. It is a new challenge to investigate such architectures deeply for different image matching algorithms and various matching strategies in order to prepare an efficient environment for medical applications.

References

1. Athmas P.M., Abbott A.L.: *Realtime Image Processing On a Custom Computing Platform.* IEEE Computer, pp. 16–24, Feb. 1995.
2. Böszörmenyi L., Kosh H., Slota R.: *PARMED — Information System For Long Distance Collaboration in Medicine.* Proc. of Intl. Conference on Software Communication Technologies, Hagenberg–Linz, Austria, pp. 157–164, April 1999.
3. Czarnul P., Krawczyk H.: *Dynamic Assignment with Process Migration in Distributed Environments.* Recent Advances in PVM and MPI, Lecture Notes in Computer Science, vol. 1697, pp. 509–516, 1999.
4. Figueiredo M.A.T., Leitao J.M.N., Jain A.K.: *Unsupervised Contour Representation and Estimation Using B-Splines and a Minimum Description Length Criteria.* IEEE Trans. on Image Processing, Vol. 9, No. 6, pp. 1075–1089, June 2000.
5. Geist, A., Beguelin, A., Dongarra, J., Jiang, W., Manchek, R., Sunderam, V.: PVM 3 user's guide and reference manual. Oak Ridge National Laboratory, Oak Ridge, Tennessee (1994)
6. Gevers T., Smeulderds A.W.M.: *PictoSeek: Combining Colour and Shape Invariant Features for Image Retrieval.* IEEE Trans. on Image Processing, Vol. 9, No. 1, January 2000.
7. Krawczyk H., Knopa R., Lipczynska K., Lipczynski M.: *Web-based Endoscopy Recommender system — ERS.* In Proc. of IEEE Intl. Conference on Parallel Computing in Electrical Engineering PARELEC 2000, August, Canada, pp. 27–30, 2000.
8. Koenen R.: *MPEG-4. Multimedia for our time.* IEEE Spectrum, pp. 26–33, Feb. 1999.
9. Maintz J.B.A., Viergever M.A.: *A survey of medical image registration.* Medical Image Analysis No. 2, pp. 1–36, 1998.
10. Okatani T., Deguchi K.: *Shape Reconstruction from an Endoscope Image by Shape from Shading Technique for a Point Light Source at the Projection Center.* Computer Vision and Image Understanding, vol.66, No.2 May, pp. 119–131, 1997.
11. Sacile R., Ruggiero C. et al.: *Collaborative Diagnosis over the Internet. Working Experience.* IEEE Internet Computing, Nov–Dec., pp. 29–37, 1999.
12. Shintani T., Oguchi M., Kitsuregawa M.: *Performance Analysis for Parallel Generalized Association Rule Mining on a Large Scale PC Cluster.* In Proc. of Euro–Par, pp. 15–25, 1999.
13. Sonka M., Hlavac V., Boyle B.: *Image Processing, Analysis and Machine Vision.* PWS Edition, USA 1998.
14. Thurman D.: *jPVM.* http://www.chmsr.gatech.edu/jPVM
15. You J., Bhattacharya P.: *A Wavelet-Based Coarse-to-Fine Image Matching Scheme in A Parallel Virtual Machine Environment.* IEEE Transactions on Image Processing, Vol.9, No.9, pp. 1547–1559, September 2000.

Computing Partial Data Cubes
for Parallel Data Warehousing Applications

Frank Dehne[1], Todd Eavis[2], and Andrew Rau-Chaplin[3]

[1] School of Computer Science
Carleton University, Ottawa, Canada K1S 5B6
frank@dehne.net, www.dehne.net
[2] Faculty of Computer Science
Dalhousie University, Halifax, NS, Canada B3H 1W5
eavis@cs.dal.ca
[3] Faculty of Computer Science
Dalhousie University, Halifax, NS, Canada B3H 1W5
arc@cs.dal.ca, www.cs.dal.ca/~arc
Corresponding Author.

Abstract. In this paper, we focus on an approach to *On-Line Analytical Processing* (OLAP) that is based on a database operator and data structure called the datacube. The *datacube* is a relational operator that is used to construct all possible views of a given data set. Efficient algorithms for computing the entire datacube – both sequentially and in parallel – have recently been proposed. However, due to space and time constraints, the assumption that all 2^d (where $d =$ dimensions) views should be computed is often not valid in practice. As a result, algorithms for computing partial datacubes are required. In this paper, we describe a parallel algorithm for computing partial datacubes and provide preliminary experimental results based on an implementation in C and MPI.

1   Introduction

As databases and data warehouses grow ever bigger there is an increasing need to explore the use of parallelism for storage, manipulation, querying, and visualization tasks. In this paper, we focus on an approach to On-Line Analytical Processing (OLAP) that is based on a database operator and data structure called the datacube [4]. Datacubes are sets of pre-computed views of selected data that are formed by aggregating values across attribute combinations (a *group-by* in database terminology) as illustrated in Figure 1. A generated datacube on d attribute values can either be complete, that is, contain all of the 2^d possible views formed by attribute combinations, or partial, that is, contain only a subset of the 2^d possible views. Although the generation of complete and partial views is related, the latter is a significantly more difficult problem. Despite this difficulty, in practice it is important to be able to generate such partial datacubes because, for high dimensional data sets (i.e., between four and

Y. Cotronis and J. Dongarra (Eds.): Euro PVM/MPI 2001, LNCS 2131, pp. 319–326, 2001.

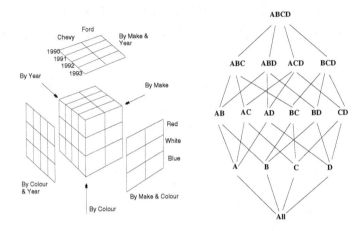

Fig. 1. An example 3 dimensional datacube and a 4 dimensional lattice. Lefthand side: An example three dimensional datacube concerning automobile data. Righthand side: The lattice corresponding to a four dimensional data cube with dimensions A, B, C and D. The lattice represents all possible attribute combinations and their relationships. The "all" node represents the aggregation of all records.

ten), a fully materialized datacube may be several hundred times larger than the original data set.

The datacube, which was introduced by Jim Gray et. al [4], has been extensively studied in the sequential setting [1,2,4,5,6,7,8] and has been shown to dramatically accelerate the visualization and query tasks associated with large information sets. To date the primary focus has been on algorithms for efficiently generating complete datacubes that reduce computation by sharing sort costs [1,7], that minimize external memory sorting by partitioning the data into memory-size segments [2,6], and that represent the views themselves as multidimensional arrays [4,8]. The basis of most of these algorithms is the idea that it is cheaper to compute views from other views rather than from starting again with the original data set. For example, in Pipesort [7] the lattice is initially augmented with both estimates for the sizes of each view and cost values giving the cost of using a view to compute its children. Then a spanning tree of the lattice is computed by a level-by-level application of minimum bipartite matching. The resulting spanning tree represents an efficient "schedule" for building the actual datacube.

Relatively little work has been done on the more difficult problem of generating partial datacubes. Given a lattice and a set of selected views that are to be generated, the challenge is in deciding which view should be computed from which other view, in order to minimize the total cost of computing the datacube. In many cases computing intermediate views that are not in the selected set, but from which several views in the selected set can be computed cheaply, will reduce the overall computation time. In [7], Sarawagi et al. suggest an approach based on augmenting the lattice with additional vertices (to represent all possible orderings of each view's attributes) and additional edges (to represent all relationships between views). Then a Minimum Steiner Tree approximation algorithm is run to identify some number of "intermediate" nodes (or so-called Steiner points) that can be added to the selected subset to "best" reduce the overall cost. An approximation algorithm is used here because the optimal Minimum Steiner Tree is NP-Complete. The intermediate nodes introduced by this method are, of course, to be drawn from the non-selected nodes in the original lattice. By adding these additional nodes, the cost of computing the selected nodes is actually reduced. Although theoretically neat this approach is not effective in practice. The problem is that the augmented lattice has far too many vertices and edges to be efficiently handled. For example, in a 6 dimensional datacube the number of vertices and edges in the augmented lattice increases by a factor of 326 and 8684 respectively, while for a 6 dimensional datacube the number of vertices and edges increase by a factor of 428 and 701,346 respectively. A 9 dimensional datacube has more than 2,000,000,000 edges. Another approach is clearly necessary.

In this paper we describe a new approach to efficiently generate partial datacubes based on a parallel version of Pipesort [3] and a new greedy algorithm to select intermediate views. We also present initial experimental results based on an implementation of our algorithm in C and MPI. The experimental results are encouraging in that they show an average reduction in computing a partial datacube of 82% over computation directly from the raw data. This reduction applies to both the sequential and parallel cases. Furthermore, the parallel version of our algorithm appears to achieve linear speedup in experiments on an eight node cluster.

2   Generating Partial Datacubes in Parallel

In the following we present a high-level outline of our coarse grained parallel partial datacube construction method. This method is based on sequential Pipesort [7] and a parallel version of Pipesort described in [3]. The key to going from these methods for computing *complete* datacubes to a method for computing *partial* datacubes is Step 2 of the following algorithm - the greedy method for computing an efficient schedule tree for the partial datacube generation problem.

A Parallel Algorithm for Generating Partial Datacubes

1. **Build a Model:** Construct a lattice for all 2^d views and estimate the size of each of the views in the lattice. To determine the cost of using a given view to directly compute its children, use its estimated size to calculate (a) the cost of scanning the view and (b) the cost of sorting it.
2. **Compute a schedule tree using the model:** Using the bipartite matching technique presented in Pipesort [7], reduce the lattice to a spanning tree

that identifies the appropriate set of prefix-ordered sort paths. Prune the
spanning tree to remove any nodes that cannot possibly be used to compute
any of the selected nodes. Run a greedy algorithm using the pruned tree to
identify useful intermediate nodes. The tree built by the greedy algorithm
contains only selected nodes and intermediate nodes and is called the schedule tree as it describes which views are best computed from which other
views.

3. **Load balance and distribute the work:** Partition the schedule tree into
   $s \times p$ sub-trees (s = oversampling ratio). Distribute the sub-trees over the p
   compute nodes. On each node use the sequential Pipesort algorithm to build
   the set of local views.

Given that finding the optimal schedule tree is NP-Complete[4], we need to
find a method that takes a manageable amount of time to find a reasonable
schedule. In computing the schedule tree we propose starting from the spanning
tree that is derived from Pipesort. Clearly there are many other approaches that
could be taken. We chose this approach for our initial try at generating partial
cubes because the Pipesort tree has proven to be effective in the generation of
complete datacubes and therefore appears to be a good starting point for a schedule for partial datacubes. This choice is indeed supported by our experimental
findings.

In the following sections we will describe exactly how the Pipesort tree is
pruned, as well as the greedy algorithm for selecting intermediate nodes/views.
For a description of how the model is built and the details of the load balancing
algorithm see [3].

The Pruning Algorithm. Before passing the Pipesort tree to the greedy
algorithm, we want to ensure that is has been pruned of any unnecessary nodes.
Quite simply, we remove any node from the tree whose attributes are not a
superset of at least one selected node. The pseudo code can be written as follows:

```
Input: Spanning tree T and Subset S
Output: Pruned (spanning) tree T

for every node i in T - S
    for all nodes j of S
        if there is no node j whose attributes are a
            subset of the attributes of i
            delete node i from T
```

The operation of this simple quadratic time algorithm is illustrated in Figure 2.

The Greedy Algorithm. The greedy algorithm takes as input a spanning tree
T of the lattice that has been pruned and a set S of selected nodes representing
those views to be materialized as part of the partial datacube. The algorithm

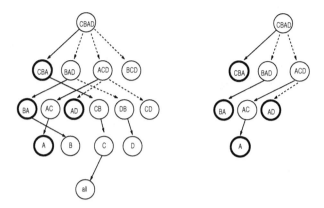

Fig. 2. Graph Pruning. Lefthand side: Spanning tree of the lattice as created by Pipesort with selected nodes in bold. Righthand side: Pruned tree.

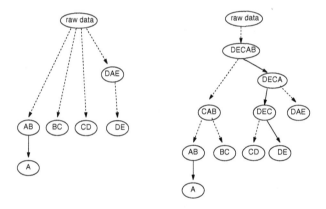

Fig. 3. The Schedule tree. Lefthand side: The initial schedule tree T' containing only selected nodes. Righthand side: An example final schedule tree T' containing both selected and intermediate nodes.

begins by assuming that its output, the schedule tree T', will consist only of the selected nodes organized into a tree based on their relative positions in T. In other words, if a selected node a is a descendant of a selected node b in T, the same relationship is true in the initial schedule tree T'. The algorithm then repetitively looks for intermediate nodes that reduce the total cost of the schedule. At each step, the node from T that most reduces the total cost of T' is added. This process continues until there are no nodes which provide a cost improvement. Figure 3 shows an example of an initial schedule tree T' for a five dimensional cube with attributes A, B, C,D, and E and selected views A, AB, BC, CD, DE, and DAE, as well as a possible final schedule tree, T'.

The pseudo code for the greedy algorithm is as follows:

Table 1. Cost reductions in five, seven, and nine dimensions.

Dim.	Partial datacube	(1) Base Cost	(2) Our Cost with no intermediate nodes	(3) Our Cost with intermediate nodes	1 vs 2	1 vs 3
5	8/32	35,898,852	23,935,211	6,203,648	44%	83%
7	15/128	95,759,942	36,353,947	14,139,249	62%	85%
9	25/512	119,813,710	49,372,844	28,372,844	59%	77%

```
Input: Spanning tree T and Subset S
Output: Schedule tree T'

Initialize T' with nodes of S
    each node of S is connected to its immediate predecessor in S
    edges are weighted accordingly

while global_benefit >= 0
    for each node i in T - S
        /* compute the benefit of including node i */
        for every node j in S
            if attributes of j are a prefix of i
                local_benefit += current_cost of j  - cost of scanning i
            else if attributes are a subset of i
            local_benefit += current_cost of j  - cost of resorting i
        local_benefit -= cost of building i
        if local_benefit > global_benefit
            global_benefit = local_benefit
            best node = i

    if global_benefit > 0
        add node i to T'
```

3   Experimental Evaluation

Our preliminary results indicate that our new greedy approach significantly reduces the time required to compute a partial datacube in parallel. In this section we examine experimentally two aspects of our algorithm: 1) how well does the greedy algorithm reduce the total cost of building the partial datacube, and 2) what speedups are observed in practice.

We first examine the reduction in total cost. Table 1 provides the results for arbitrarily selected partial datacubes in five, seven, and nine dimensions. Each result represents the average over five experiments. In each row we see three graph costs. The first is the cost of computing each view in the partial datacube directly from the raw data set. The second cost is that obtained by our algorithm without the addition of intermediate nodes. Finally, the third column shows the cost obtained by our algorithm when utilizing intermediate nodes. Columns four and five show the percentage reductions in cost our algorithm obtains. Note that our algorithm for generating schedules for the computation of partial datacubes reduces the cost of such computation by between 77% and 85% over a range of

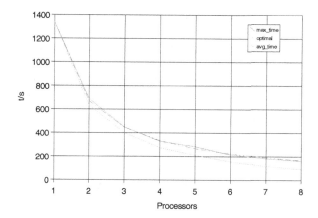

Fig. 4. Running Time In Seconds As A Function Of The Number Of Processors. (Fixed Parameters: Data Size = 1,000,000 Rows. Dimensions = 7. Experiments Per Data Point = 5.)

test cases. It appears that the algorithm works best when the number of selected views is not too small. This is what one might expect given that when there are only a small number of selected views, there is little to be gained by introducing intermediate nodes. More extensive experimental results will be reported on in the final version of this paper.

Figure 3 provides a graphical illustration of the algorithm's benefit. The image depicts a "before and after" scenario for a five dimensional lattice and associated partial datacube (this was an actual test case). On the left we see a spanning tree containing only the selected nodes (constructed during the initialization process in the greedy algorithm). On the right we have the final result - a new spanning tree with four additional nodes. In this case, the tree was reduced in size from 32,354,450 to 5,567,920 for an 83% reduction in total cost.

The following experiments were carried out on a very modest parallel hardware platform, consisting of a front-end machine plus 8 compute processors in a cluster. These processors were 166 MHZ Pentiums with 2G IDE hard drives and 32 MB of RAM. The processors were running LINUX and were connected via a 100 Mbit Fast Ethernet switch with full wire speed on all ports.

Figure 4 shows the running time observed as a function of the number of processors used. There are three curves shown. The *runtime* curve shows the time taken by the slowest processor (i.e., the processor that received the largest workload). The second curve shows the *average time* taken by the processors. The time taken by the front-end machine to compute the model and schedule and distribute the work among the compute nodes was insignificant. The *theoretical optimum* curve shown in Figure 4 is the sequential Pipesort time divided by the number of processors used. Note that, these experiments were performed with schedule trees for complete datacubes rather than for partial datacubes but we expect the results to hold as these trees have very similar properties.

One can observe that the *runtime* obtained by our code and the *theoretical optimum* are essentially identical. Interestingly, the *average time* curve is always below the *theoretical optimum* curve, and even the *runtime* curve is sometimes below the *theoretical optimum* curve. One would have expected that the *runtime* curve would always be above the *theoretical optimum* curve. We believe that this *superlinear speedup* is caused by another effect which benefits our parallel method: improved I/O.

4   Conclusions

As data warehouses continue to grow in both size and complexity, so too does the need for effective parallel OLAP methods. In this paper we have discussed the design and implementation of an algorithm for the construction of partial datacubes. It was based on the construction of a schedule tree by a greedy algorithm that identifies additional intermediate nodes/views whose computation reduces the time to compute the partial datacube. Our preliminary results are very encouraging and we are currently investigating other related approaches.

References

1. S. Agarwal, R. Agrawal, P. Deshpande, A. Gupta, J. Naughton, R. Ramakrishnan, and S. Sarawagi. On the computation of multidimensional aggregates. *Proceedings of the 22nd International VLDB Conference*, pages 506–521, 1996.
2. K. Beyer and R. Ramakrishnan. Bottom-up computation of sparse and iceberg cubes. *Proceedings of the 1999 ACM SIGMOD Conference*, pages 359–370, 1999.
3. F. Dehne, T. Eavis, S. Hambrusch, and A. Rau-Chaplin. Parallelizing the datacube. *International Conference on Database Theory*, 2001.
4. J. Gray, S. Chaudhuri, A. Bosworth, A. Layman, D. Reichart, M. Venkatrao, F. Pellow, and H. Pirahesh. Data cube: A relational aggregation operator generalizing group-by, cross-tab, and sub-totals. *J. Data Mining and Knowledge Discovery*, 1(1):29–53, April 1997.
5. V. Harinarayan, A. Rajaraman, and J. Ullman. Implementing data cubes. *Proceedings of the 1996 ACM SIGMOD Conference*, pages 205–216, 1996.
6. K. Ross and D. Srivastava. Fast computation of sparse data cubes. *Proceedings of the 23rd VLDB Conference*, pages 116–125, 1997.
7. S. Sarawagi, R. Agrawal, and A.Gupta. On computing the data cube. Technical Report RJ10026, IBM Almaden Research Center, San Jose, California, 1996.
8. Y. Zhao, P. Deshpande, and J. Naughton. An array-based algorithm for simultaneous multi-dimensional aggregates. *Proceedings of the 1997 ACM SIGMOD Conference*, pages 159–170, 1997.

PDES: A Case Study Using the Switch Time Warp*

Remo Suppi, Fernando Cores, and Emilio Luque

Dept. of Computer Science, University Autonoma of Barcelona,
08193, Bellaterra, Spain
{Remo.Suppi,Emilio.Luque}@uab.es, Fernando@aows10.uab.es

Abstract. The present work outlines the results of a technique for improving Parallel Discrete Event Simulation (PDES) using optimistic protocols on a real physical system. This technique, designated Switch Time Warp (STW) is developed on PVM and is based on the research carried out by the authors at the *Universitat Autònoma de Barcelona*. The STW is used to limit excessive optimism in the Time Warp method, and has a negative effect on optimistic PDES. The idea of the STW consists of dynamically adapting the simulation speeds of different application processes. This change is accomplished through the priority's agent of the operating system and is based on decisions taken by the heuristic control of the STW.

1. Introduction

In recent years, simulation has been extending toward fields in which it is necessary to handle large information volumes and resources. A consequence of this expansion is that the resources necessary for large simulations are greater than the performance offered by monoprocessor systems. The new simulation systems need ever greater computing capacity, memory and secondary storage (for example: simulation of telecommunications nets or the design of VLSI circuits). Because of this, PDES is a useful tool (and indispensable in some instances) in providing response to these problems within an acceptable time. There are two main objectives to PDES: to reduce simulation execution time of the simulation, and to increase the potential dimension of the problem to be simulated. Moreover, PDES is justified economically by the possibility of distributing tasks between interconnected low cost machines.

1.1 Optimistic PDES

The most referenced optimistic simulation method is Time Warp (TW), which was defined by Jefferson and Sowizral [1,2]. This mechanism is based on logical processes (LP) for distributing the simulation and the utilization of messages for synchronization. In optimistic protocols, each LP executes the events as soon as these are avail-

---

* This work has been supported by the CICYT under contract TIC98-0433

Y. Cotronis and J. Dongarra (Eds.): Euro PVM/MPI 2001, LNCS 2131, pp. 327–334, 2001.

able. This processing order can be different from the processing order produced in the real physical system. Some of these events processed outside of order will have to be canceled. The TW algorithm provides a mechanism (rollback) to solve these causality errors. The TW uses a global virtual time -GVT- (this is the smallest time taken by all messages in the system) to guarantee the maximum possible rollback in any LP. The main TW problem is excess of optimism. This excess optimism emerges when one or more LPs advance their local virtual time -LVT- with respect to the GVT in an excessive form, generating a higher number of rollbacks and reducing simulation performance. In the literature, there are several methods for controlling LP optimism in the TW. These methods generally attempt to block the most optimistic LPs from releasing resources that would be spent on incorrect calculation. These resources will be reassigned to the slowest LPs in order to advance their simulation. Amongst such methods there are the following: Moving Time Windows [3], the Adaptive TW concurrency control algorithm [4] and Warped (a TW simulator that includes several methods) [5].

1.2   Switch Time Warp (STW): Our Proposal for Controlling Optimism

Our proposal is based on an adaptive technique for optimism control on optimistic PDES. This proposal is based the change of LP priority, dynamically benefiting those processes that accomplish useful work and penalizing processes that continually accomplish rollbacks. Our proposal has two main objectives: to reduce the non-balanced execution that could emerge between LPs, and to control LP optimism. The first objective will reduce rollback size, reducing the existing deviations between LVT and reducing the memory size required for causality error recovery. The second objective is to reduce the probability of rollbacks being produced. The STW mechanism has been developed on PVM and a detailed description of the algorithm can be found in [6,7].

2.   A Case Study: Ecological System Simulation

In order to validate the proposed technique, a model was designed and a real physical system simulator with a high degree of complexity was developed. The results provided by the STW simulation were compared with the original technique (TW). The model selected is an ecological prey-predator physical system. The selection of this physical system was guided by the following criteria:
- High model complexity to allow an in-depth study of the PDES simulation.
- A physical system model that allows a distribution of the simulation domain between a variable numbers of sub-domains.
- Capacity to easily modify the simulation parameters such as:
  1. The number of logical processes.
  2. The interconnection graph of the LPs.
  3. The size of the problem to be simulated and LP workload.
  4. The external event frequency sent to other processes.

2.1   Model of the Prey-Predator System

The model will be used as base of the ecosystem representation that, combined with the input data and its evolution, could determine the behavior of a given ecosystem and population of a species in general terms. The predator has an inhibiting effect on the growth of the prey, and the prey causes a growth effect (acceleration) on the growth of the predator. An initial approximation to the model is to consider the eco-system without aging. The simplest equation to describe the growth of a population is **dx/dt = rx**, where x is the population density at time t and r is a constant. The solution to this equation is $x = x0\ e^{rt}$, where x0 is the density to t=0. This equation can only give real values for a limited period, since an increase in population will imply that the resources are finished. In this sense, the population can decrease to other stationary values, oscillate (regularly or not) or decrease until it is extinguished. A more ade-quate equation for describing this ecosystem behavior is the logistic equation: **dx/dt = ax-bx2**. In this equation, when x is small, the equation is reduced to the first equation, and growth is exponential; when t increases, x tends to a stationary value without oscillations. A more accurate and exact model is presented by the Volterra equations that describe the interaction between prey with density x and predators with density y:

$$\textbf{dx/dt = ax-bx2-cxy} \qquad\qquad \textbf{dy/dt = - ey+c'xy} \qquad\qquad (1)$$

In these equations: with the absence of predators, the prey corresponds to the logis-tic equation; the speed at which the prey is eaten is proportional to the product of the prey-predator densities. If the prey is limited only by the predator (b=0), the prey has an exponential growth. If it is necessary to consider that a given number of prey (xr) finds refuge and makes itself inaccessible to the predator, the equations are:

$$\textbf{dx/dt = ax-cy (x-xr)} \qquad\qquad \textbf{dy/dt = -ey+c'y (x-xr)} \qquad\qquad (2)$$

The prey-predator system with aging and discrete equations (Xn, Yn are the prey and predator densities) in year n, are: $X_{n+1}=RXn\ Y_{n+1}=rY_n$ where R and r will be func-tions of Xn and Y_n, respectively. If we consider that the predator is only limited by prey numbers:

$$\textbf{X}_{n+1}= \textbf{aX}_n - \textbf{CX}_n\textbf{Y}_n \qquad\qquad \textbf{Y}_{n+1}= \textbf{-eY}_n + \textbf{C'X}_n\textbf{Y}_n \qquad\qquad (3)$$

These equations are then equivalent to the Volterra equations without feedback [8].

Our implementation of the ecosystem model will use the last series of equations (3) under the following conditions:

- The habitat is split into a fixed number of cells.
- Within these cells, each population can be treated as populations without spatial separation.
- Between neighboring cells, migration is possible and therefore immigration and emigration have effects on the size of the population in similar proportions to re-production or death in the cell.
- The effects of migration are immediate (small migration time compared with the life cycle).
- The environment is uniform (all cells have similar characteristics).

- Prey migration will depend on two factors: a) predator density; b) an excessive prey population can imply a reduction of available resources, thereby causing migration.

Therefore, this ecological prey-predator model with aging and migration will be our model for PDES and STW-TW simulation.

2.2   Prey-Predator PDES Model Implementation

The territory of the physical system will be split into cells where the previous equations will be valid. These cells will be modeled through their position in the global territory, and the populations will be modelled on the prey and predators in the cell. PDES simulation of the global territory will be accomplished by simulating the individual evolution of different populations in each cell, where this evolution will be modeled through previous equations (3). Each cell in the territory is related to other cells through migration. The quantity and destination of this migration will depend on cell populations.

The distribution of the physical system will be accomplished through the division of the simulation space, so that each one of the objects (LP) must be capable of simulating the complete model, but on a limited simulation space. In this case, the simulation space is represented by the territory of the ecological system to be simulated. The migration option presents considerable advantages in terms of simulation, because it allows the creating of simulations with a variable computing volume. The computing volume will depend on the territory size to be simulated, and process number will depend on the territory partition and assignment to the different logical processes. The following migratory actions will be taken in order to simplify the study: migration in every direction, or only one address (up, down, left and right). Migration probability will depend on the cell characteristics and on the number of neighboring cells.

2.3   Performance of the STW and TW with Different Communication Paths

The first step to the simulation runs is the tuning of the STW algorithm control heuristic in order to fix the necessary parameters, according to a given LP distribution. For this case, the most unfavorable distribution possible for an optimistic simulation is used, allowing migration to form only one cycle. With these values, the next step in the simulation process is to execute simulation runs for TW and STW, for different communication paths. All these tests were accomplished through using simulations with size territories of 40 x 40 cells, distributed in a pool of 12 processors (SunSparc Workstations).

For this study, four different topologies were used: pipeline, tree, ring and all-with-all. The most interesting topologies will be the ring and all-with-all, since they better characterize the behavior of most simulation applications, particularly where PDES simulation techniques have certain problems.

Pipeline

From the STW point of view, the pipeline topology represents an optimum graph. With some adjustment to LP simulation speeds, a simulation with very few rollbacks is allowed. This migration policy allows the simulation process to be balanced, and to obtain better performance with respect to consumed CPU time and simulation time. Furthermore, such a communication path shows good response to situations of dynamic change in simulation behavior. The topology can be used to identify the maximum STW performance obtained under a set of specific conditions.

Figure 1 shows improvements of 70% in the CPU time, 65% in simulation time and 80% in rollback number generated in large simulations (GVT > 10000). Another interesting characteristic that can be taken from these results is that STW obtains better results than TW when the problem size is increased.

Tree

The tree topology is very similar to the previous one, and allows a considerable reduction in rollback number (if LP simulation speeds are tuned correctly). The main difference in this topology is that it has two independent communication paths and it is necessary to synchronize the control heuristic for both paths.

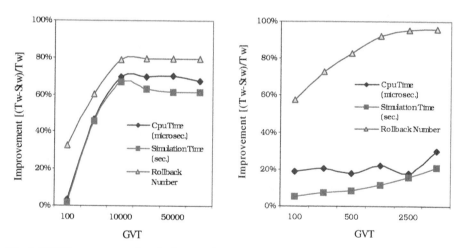

Fig. 1. STW Improvement for pipeline topology **Fig. 2.** STW Improvement for tree topology

The results obtained (fig. 2) show a spectacular reduction in the number of rollbacks, (in some cases, 95%), which does not correspond to a similar reduction in simulation time. This effect emerges when the STW attempts to reduce the rollbacks generated by the two tree communication paths, generating an imbalance in LP priorities, and therefore in the quantity of the CPU that is received.

Ring

The ring topology is especially interesting for our STW evaluation, as it models a classic parallel and distributed simulation behavior: logical communication loops between LPs. This case is the most unfavorable, since it can produce rollback chains. However, the results obtained (fig. 3) show, as with the previous cases, better STW performance with respect to the Time Warp.

Comparing the results, we can see that an excessive difference between reduction in rollback number, simulation time and CPU time does not appear in the ring topology.. The STW performance for this topology is more balanced for the different categories (29, 15% and 26%, respectively).

All with All

This topology is possibly the most realistic communication path from the point of view of the physical model, and is one that would most likely give greater problems to the STW optimism control policy. This topology allows all the logical processes to exchange events with all remaining LPs. It shows us the true potential of the STW heuristic in controlling simulation optimism.

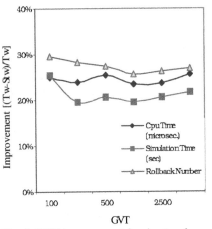
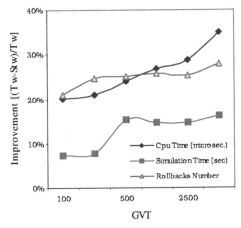

Fig. 3. STW improvement for ring topology   **Fig. 4.** STW improvement for all-with-all topology

It can be observed in the results (fig. 4) that improvement with respect to CPU time and rollback reduction is 26% and 30%, respectively, while the improvement in execution time is reduced to 18%.

Figure 5 shows that, for the three last topologies, STW has better execution times than Time Warp. In addition, it can be verified that while the GVT (simulation work) is increased, the performance obtained by the STW also increases (see fig. 1). Figure 6 shows rollback numbers for STW and TW. In all cases, STW reduces simulation optimism, and therefore a smaller rollback number is produced.

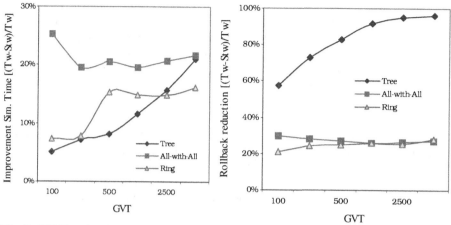

Fig. 5. STW Speedup for different toplogies   **Fig. 6.** STW Rollback Reduction

Figure 5 shows speedup obtained by the STW with respect to Time Warp (from 5% to 25%). However, this percentage grows when the final GVT is greater (see fig. 1). For the case of the pipeline topology with GVT=10000, improvement of up to 70% was obtained.

As conclusions to the STW validation, we would mention:

1. STW improves TW performance for all topologies: As has been demonstrated in the previous figures, STW improves the simulation result independently of LP communications paths. This characteristic is very important in terms of optimism control methods, because LP communication paths depend only on the simulation tool and physical model, and this can vary considerably from one application to other.

2. The reduction in rollback number does not necessarily imply an improvement in simulation execution time: an interesting conclusion from the accomplished tests is that a reduction does not imply an improvement of the same order of magnitude in simulation time. This effect is due to a non-balanced LP execution, as a consequence of priority management.

3. STW improves large simulation applications: performance obtained by STW is greater when the simulation application is greater; this is because STW heuristic control runs better with higher data quantity .

3.   Conclusions

STW mechanism controls excessive optimism in the optimistic PDES (as was verified in the previous real simulation environment). The mechanism allows the execution of optimistic PDES, reducing the high cost of state management (memory) and rollbacks.

The overhead introduced by STW is minimal and is fixed at about 2% of simulation execution time. This additional load is not very important since, as has been proved, STW obtains an improvement of 20% in most of the accomplished tests.

STW integration within a distributed simulation system is very easy. STW performance depends on the number of processes and their assignment within the distributed architecture.

The mechanism of priority management through the O.S allows better STW integration.

STW accomplishes a control of coarse grain optimism (repeated in time) and has certain problems with fine grain optimism (small interval optimism).

As future work we we would point to:
- The study of how STW & system load affects OS priority management.
- The study of STW behavior in terms of the number of processes assigned to each processor, and the analysis of load-balancing policies.
- The study of STW behavior with balanced simulations and the design of a new control heuristic for this case.

References

1. Jefferson, D., Sowizral, H. Fast Concurrent Simulation Using the Time Warp Mechanism. Distributed Simulation. SCS (1985) 63-69
2. Jefferson, D. Vistual Time. ACM Transactions on Programing. Languages and Systems. 7(3) (1985), 404-425
3. Sokol, L., Briscoe, P., Wieland. A. MTW: A Strategy for Scheduling Discrete Simulation Events for Concurrent Execution. SCS Multiconference on Distributed Simulation 19, 3. SCS. 1988. 34-42.
4. Ball, D., and Hoyt, S. The Adaptive Time-Warp Concurrency Control Algorithm. SCS 22 1. 1990. 174-177.
5. Martin, D., McBraayer, T., Radhakrishnan, R., Wilsey, A. WARPED, a Time Warp Parallel Discrete Event Simulator. Dept. of ECECS. University of Cincinnati, Cincinnati, OH, 1995.
6. Suppi, R., Cores, F., Luque, E. Improving Optimistic PDES in PVM environments. Lecture Notes in Computer Science 1908. EuroPVM-MPI2000. 2000. 304-312.
7. Serrano, M., Suppi, R., Luque, E. Parallel Discrete Event Simulation. STW: Switch Time Warp, Caos, University Autonoma of Barcelona, http://pirdi.uab.es/document/pirdi11.htm, 1999.
8. Chapman, J., Reiss, M. Ecology. Principles and Applications. 2nd edition. Cambridge University Press. 1999.

Application of MPI in Displacement Based Multilevel Structural Optimization

Craig L. Plunkett[1], Alfred G. Striz[1], and J. Sobieszczanski-Sobieski[2]

[1] University of Oklahoma, School of Aerospace and Mechanical Engineering,
Norman, Oklahoma 73019-1052
striz@ou.edu
[2] NASA Langley Research Center, Mailstop 139, Hampton, Virginia 23681-0001
j.sobieski@LaRC.NASA.gov

Abstract. The weight optimization of trusses using **D**isplacement based **M**ultilevel **S**tructural **O**ptimization (DMSO)[1-4] is investigated. In this approach, the optimization task is subdivided into a single system and multiple sub-systems level optimizations. The approach is considered to be efficient for large structures, since parallel computing can be utilized in the different optimization levels. A 240-element truss was tested for efficiency improvement in DMSO on both, a network of SUN workstations using the MPICH implementation of the Message Passing Interface (MPI) and on a faster Beowulf cluster using the LAM implementation of MPI. Parallel processing was applied to the subsystems level, where the derivative verification feature of the optimizer NPSOL was first used in the optimizations. Even with parallelization, this resulted in large runtimes. Then, the optimizations were repeated without derivative verification, and the results were compared. They were consistent and showed no need for derivative verification, giving a large increase in efficiency in the DMSO algorithm. They also showed that very little improvement in performance was obtained by parallelization once the derivative verification feature in NPSOL was turned off.

1   Introduction

Early uses of structural optimization were endeavored during World War II, when the need for high performance aircraft led to increased research in this area. Today, with the use of modern computers, structural optimization has become a more important consideration in structural design. For static structures, it is of interest to create an optimized structure. This produces a minimum weight structure and, thus, a minimum material structure, which lowers the cost of producing the structure. For mobile structures, an optimal design not only helps to lower the cost of building the structure, it also helps to lower the operating cost. Therefore, structural optimization continues to be an important area of research in transportation. In an effort to design the lightest and most efficient structure for a given task and vehicle, constraints from many different disciplines must be considered, such as aerodynamics, controls, cost, etc. The development of a design that simultaneously satisfies all of these criteria is often

Y. Cotronis and J. Dongarra (Eds.): Euro PVM/MPI 2001, LNCS 2131, pp. 335–343, 2001.

com-putationally complex. It is here that **M**ultidisciplinary **D**esign **O**ptimization (MDO)[2,3] can help by providing mathematically based design tools to obtain a minimum weight structure that satisfies the multifaceted constraints of multiple disciplines.

2   Displacement Based Multilevel Structural Optimization (DMSO)

One approach to improve the MDO of vehicles through more efficient structural optimization is Displacement Based Multilevel Structural Optimization (DMSO). In this approach, the optimization task is subdivided into a single system and multiple subsystems level optimizations. The system level optimization minimizes the load unbalance resulting from the use of displacement functions to represent the displacements of the structure. Here, the function coefficients are the design variables of the system level. The system level can also be solved using the displacements themselves as design variables in the optimization, or by replacing the system level optimization with a standard finite element analysis using, e.g., Gauss elimination. All approaches ensure that the calculated loads match the applied loads. In the subsystems level, the weight of the structure is minimized using the element dimensions as design variables.

3   Processing

A parallel computation model[5] is a conceptual view of the types of parallel operations that are available to a program. For the present approach, the message-passing model was used. One advantage of the message-passing model is that it fits well on separate processors that are connected to a communications network. Therefore, it can be used on super-computers as well as on networks of workstations or PCs.

3.1   MPI Implementations

In this project, the **M**essage **P**assing **I**nterface (MPI)[5] was chosen for the required parallel communications library. MPI is attractive since it can utilize like machines as a parallel network. Its only drawback is that, to date, most available implementations do not allow for its use on heterogeneous networks, since they were generally designed for use on specific computers. An effort is underway to produce an interoperable MPI standard, which would allow MPI implementations from different vendors to inter-operate and to run on heterogeneous computer installations.

MPICH 1.2.1, an implementation of MPI, was chosen for this research because it could easily be implemented on the four SUN Ultra10 SPARC workstations available in the University of Oklahoma's Computational Mechanics Research Laboratory which is associated with the OU Center for Engineering Optimization. These processors run at 300 MHz with 512 MB of memory each. They are connected through a University LAN on a local subnet.

Further, the LAM 6.2b implementation of MPI was utilized on the Pentium processors of the Beowulf cluster at the Institute for Computer Applications in Science and Engineering (ICASE) at NASA Langley Research Center, where the first author spent a summer. At the time, this cluster was front-ended by a dual CPU 400 MHz Pentium II server with dual CPU 500 MHz Pentium III file servers. It had 32 single CPU 400 MHz Pentium II nodes and 16 dual CPU 500 MHz Pentium III nodes for a total of 64 processors. The Pentium IIs had 384 MB of 100 MHz RAM and 6.5 GB hard drives, each. The Pentium IIIs had 512 MB of 100 MHz RAM and 14.4 GB hard drives, each. Only Pentium II nodes were used in this project due to better availability.

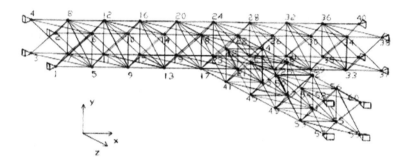

Fig. 1. 240 Bar Three-Dimensional Truss Model

4   240 Element Truss Optimization

The model studied was a large, three-dimensional truss structure, consisting of 240-elements and 60 nodes. It is shown in Figure 1. This model was specifically chosen to investigate any potential efficiency improvements obtainable in DMSO from the application of parallel processing. For this purpose, the subsystems level optimizations had to be easily divisible into sections. Since the weight of each element in the truss only depends on the element force and the element dimensions, the weight minimization of each element is a separate optimization problem. Therefore, the truss can be divided into any number of arbitrary element groupings. For the present investigation, it was divided into a number of sections equal to the number of processors used, each with an equal number of elements. The elements were selected into the sections only by order of the element numbers.

4.1   Optimization with Derivative Verification

Parallel processing in the form of MPICH on the SUN workstations and LAM on the Beowulf cluster as described above was utilized in the sub-systems level optimizations. All programs were coded in FORTRAN77. NPSOL[6] was the optimizer. The code utilizes the Sequential Quadratic Programming (SQP) algorithm in the optimization process. By default, NPSOL checks the user provided gradients by finite difference approximations. This feature was left operable for the initial set of runs. Also, for the 8 processors runs on the SUN network, 2 virtual processors per actual computer were used.

All cases converged to the same element dimensions and structural weight. The run-time improvements achieved by parallel processing on the SUN and Beowulf systems are given in Table 1. The second and third columns state the *average individual sub-systems level optimization runtimes* for the Sun network and the Beowulf cluster, respectively. The fourth and fifth column give the respective *times to converge the entire optimization* for the case where the displacement function coefficient are used as the design variables.

Table 1. Runtimes Observed in 240-Element Truss Optimization
(Displacement Function Coefficients as Design Variables, Derivative Verification On)

Number of Processors	Avg. Individual Sub-Systems Level Runtimes SUN	Avg. Individual Sub-Systems Level Runtimes Beowulf	Time for Total Optimization (hr) SUN	Time for Total Optimization (hr) Beowulf
Sequential	1.35 hr	28.69 min	32.72	8.89
1	1.36 hr	29.17 min	32.76	9.38
2	4.54 min	1.05 min	16.12	3.10
3	55.3 sec	12.48 sec	15.02	2.91
4	18.1 sec	4.03 sec	15.00	2.87
5	-	1.56 sec	-	2.86
6	-	0.68 sec	-	2.84
8	2.71 sec	0.24 sec	14.83	2.86
10	-	0.23 sec	-	2.94

For this case, the *average individual subsystems level optimization runtimes* dropped significantly with an increase in the number of parallel processes, specifically, from 1.35 hr for 1 processor to 2.71 sec for 8 processors for the SUN network and from 28.69 min to 0.235 sec for the Beowulf cluster. The *overall optimization time* decreased by > 50% for 8 parallel processors for the SUN system and by about 70% for the Beowulf cluster. Overall, the Beowulf cluster experienced considerably shorter runtimes than the SUN network for the overall optimization, by 70 - 80 %. For both systems, the time to completion stayed about equal beyond 3 processors. This would imply that the system level optimization is driving the execution time, and any further increase in the subsystems parallelism beyond 3 processors will not contribute significantly.

In order to investigate the influence of the respective design approach on the *average individual system level runtimes* and, thus, on the overall optimization runtimes, three different cases were run on both parallel systems: the displacement coefficients as design variables, the displacements as design variables, and a simple Gauss elimination analysis. The SUN results are given in column 2 of Table 2 and the Beowulf results in column three. For the 240-element problem investigated here, the simple Gauss elimination approach gave by far the fastest results. Again, the Beowulf cluster outperformed the SUN network by similar margins as in the previous case.

Table 2. Average Individual System Level Runtimes for Various Design Approaches (Derivative Verification On)

System Level Formulation	Avg. Runtime SUN	Avg. Runtime Beowulf
1a. Displacement Function Coefficients (Sequential)	1.16 hr	12.35 min
1b. Displacement Function Coefficients (Parallel)	1.15 hr	13.14 min
2. Displacements	2.71 min	28.60 sec
3. Gauss Elimination	0.292 sec	0.066 sec

The equivalent *overall runtimes* for the design approaches 2 and 3 are presented in Tables 3 and 4 with SUN and Beowulf comparisons in each table. Parallel processing again provided tremendous improvements. It can be seen that both cases ran faster than the displacement function case. The Gauss elimination approach showed the best overall performance. Again, the Beowulf cluster outperformed the SUN network by margins of 3:1 to 10:1. For the optimization case with the displacements as design variables, there was no real improvement in the overall runtime when more than about 4 processors were used. Again, the system level optimization was driving the code's performance (see Table 2). When a Gauss elimination analysis was used to solve the system level, however, continuous improvement in overall runtime could be seen for an increase in processors, since the analyses were considerably faster than the equivalent optimizations for the problem size under discussion. Only for very large problems requiring the solution of very large matrices is this trend expected to reverse.

4.2   Optimization with Derivative Verification

Even with parallel processing, however, the runtimes were prohibitively long for all approaches except the Gauss elimination for this relatively simple problem. Thus, the derivative verification feature in NPSOL was turned off to see if efficiency could be improved. The resulting code was run for the different system level design-variable alternatives on the SUN network and the Beowulf cluster. The *average individual system level runtimes* without derivative verification for the different system level formulations are given in Table 5. The SUN results are given in column 2 of this table and the Beowulf results in column three. All previous trends hold. The displacement function coefficient approach represents the most complex formulation and requires the largest execution times. The displacement formulation is more than an order of magnitude faster. The Gauss elimination analysis gives essentially the same numbers as before since derivative verification does not apply for this approach. It is by far the fastest approach for this problem size. Again, the Beowulf results were faster by factors of 5 to 6.

Table 3. Runtimes Observed in 240-Element Truss Optimization (Displacements as Design Variables, Derivative Verification On)

Number of Processors	Time for Total Optimization SUN	Time for Total Optimization Beowulf
1	18.22 hr	6.60 hr
2	1.56 hr	20.02 min
3	49.83 min	9.02 min
4	39.86 min	7.01 min
5	-	6.41 min
6	-	6.36 min
8	35.87 min	6.25 min
10	-	6.13 min

Table 4. Runtimes Observed in 240-Element Truss Optimization (Gauss Elimination for System Level, Derivative Verification On)

Number of Processors	Time to Complete Optimization SUN	Time to Complete Optimization Beowulf
1	17.66 hr	5.93 hr
2	59.08 min	13.21 min
3	12.19 min	2.64 min
4	4.05 min	52.42 sec
5	-	21.28 sec
6	-	10.25 sec
8	44.62 sec	4.61 sec
10	-	3.11 sec

For the displacement function coefficient formulation without derivative verification, the *average individual sub-systems level optimization runtimes* together with the respective *overall times to converge* for the entire optimization are given in Table 6. The Sun network results are displayed in columns 2 and 4 and those for the Beowulf cluster in columns 3 and 5, respectively. It is clear from the results that, without derivative verification, the system level optimizations were driving the overall performance of the approach. Although the subsystems level optimization runtimes still decreased with increasing parallelization, they had become so short that there was virtually no improvement from parallel processing on the overall optimizations. The SUN network took consistently 6 times longer to finish the optimization than the Beowulf cluster.

Table 5. Average Individual System Level Runtimes for Various Design Approaches
(No Derivative Verification)

System Level Formulation	Average Runtime SUN	Average Runtime Beowulf
1a. Displacement Function Coefficients (Original)	31.39 min	4.59 min
1b. Displacement Function Coefficients (Parallel Subsystems)	30.34 min	5.11 min
2. Displacements	2.08 min	19.80 sec
3. Gauss Elimination	0.300 sec	0.067 sec

Table 6. Runtimes Observed in 240-Element Truss Optimization
(Displacement Function Coefficients as Design Variables, No Derivative Verification)

Number of Processors	Avg. Individual Sub-Systems Level Runtimes (sec) - SUN	Avg. Individual Sub-Systems Level Runtimes (sec) - Beowulf	Time to Complete Optimization (hr) - SUN	Time to Complete Optimization (hr) – Beowulf
Sequential	17.45	2.27	6.87	1.0
1	19.52	2.85	6.91	1.12
2	2.64	0.332	6.68	1.12
3	0.989	0.148	6.56	1.11
4	0.539	0.073	6.44	1.11
5	-	0.058	-	1.11
6	-	0.043	-	1.11
8	0.295	0.038	6.39	1.10
10	-	0.022	-	1.10

The results for the displacement formulation showed very similar behavior. Here, the *overall optimization runtimes* dropped from 32 min to about 26 min for the SUN network cases as the number of processors increased from 1 to 8. The equivalent runtimes dropped from 4.4 min to 4.3 min for the Beowulf cluster runs, as the number of processors increased from 1 to 10.

For the case of the Gauss elimination analysis in the system level, parallelization did provide an advantage. For the SUN network, the *overall optimization runtimes* dropped from 3.84 min for 1 processor to 11.77 sec for 8 processors. For the Beowulf cluster, the drop ranged from 37.5 sec for 1 processor to 1.9 sec for 10 processors.

Even for this case, however, there was very little improvement beyond the use of 4 processors.

Overall, turning off the derivative verification feature in NPSOL provided anywhere from a 27.9% decrease in overall runtime for the SUN-network 8-processor case with displacements as design variables to a 99.8 % decrease for the Beowulf-cluster 1-processor case with Gauss elimination. This implies that parallelization gives improved results only for the optimization cases with derivative verification but contributes little to those without it for the present problem. It can be surmised, however, that a possible parallelization of the system level optimizations would improve overall performance

5   Continuing Research

The results of the truss optimizations indicate that parallel processing can be applied to the DMSO methodology as a means of increasing optimization efficiency, depending on the respective optimization formulation. Thus, we will first investigate the possible parallelization of the different system level optimization approaches.

Then, since more and more design and optimization work is expected to occur in distributed computational environments, we will attempt to perform DMSO on computers connected through the Internet as a parallel network. This will include the SUN workstations at OU and computers at other national and international institutions. To test how the connection speed between remote computers affects the overall optimization efficiency, we will also run these connections through Internet2. Finally, we will examine the feasibility of having a user at a secondary site interact with the program. This will consist of having the remote user change a problem parameter.

Acknowledgments

The second author would like to express his appreciation for sponsorship received from NASA Headquarters under EPSCoR Grant NCC5-171 and from the State Regents of Oklahoma under Grant NCHK000069. Also, thanks go to ICASE at NASA Langley Research Center for letting us use the Beowulf cluster and to Brian Cremeans for keeping our SUN network running smoothly.

References

1. A.G. Striz, T. Srivastava, and J. Sobieszczanski-Sobieski, "An Efficient Methodology for Structural Optimization", in: *Structural Optimisation,* Proceedings of the ACSO'98 - Australasian Conference on Structural Optimization, Sidney, Australia, February 11-13, 1998, Oxbridge Press, 1998, Victoria, Australia, pp. 259-266.
2. A.G. Striz, S. Sharma, T. Srivastava, and J. Sobieszczanski-Sobieski, "Displacement Based Multilevel Structural Optimization: Beams, Trusses, and Frames", Proceedings of the 7th AIAA/USAF/NASA/ISSMO Symposium on Multidisciplinary Analysis and Optimization, St. Louis, Missouri, September 2-4, 1998, pp. 670-680.
3. Missoum, S., Hernandez, P., Gürdal, Z., and Guillot, J., "A Displacement-based Optimization for Truss Structures Subjected to Static and Dynamics Constraints", Proceedings of the 7th AIAA/USAF/NASA/ISSMO Symposium on Multi-disciplinary Analysis and Optimization, St. Louis, Missouri, September 2-4, 1998, pp. 681-690.

4. A.G. Striz, C. Plunkett, and J. Sobieszczanski-Sobieski, "Parallel Processing on a Variant of Displacement Based Multilevel Structural Optimization," AIAA-99-1301-wip, 40th AIAA/ASME/ASCE/AHS/ASC Structures, Structural Dynamics, and Materials Conference, St. Louis, Missouri, April 1999.
5. *DOT User's Manual (4.20)*, Vanderplaats Research and Development, Inc., Colorado Springs, Colorado, 1995.
6. C.L. Plunkett, "New Developments in Displacement Based Multilevel Structural Optimization", M.S. Thesis, University of Oklahoma, Norman, Oklahoma, 2001.
7. P.E. Gill, W. Murray, M. Saunders, and M. Wright, *User's Guide for NPSOL (Version 4.0)*, Technical Report SOL 86-2, January 1986, Stanford University, Stanford, California.
8. W. Gropp, E. Lusk, and A. Skjellum, Using MPI: Portable Parallel Programming with the Message-Passing Interface, The MIT Press, Cambridge, Massachusetts, 1994.

Parallelization of Characteristics Solvers
for 3D Neutron Transport

Guang Jun Wu and Robert Roy

École Polytechnique de Montréal,
P.O. Box 6079, Station Centre-Ville, Montréal H3C 3A7, Canada
`wu@meca.polymtl.ca`, `roy@info.polymtl.ca`

Abstract. In this paper, recent advances in parallel software development for solving neutron transport problems are presented. Following neutron paths along their characteristics, these solvers use ray tracing techniques to collect the local angular flux components. Due to the excessive number of tracks in the demanding context of 3D large-scale calculations, reliable acceleration techniques are shown in order to obtain fast iterative solvers. The load balancing strategies include a global round-robin distribution of tracks, an approach where we forecast the calculation load implied by each track length and a macro-band decomposition where tracks crossing the same regions are grouped together. The performance of the PVM implementations for these various track distributions are analyzed for realistic applications.

1 Introduction

In recent years, the standard techniques for the deterministic solution of the multi-group transport equation have drastically changed. It is not still possible to solve the transport equation over the complex geometry of a whole power reactor core without various approximations. However, the exponential growth in capacity and power of computer resources enables the nuclear engineering community to enhance their computational models and to explore the limits of older models. Day-to-day simulations of three-dimensional effects for various critical configurations are now done using new parallel algorithms, and the activity of benchmarking these simulations either scrutinized in light of older approaches or in comparison with stochastic-based results needs an on-going industrial support. Here, we will resume one of the most flexible and well-organized algorithms known to perform such 3D state-of-the-art computations.

The simplified static transport equation for obtaining a flux ϕ from a source q is given by:

$$\boldsymbol{\Omega} \cdot \boldsymbol{\nabla}\Phi(\boldsymbol{r}, \boldsymbol{\Omega}, E) + \Sigma_t(\boldsymbol{r}, E)\Phi(\boldsymbol{r}, \boldsymbol{\Omega}, E) = Q(\boldsymbol{r}, \boldsymbol{\Omega}, E) \ , \tag{1}$$

where $\boldsymbol{r}$ is a spatial point in the domain D, $\boldsymbol{\Omega}$ is a solid angle and E is the energy. The parallelization obtained after distributing the energy variable in multi-group solvers, as well as the usual spatial domain decomposition methods, have already

Y. Cotronis and J. Dongarra (Eds.): Euro PVM/MPI 2001, LNCS 2131, pp. 344–351, 2001.

been developed a few years ago [1], and the spectrum of applications of similar techniques has been extended to other related fields such as radiography [2]. The aim of the current article is to explain the newest approaches based on the characteristics formulation of (1).

The recent interest in the development of characteristics solvers is sustained by many interesting features: [3]

1. As for collision probability and Monte Carlo methods, the characteristics formulation can be implemented in the context of general geometry, providing great flexibility with regard to the geometric domains that can be treated;
2. Because the aim of these solvers is basically to compute the angular flux on certain directions, the spectrum of applications can be easily extended to many closely related fields (shielding calculations, radiography, photon or charged-particle transport on complicated geometries);
3. The scalar flux (or other angular modes, such as neutron currents) is recovered by averaging all contributions of the angular flux in a region, thus the memory requirements are much less than with standard collision probability solvers where huge matrices have to be defined to collect region-to-region contributions.

This explains why nowadays development of transport methods for lattice or whole-core heterogeneous calculations has been focused around the differential formulation rather than the integral one. Within the framework of large scale deterministic solvers for particle calculations, the parallelization of characteristics solvers sustains the most active research groups all over the world.

Recently, domain decomposition techniques have been applied to large scale parallel transport calculations using the method of characteristics. The spatial decomposition of multi-assembly problems where neutron paths are directly linked together when crossing assemblies exhibit limited speedup (5.3 on 8 processors) on a shared memory Sun Enterprise4000 because of the tight coupling between the assemblies.[4] The angular decomposition technique, where directions are distributed among a set of processors, has also been implemented and tested on a COTS cluster composed of Pentium II 300MHz with 100Mbps bandwidth.[5] Speedups obtained using this angular decomposition are of the order of 3.7 for 4 processors, and 6.8 for 8 processors. These techniques were restricted to two-dimensional problems, but they give the trend that we are looking for when we apply our characteristics method to three-dimensional geometries.

2   Acceleration of Characteristics Solvers

2.1   Ray Tracing

Neutron paths are straight lines, and we generally neglect neutron/neutron interactions. Statistical solvers attempt to generate neutron histories using the following approximations: a neutron will collide at a certain point following the attenuation probability law in materials, and scattering direction after this collision is also randomly determined. In deterministic characteristics solvers, the ray tracing is quite different from what can be found in statistical solvers or in computer graphics applications. In these solvers, no secondary ray is followed; however, the primary rays must cover all angular direction in the geometric domain.

A characteristics line T (tracking line) is defined by its orientation (solid angle Ω) along with a reference starting point p for the line. A quadrature set of solid angles is selected in order to preserve the even flux moments and that the starting point p is chosen by scanning the plane π_Ω perpendicular to the selected direction Ω. In order to cover accurately a 3D domain with reflected boundary conditions (such as a Candu supercell, see Ref. [6]) we need $\approx 10^3 - 10^6$ such tracks depending on the spatial mesh dimensions.

For a chosen line $T = (\Omega, p)$, the collection of segment lengths L_k and numbers N_k for each region encountered along the line must be calculated (and can eventually be recorded in sequential binary files). Assuming isotropic input current at the external boundary and isotropic sources, a recursive calculation is necessary to compute the local angular flux along line T; this recursion is based on the following simple attenuation equation:

$$\phi_k = \phi_{k-1} e^{-\tau_k} + \frac{q_k}{\sigma_k}(1 - e^{-\tau_k}) \, , \tag{2}$$

where local sources are $q_k = \frac{Q_{N_k}}{4\pi}$, local total cross sections are $\sigma_k = \Sigma_{N_k}$ and the total optical path when crossing the region N_k are $\tau_k = \Sigma_{N_k} \times L_k$. Reciprocity relations allow us to solve concurrently for direction Ω and for the inverse direction $-\Omega$ using the same line; however, the input currents are not the same at both ends. The scalar flux (i.e. the isotropic component of the angular flux) in region j can be recovered by a reductive operation over all lines:

$$\Phi_j = -\frac{1}{4\pi \Sigma_j V_j} \sum_T w_T \sum_k \delta_{jN_k} \Delta\phi_k + \frac{Q_j}{\Sigma_j} \, , \tag{3}$$

where δ is the Kronecker symbol and $\Delta\phi_k = \phi_k - \phi_{k-1}$. Equations (2) and (3) are the basis of all characteristics solvers which then works the following iterative way:

1. Guess new fission sources and incoming currents (called the outer loop).
2. Guess scattering sources (called the inner loop).
3. Compute the flux map for each energy group:
   (a) compute local solutions for each characteristics line;
   (b) apply local acceleration techniques;
   (c) perform a reductive sum of all contributions to the flux moments;
   (d) apply global inner acceleration techniques.
4. If flux map are not converged, go to 2.
5. Apply global outer acceleration techniques.
6. Compute critical factors for next neutron generation.
7. If not converged, go to 1.

Once the characteristics lines have been generated, the local solutions can be computed concurrently. The global solution is a succession of calls to these local solvers, and the sequential CPU time depends on the number of lines multiplied by the mean calculation time on one line and the number of iterations which can be reduced by various acceleration techniques.

2.2 Calculation Load on a Characteristics Line

The number of floating point operations necessary for computing the local angular fluxes on a single track containing K segment lengths gives us an estimate of the calculation load:

$$t_T = 2t_\times + 2t_+ + K(t_e + 5t_\times + 7t_+) \ , \tag{4}$$

where $t_T, t_+, t_\times$ and t_e are the total time for the track, the time for an addition, a multiplication and an exponential evaluation. As the dominant term is the exponential evaluation, our 3D characteristics solver can use linear tabulations for evaluating the exponential terms. For strong absorbers or simply for the most accurate results, the code can switch to exact exponential evaluations once convergence of the iterative scheme is sufficient. In all cases, the calculation load is a linear function of the number of segments; this behavior will be important when explaining our load balancing algorithms.

2.3 Reducing the Number of Iterations

A local acceleration technique to reduce the number of iterations in multi-group characteristics solver is the so called *self-collision rebalancing technique*.[6,7] Linear in the number of regions, this technique requires a one-step *sweep* of all tracks in order to collect the local collision terms of each region to itself. These collision probabilities are kept in an array of the same size as an usual flux map, and are then used at every iteration to rebalance the flux. Other inner/outer rebalancing schemes and variational acceleration with dynamically-computed relaxation factors are also applied.

2.4 Reducing the Number of Tracks

An important requirement to obtain accurate transport solutions is to preserve the particle conservation laws in the domain. In order to conserve sources, the segment lengths are renormalized for each angle so that the analytic values of volumes for all regions are the same as the numerical values obtained by integration. In fact, the number of characteristics lines necessary to obtain an accurate covering (with normalization factors near one) can be huge. As explained earlier, the ray tracing generally proceeds by covering uniformly the plane perpendicular to the direction. A lot of these tracks will be very similar one from another, so that some techniques were designed for reducing their number.

Table 1. Mean number of characteristics line per processor.

Density (lines/cm^2)	Type of merging	Number of processors		
		1	4	16
2.5	No merge	68391	17098	4274
	TMT-1	29446	7362	1840
	TMT-2	5703	1426	356
10.0	No merge	275262	68816	17203
	TMT-1	69801	17450	4363
	TMT-2	7092	1773	443

Track Merging Technique. Assume that two successive tracking lines cross exactly the same regions with the same direction. Segment lengths can be slightly different: for a reference length L, one track may have length $L + \epsilon$ while the other has $L - \epsilon$. It was shown that the two tracks can be merged together with $O(\epsilon^2)$-order of error on the local angular flux. Moreover, this can be done without changing the accuracy of the numerical volumes. Two different forms of track merging are available in our characteristics solver:[6]

 - TMT-1: lines can be merged at the same time as they are generated by scanning one dimension of the perpendicular line;
 - TMT-2: lines are sorted and reduction is done on both dimensions of the perpendicular plane.

Table 1 shows that the more the density of lines is increased, the more the tracking merge options are efficient.

Macro-band Grouping. Another step in attempting to group together characteristics lines having the same behavior is the *macro-band grouping.*[7] Using the concept of a macro-band, we identify tracks specifically by the region numbers that they are crossing. This classification encompasses two phase attributes: the spatial regions that are crossed plus the set of directions that allow this crossing sequence. Once the finite set of these attributes has been identified, it is possible to define an almost perfectly vectorized process for each such macro-band.

3   Performance of Message-Passing Algorithms

The MCI solver have been included as a module of the DRAGON lattice cell code. This solver uses an hostless model based on the PVM library.[8] Parallel calculations were performed on a non-dedicated cluster of Pentium-III 450MHz limited by 100Mbps Ethernet cards. Since this laboratory was also available for login to undergraduate students, the tests were run several times (mostly during weekends and Christmas holidays), so that workloads are not affected by sequential jobs. The total number of tracks N_T was distributed on the processors. The computation time on processor p is:

$$t_{\text{comp},p} = N_G n_p [2t_\times + 2t_+ + \bar{K}_p(t_e + 5t_\times + 7t_+)] \tag{5}$$

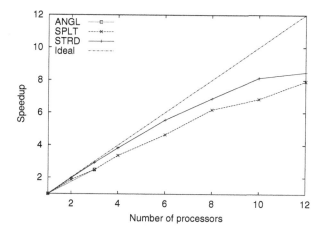

Fig. 1. MCI speedups with TMT1 merging (69801 tracks).

where $\bar{K}_p$ is the mean number of segments for the n_p characteristics and N_G is the total number of energy groups. At each iteration step, a multicasting communication process is used to recover the partial contributions to the angular fluxes; each processor has its own copy of the flux and source arrays with the same consistent ordering for unknowns (regions and energies). One interesting feature of this multicasting process is that it is independent of the number of tracks and it requires a total communication time of:

$$t_{\text{comm}} = N_P(N_P - 1)[t_{\text{startup}} + N_R N_G t_{\text{data}}] \tag{6}$$

to complete any inner-loop iteration on N_P processors. The 3D problem is composed of two horizontal bundles (a bundle is a cluster arrangement of rods containing Uranium oxide, the standard nuclear fuel used in nuclear reactors) with a vertical absorber rod at the center of the assembly. There is a 1/8 symmetry and angular quadratures EQ_2 and EQ_4 were used. Dimensions are $N_G = 89$ (number of groups), $N_R = 48$ (number of regions), and the number of iteration to converge is ≈ 20 for all cases described in this section.

3.1   Statistical Load Balancing

These options are statistical in that they rely on the uniform distribution of number segments when grouping tracks. Three such options have been developed:

- ANGL: all tracks of the same direction are grouped together; the number of directions must be equal to the number of processors;
- SPLT: subsets of N_T/N_P tracks respect the order in which these would be sequentially generated;
- STRD: each track is given in a round-robin fashion to each processor, so that track i is on processor $i \bmod p$.

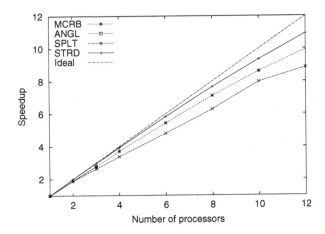

Fig. 2. MCI speedups with no track merging (275262 tracks).

Figure 1 gives the speedups observed when using these options with up to 12 processors. The last option (STRD) seems the most efficient to statistically preserve the load balance; as the TMT1 merging technique was used, the number of tracks is low so the efficiency is still low.

3.2 Deterministic Load Balancing

A single option was coded using a deterministic approach:

– MCRB: the number of segments of each track is taken into account as a weighting factor for distributing tracks; each processor receives an equivalent total weight based on the calculation load.

This represents the most uniform load balancing. Figure 2 shows that speedups are nearly linear for this deterministic balancing and with the statistical STRD option. The MCRB option is still under investigation, and the dispatching algorithm based on (5) does not seem to provide the best equilibrium. In this last case, there was no track merging, so the number of tracks is large enough to give us a good efficiency of the processors. Figure 3 gives the computational and communication times for 12 processors in the STRD option.

4   Conclusion and Perspective

Characteristics solvers now offer a fast and accurate solution to various transport problems. Implemented with efficient parallel computation algorithms, this approach can also handle heterogeneous large scale problems. The decrease in the computational time strongly depends on the number of tracks in the domain. When this number is significantly larger than the number of regions, the computing time is dominant and parallelism is efficient.

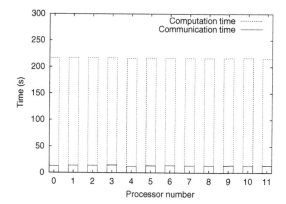

Fig. 3.   Computation and communication times for 12 processors calculation (STRD option, no track merging).

Next step would be to introduce modular ray tracing schemes based on pattern recognition for the macro-bands. Calculation along a characteristics is so simple that mobile agents could be defined to compute on a very limited view of the whole-core system, allowing these solvers to evolve on grid computers.

Acknowledgment. This work has been carried out partly with the help of a grant from the Natural Science and Engineering Council of Canada.

References

1. Qaddouri, A., Roy, R., Mayrand, M., Goulard, B., "Collision Probability Calculation and Multigroup Flux Solvers Using PVM", Nucl. Sci. Eng. **123**, 392 (1996).
2. Inanc, F., Vasiliu, B., Turner, D., "Parallel Implementation of the Integral Transport Equation-Based Radiography Simulation Code,", Nucl. Sci. Eng. **137**, 173 (2001).
3. Roy, R. "The Cyclic Characteristics Method with Anisotropic Scattering," Proc. Mathematics and Computation ANS-M&C'99, September 1999.
4. Kosaka, S., and Saji, E., "The Characteristics Transport Calculation for a Multi-Assembly System using Neutron Path Linking Technique," Proc. Mathematics and Computation ANS-M&C'99, September 1999.
5. Lee, G.S., Cho, N.Z., and Hong, S.G., "Acceleration and Parallelization of the Method of Characteristics for Lattice and Whole-Core Heterogeneous Calculations," Proc. of PHYSOR 2000, may 2000.
6. Wu, G.J. and Roy, R., " A New Characteristics Algorithm for 3D Transport Calculations," Proc. 19th Canadian Nucl. Soc., October 1998; see also "Self-Collision Rebalancing Technique for the MCI Characteristics Solvers," Proc. 20th Canadian Nucl. Soc., September 1999; see also "New Development of the 3D Characteristics Solver in DRAGON," Proc. 21st Canadian Nucl. Soc., October 2000.
7. Wu, G.J., Ph.D. dissertation, École Polytechnique de Montréal, June 2001.
8. Geist, A., Beguelin, A., Dongarra, J., Jiang, W., Manchek, R., Sunderam, V., "PVM 3 User's Guide and Reference Manual", Report ORNL/TM-12187, Oak Ridge National Laboratory (1994).

Using a Network of Workstations
to Enhance Database Query Processing Performance

Mohammed Al Haddad and Jerome Robinson

Department of Computer Science, University of Essex,
Wivenhoe Park, Colchester,
CO4 3SQ, United Kingdom
mjalha@essex.co.uk

Abstract. Query processing in database systems may be improved by applying parallel processing techniques. One reason for improving query response time is to support the increased number queries when databases are made accessible from the Internet.

1. Introduction

Applying parallel processing techniques, like Parallel Query Processing in database systems, will improve the database query answering time, and hence the overall response time of a query. The need for this improvement has become apparent due to the increasing size of the relational database as well as the support of high level query languages like SQL which allows users to present complex queries.

Commercially available Parallel Processing servers are expensive systems and do not present a viable solution for small size businesses, therefore we are interested in trying to find alternative parallel processing methods. Such method as described in this paper is by the utilization of a network of workstations. It has been observed that, up to 80% of workstations are idle depending on the time of the day [1].

Parallel Virtual Machine is software that allows utilization of networked workstations as a single computational resource.

We present in this paper the effective use of Parallel Virtual Machine in enhancing the performance of Parallel Query Processing, with the use of a proposed Parallel Query Interface PQI, which is explained in section 4.2. We also demonstrate the cost effectiveness of the proposed Expandable Server Architecture ESA, which uses the shared nothing architecture. The shared nothing architecture is a relatively straightforward to implement, and more importantly has demonstrated both scalability and reliability. It has also proved to be a cheap way of connecting processors to build parallel database system that could be affordable to small businesses in contrast to the high cost of the most commercially available parallel database systems.

2. Parallel Architecture

Parallel database systems constitute a combination of database management and parallel systems to increase performance and availability. Parallel systems provide much better price / performance than their mainframe counterparts. These advantages

Y. Cotronis and J. Dongarra (Eds.): Euro PVM/MPI 2001, LNCS 2131, pp. 352–359, 2001.

could be categorised as follows: ***high performance*** - obtained by operating system support, parallelism and load balancing, ***high availability*** - obtained by data replication, and ***Extensibility*** – obtained by smooth expansion of processing and storage power to the system [5].

2.1   Parallel Database System

A parallel database system can be built by connecting together several processors, memory units, and disk devices through a network [2]. Depending upon how the different components are interconnected they can be categorized under three main classes [3]: shared-nothing, shared-disk, and shared-everything. Each architecture consists of processors, memory units, disk units, local buses, and a global interconnections network. In the experiments described in this paper, the shared-nothing architecture is used, where all processors are connected via Ethernet interconnection network.

Because none of the components in the shared-nothing architecture are shared, the need for a complex interconnection network is removed. In contrast, the shared-memory and shared disk systems needs a powerful interconnection network because it has to transmit large quantities of data. The shared-everything and shared-disk architectures are not suitable for large systems due to the interconnection bandwidth limitations. The shared-nothing architecture is popular for its scalability and reliability.

3.   Parallel DBMS Techniques

A successful implementation of parallel architecture heaviliy relies on the parallel techniques used. This research mainly focuses on two main components, namely *partitioning* and *query parallelism techniques*.

3.1   Partitioning

In relational database schema, relations (Tables of data items) can be fragmented into a number of parts stored on different disks associated with different processors. The three most common ways of partitioning the data are: ***Hash partitioning:*** This method is suited for sequential and associative access as the tuples will be placed amongst different fragments using some hash functions mapped on an attribute or attributes of the relation. The advantage of this strategy is that the scan will be directed to only one workstation's disk instead of all of them. ***Range Partitioning (clustering):*** Tuples with the same value of attribute and are frequently accessed are placed together. This method suits sequential and associated access, but clustering can lead to a few problems during that access, such as data skew. ***Round Robin:*** Tuples are distributed in a round robin fashion. The round robin strategy works well when majority of the queries involve the whole relation to be accessed sequentially. But, bottleneck may be created when tuples have a common value for a particular attribute.

Relational data base schema has a number of tables, in this experiment the partitioning was created by allocating each table into a different node, as can be seen in section 4.1.

3.2   Query Parallelism

Query parallelism can be obtained by two main approaches, inter-query parallelism and intra-query Parallelism. Inter-query parallelism enables the parallel execution of multiple queries generated by concurrent transactions, and aims to increase the transactional throughput, while intra-query Parallelism aims to decrease response time. Intra-query parallelism could be composed by using *inter-operator* which is obtained by executing in parallel several operators of the query tree on several processors, and *intra-operator,* where the same operator is executed by many processors, each one working on a subset of the data.

In our experiment, a Parallel Query Interface (PQI) has been designed to facilitate the use of both of the approaches explained above, where a series of user's queries can be served at the same time. Each query can then be decomposed to sub-queries and executed concurrently, as explained in section 4.2

4   Experimental Environment and Results

In this experiment, Parallel Query Interface PQI was designed, and applied on the proposed Expandable Server Architecture. Different scenarios were produced (Central database on a single workstation with PQI and without PQI) in order to measure the performance of the query.

In this experiment, the TPC-D benchmark databases sets and their queries were used as well POSTGRES. The complexity of these queries are explained in [4]. POSTGRES [8] is an extended relational database system still being developed at Berkeley. It is well suited for handling massive amounts of data, it also supports large objects that allow attributes to span multiple pages and contains a generalized storage structure that supports huge capacity storage devices as tertiary memory [9].

The commercial parallel systems are very expensive, thus in this experiment a *Virtual Parallel Machine* (PVM) is used. This provides a cost-effective solution for small businesses.

In this experiment, a group of six workstations PIII 450 with 128 MB RAM, 20 GB hard disk linked with 100 MB/s network are connected via a LAN, which provides better cost/performance.

4.1   Expandable Server Architecture (ESA)

The fundamental objective of database technology is to retrieve the data in as short a time as possible. This Architecture has been designed to accomplish that by utilizing the resources of any Local Area Network (LAN) such as a small business. This means that we are making use of all the workstations that are connected in the LAN and saving the small business, for example, from buying a multi processor server machine along with all its software tools. Using some or all those available resources in LAN, PVM is responsible for connecting them in order to perform any task in the best possible way. Performing any query in those selected resources in done using Parallel Query Interface (PQI), which uses a parallel mechanism to increase the performance and efficiency of retrieving the query. Also, by expanding the server, i.e., dividing up the database and spreading it over the LAN, we are speeding up the retrieval by accessing the data using the parallel mechanism for users' queries. Instead of loading up the main server and loosing time accordingly while retrieving any data.

This method reduces the communication overhead. As a result, there is a big performance benefit from reduced transaction latency and server workload. Our system promises much higher flexibility in distribution of workload among nodes because any node can access the data at equal cost.

4.2   Parallel Query Interface (PQI)

The main function of Query Processing is to transform a high-level declarative query into an equivalent lower-level procedural query. The transformation must achieve both correctness and efficiency [5]. The well-defined mapping from relational calculus to relational algebra makes the transformation correct, efficient and easy, but producing an efficient execution strategy is more involved. The lower-level query actually implements the execution strategy for the query. Since each equivalent execution strategy can lead to a very different consumption of computer resources, the main difficulty is to select the execution strategy that minimises resources consumption. On the other hand, our *Expandable Server Architecture ESA* and *Parallel Query Interface PQI* (Coordinator Process) has all the relations allocated in different sites, we simply send each sub-query into the corresponding site. The PQI algorithm used by the ESA is given in [16]. When a new user's query arrives, an arbitrary processing node receives it and becomes the coordinator in charge of optimization and supervision for this query. The coordinator first determines the degree of parallelism for the query by decomposing the query into sub-queries according to the join predicate and selection predicate. By determining the number of *Processing Nodes* (PNs) for scan and join as well as the number of disks that hold the buckets that are derived, and through message passing each operator can process the output of the previous one without delay.

The number of buckets is computed based on attribute value that has been retrieved from the corresponding disk and stored on the *table_info* at run time. When the buckets are processed separately, the coordinator will use the information from the *table_info* to command the PN, which holds the small buckets to sort them by Quicksort algorithm. It will then send them to the according process to be joined, using a binary merge algorithm, according to the join predicate, see figure 1.
When the size of the buckets exceeds the main memory of any PN, a delay will be caused in the processing of the query, as can be observed from section 5.

4.3   Experimental Results

In this experiment, query Q2 in the TPC-D benchmark, which is a correlated sub-query based on a 5-table join in both outer query and inner query [4], was performed in three different environments: ESA, Single workstation with imbedded Postgresql and without support of PQI and single workstation with support of PQI.

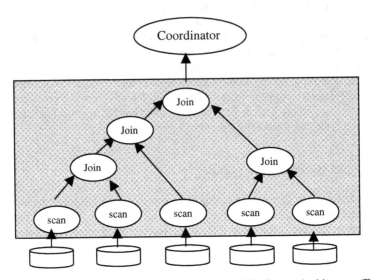

Fig. 1. Parallel Query Interface (PQI) applied in Expandable Server Architecture (ESA)

In order to effectively optimize queries in a distributed environment, it is necessary to have a reasonably accurate model that estimates the response time. Where the response time of a query is defined to be the elapsed time from the initiation of query execution until the time that the last tuple of the query result is displayed at the user's site. As can be seen from figure 2, the elapsed time obtained from ESA is 88 sec. while at the workstation with support of PQI it was 177 sec. and at the workstation without support of PQI it was 236sec.

In the ESA environment, the sub-queries were executed in parallel, which saved considerable time in comparison with the other environments. In the case where one workstation was used with the support of PQI, the sub-queries were executed sequentially and the time of each execution was added up to give the overall response time of the main query. Finally, the case of one workstation without the support of PQI, the time was spent on the query process trying to find the best strategy plan for the execution of the query and on the sequential access of the data.

5   Parallel Virtual Machine (PVM)

The PVM communication model assumes that any task can send a message to any other PVM task and that there is no limit to the size or number of such messages. While all hosts have physical memory limitations that limits potential buffer space, the communication model does not restrict itself to a particular machine's limitations and assumes sufficient memory is available.

The PVM communication model provides asynchronous blocking send, asynchronous blocking receive, and non-blocking receive functions. A blocking send returns as soon as the send buffer is free for reuse, and an asynchronous send does not depend on the receiver calling a matching receive before the send can return. There are options in PVM 3 that request that data be transferred directly from task to task. In this case, if the message is large, the sender may block until the receiver has called a matching receive. A non-blocking receive immediately returns with either the data or a flag that the data has not arrived, while a blocking receive returns only when the data is in the receive buffer.

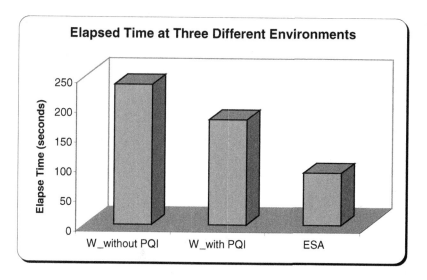

Fig. 2. Elapsed Time at three different environments; ESA, Workstation with support of PQI and Workstation without support of PQI

Message buffers are allocated dynamically. Therefore, the maximum message size that can be sent or received is limited only by the amount of available memory on a given host. There is only limited flow control built into PVM 3.3. PVM may give the user a can't get memory error when the sum of incoming messages exceeds the available memory, but PVM does not tell other tasks to stop sending to this host [6]. Figure 3, which represents the sending time for a range of data sets over different number of hosts. It shows that the curves start off at a large sending time when the data is sent to one host and then the time reduces as the number of hosts increases. The curves eventually level off when the data are sent to about 5 hosts. The reason for this phenomenon is that the size of the data when distributed over 5 machines fits totally in the buffer of the machines. This means time spent on page swapping, when large sets of data are dealt with, is saved and there is less traffic to each machine. This complies with the way PVM communication model operates as explained above.

6.   Conclusion

In this paper, PVM has clearly demonstrated its ability to use networked workstations as a single computational resource, hence exploiting parallelism.

Experiments measuring the scalability of PVM show that there is a limitation to the size of data that could be sent to one workstation. This limitation is caused by the restricted size of memory that could hold the data, which will increase the sending time as page swapping will take place. The sending time is reduced as the number of workstations is increased and the size of data per machine is decreased, which allows the data to be distributed and dealt with in a shorter time.

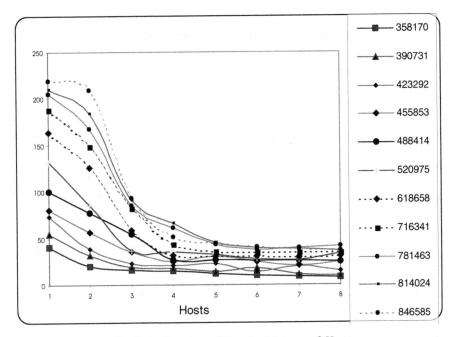

Fig. 3. Sending time of 11 sets of data over 8 Hosts

Moreover, PVM shows significant improvement in performance by applying the proposed designed Parallel Query Interface PQI in various scenarios, namely Expandable Server Architecture ESA, a central database (in a single workstation) with support of PQI, and a central database (in a single workstation) without support of PQI. The experimental results clearly show that query performance seems to work best with ESA. The Expandable Server Architecture (ESA) uses the shared-nothing architecture that is popular for its scalability and reliability in connecting its processors. It is a cheap way of building a parallel database system that could be affordable to small businesses in contrast to the high cost of most commercially available parallel database systems. The proposed PQI in this paper suggests several directions for future research, utilization of more workstations using PVM, enabling better load balancing and distributed partitioning.

Due to space restriction of this paper, details of it can be seen [16].

References

1. M. Mutka and M. Livny.: The Available Capacity of a Privately Owned Workstation Environment. Performance Evaluation, Vol. 12, No. 4 (July 1991) pp. 269-284
2. Sivarama P. Dandamudi and Gautam Jain.: Architectures for Parallel Query Processing on Networks of Workstations. Int. Conf. Parallel and Distributed Computing Systems, 1997.
3. D. Dewitt and J. Gray.:Parallel Database Systems: The Future of High Performance Database System Comm. ACM, Vol. 35, No. 6 (June 1992) pp. 85-98
4. Carrie Ballinger. Relevance of the TPC-D Benchmark Queries: The Questions You Ask Every Day. NCR Parallel System, www.tpc.org/articals/TPCDart1.97.html (1997)
5. M. Tamer Ozsu, Patrick Valduriez.: Principles of Distributed Database System. Second edition, ISBN 0-13-659707-6, Prentice Hall, 200.
6. A Geist, A Beguelin, J Dongarra, W Jiang, R Manchek and V Sunderam (1994), PVM.: Parallel Virtual Machine A users' guide and tutorial for Networked Parallel Computing.ed. J Kowalik, MIT Press (1994). Also available on-line:
http://www.netlib.org/pvm3/book/pvm-book.html
7. Michael J. Franklin, Bjorn Thor Jonsson and Donald Kossmann.: Performance Tradeoffs for Client-Server Query Processing, SIGMOD Conf. 1996: 149-160.
8. Michael Stonebraker and Greg Kemnitz.: The POSTGRES next generation database management system. Communications of ACM, 34 (1991).
9. Michael Allen Olson.: Extending the POSTGRES database system to manage tertiary storage. Master's thesis, University of California, Berkeley (1992)
10. Claire Mosher.: Postgres Reference Manual, version 4. Electronics Research Laboratory, University of California, Berkeley, CA-94720 (1992) No. UCB/ERL M92/85
11. S. Ganguly, A. Gerasoulis, and W. Wang.: Partitioning Pipelines with Communication Costs. Proc. 6th Intl. Conf. on Information Systems and Data Management (CISMOD'95), Bombay, India (November 1995)
12. M. Mehta and D.J.DeWitt.: Managing Intra-operator Parallelism in Parallel Database Systems. Proc. 21st Intl. Conference on Very Large Data Bases, 1995.
13. A.N. Wilschut, J. Flokstra, and P. M. G. Apers.: Parallel Evaluation of Multi-join Queries. Proc. 1995 ACM SIGMOD Intl. Conference on Management of Data, San Jose, California (May 1995)
14. Rahm, E.: Dynamic Load Balancing in Parallel Database Systems. Proc. EURO-PAR 96, Lyon (1996), Springer-Verlag, Lecture Notes in Computer Science 1123, S.37-52, 1996.
15. Scheneider, D. A., DeWitt, D. J.: A Performance Evaluation of Four Parallel Join Algorithms in a Shared-Nothing Multiprocessor Environmnet. Proc. ACM SIGMOD, Portland (1989)
16. Mohammed Alhaddad, Jerome Robinson, "Using a Network of Workstations to Enhance Database Query Processing Performance", Computer Science Department, Technical Report, CSM-342, June 2001.
ftp://ftp.essex.ac.uk/pub/csc/technical-reports/CSM-342.ps.gz

Towards a Portable,
Fast Parallel AP³M-SPH Code: HYDRA_MPI

Gavin J. Pringle[1], Steven P. Booth[1], Hugh M.P. Couchman[2],
Frazer R. Pearce[3], and Alan D. Simpson[1]

[1] EPCC, Edinburgh University, Edinburgh, UK
{G.Pringle,S.Booth,A.Simpson}@epcc.ed.ac.uk
[2] Department of Physics & Astronomy, McMaster University, Hamilton, Canada
Couchman@physics.mcmaster.ca
[3] Department of Physics & Astronomy, University of Nottingham, Nottingham, UK
Frazer.Pearce@nottingham.ac.uk

Abstract. HYDRA_MPI is a portable parallel N-body solver, based on
the adaptive P³M algorithm. This Fortran90 code is parallelised using
a non-trivial task-farm and two domain decompositions: a 2D cycle of
blocks and a slab distribution, using both MPI-1.1 and MPI-2 communications routines. Specifically, MPI_Put and MPI_Get are employed extensively in association with the communication epochs MPI_Fence and
MPI_Lock/MPI_Unlock. The 1D FFTW is employed. We intend to extend
the use of HYDRA_MPI to cosmological simulations that include Smoothed
Particle Hydrodynamics.

1 Introduction

The HYDRA N-body particle code [1] is employed in simulations of cosmic structure. It evolves an ensemble of particles, representing both a collisionless dark
matter phase and gas, under their mutual gravitational field. The gas phase is
modelled using Smoothed Particle Hydrodynamics (SPH). At the heart of this
simulation method lies the N-body problem for very large N.

Consider an ensemble of N particles, each with an associated mass. The
acceleration on each particle may be determined, via Newton's equations of
motion, in terms of the mass and position of every other particle (and, possibly,
periodic images), thus the overall computation is of $O(N^2)$. This is prohibitively
expensive for large N unless we employ a fast N-body solver. One such solver,
the AP³M method, lies at the heart of of HYDRA and is of $O(N \log N)$.

This decrease in complexity is achieved by separating the force contributions
to near- and far-range forces. The near-range forces are calculated using the
direct summation method. This is hereafter referred to as the particle-particle,
or PP, interaction. The far-range forces are approximated via a Fast Fourier
Transform, or FFT, wherein the masses are interpolated onto a regular mesh,
which is then convolved with an appropriate Green's function. The accelerations
are then calculated, in terms of this mesh, at each particle. This is hereafter
referred to as the particle-mesh, or PM, interaction. The combination of the

Y. Cotronis and J. Dongarra (Eds.): Euro PVM/MPI 2001, LNCS 2131, pp. 360–369, 2001.

two produces the PPPM, or P^{3}M, algorithm which was developed in the plasma physics community [4].

The P^{3}M algorithm is highly efficient for smooth distributions, however, the more clustered the distribution, the further the performance degrades as an increasing number of neighbours comes within the fixed short-range cutoff. These large density contrasts, a result of the gravitational instability, are frequently less severe in other fields, such as plasma physics. This shortcoming is eliminated by the introduction of high resolution, adaptive sub-meshes which are recursively placed in regions of strong particle clustering.

Within each sub-mesh, a smaller P^{3}M calculation is performed, thus propagating the efficiency of the FFT-based calculation to successively smaller scales. This technique is known as the adaptive P^{3}M or AP^{3}M, algorithm [2]. The AP^{3}M algorithm alone is used to calculate the accelerations, due to gravity, of N bodies. It is readily extended to include hydrodynamic forces using Smoothed Particle Hydrodynamics, or SPH [6].

This HYDRA code has been parallelised [7], using a data-parallel paradigm written for the Cray T3D using the CRAFT communications harness. A portable data-parallel version hydra is parallelised using OpenMP [8]. Further, there exists a message-passing version written for the Cray T3D using the SHMEM communications harness [5], however, this version does not support adaptivity.

This paper describes a new portable parallel version of HYDRA, namely HYDRA_MPI, based on [5], but written in Fortran90 with routines from the MPI-1.1 and MPI-2 communication libraries. Firstly, the SHMEM code was converted to MPI (see Sect. 7). Then the communications were adapted to allow for both periodic and isolated boundary conditions. Lastly, new code was added to cater for adaptivity. SPH was removed from the code so that we may now utilise the antisymmetry of the gravitational pairwise force, which reduces the PP computation by an approximate factor of 2. SPH will be reinstated in the near future.

The layout of this paper is as follows. Firstly, we describe the domain decomposition, then discuss the parallelisation of the PP and PM algorithms. The topic of parallel refinements is then introduced with a brief discussion of two types of refinements. The method of performing the AP^{3}M algorithm within a refinement is then discussed. Each refinement is processed within a task-farm, and this is presented along with details on how the associated particles' information is loaded and unloaded into new AP^{3}M cycles. Finally, we discuss the use of MPI-2 single-sided communication routines.

2   Domain Decomposition

The cubic simulation volume is divided into an array of 'chaining cells' which is used to delineate the cutoff between the near- and far-fields. These cells are distributed in the 2nd and 3rd dimensions over cyclic blocks of processors, forming a 2-dimensional cuboid. This is shown for a simple case below in figure 1 for 4 processors and a block size of 4 neighbour chaining cells. The distribution in the 1st dimension is degenerate - a full 3-dimensional decomposition was shown to be unnecessary in [5].

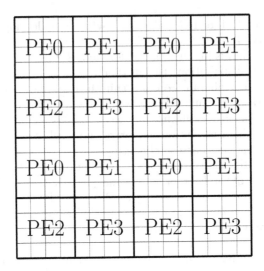

Fig. 1. Schematic of the domain decomposition in cyclic blocks for a chaining cell size of 16, a block size of 4 and a 2×2 logical processor grid

Whilst a regular domain decomposition would retain data-locality, the particles typically have a highly non-uniform distribution, thus this 'cyclic-block' domain decomposition improves load balance.

3   The Parallel AP³M Building Blocks

3.1   Parallel PP Algorithm

The short-range gravitational forces are calculated by performing the explicit particle-particle interactions up to a range of 1 chaining cell. The parallel PP algorithm employs 'ghost-cells', surrounding the local chaining cells, to contain the particles of neighbouring chaining cells which reside on other processors. The number of particles within the ghost-cells is unknown at every time step, therefore, the off-processor particle information is read for every relevant chaining cell. This information is then utilised to retrieve only the necessary particles.

Before any PP work begins, refinements may be placed if doing so reduces the cost of the PP cycle. If this is the case, then the acceleration is computed only on a sub-set of the particle ensemble, where the rules of exclusion depend on the type of refinement employed (see Sect. 4). Computing the acceleration on the excluded particles is deferred to the refinement stage.

3.2   Parallel PM Algorithm

Long-range gravitational forces are calculated using a particle-mesh method using FFTs. The widths of the 'FFT-mesh' and the chaining cells are related by a

constant of proportionality determined by the desired accuracy of the force calculation. The FFT-mesh and the chaining cell mesh are aligned at the boundaries,
with the size of the chaining cells being increased to force this alignment (this
can be done with no loss of accuracy, but at the expense of added PP work).

Within the PM algorithm, the mass of the particles is interpolated onto the
FFT-mesh, which is then redistributed from 2D to 1D across the processors.

The code described in [5] utilised a convolution routine which assumed periodic boundary conditions. A new parallel convolution routine has been written
which can compute both periodic and isolated boundary conditions. As in [5],
this involves building a 3D FFT out of 1D FFTs and packing off-processor *Real*
data into a local *Complex* data-structure. Further, a new, parallelised Green's
function, for isolated boundary conditions, has been added. This new Green's
function needs to be highly efficient as a different Green's function is required for
each refinement. It is computed in Fourier space utilising low frequency aliases
and its inherent 48-fold cubic symmetry.

A proprietary 1D FFT is employed in [5] which restricted the choice of FFT-
mesh to a power of 2. However, in HYDRA_MPI, this routine has been replaced
by the efficient and portable FFTW [9]. This dramatically extends the choice
of FFT-mesh which, in turn, allows for a wider and more optimal choice of
refinements.

Lastly, the convolved mesh must be redistributed from 1D back to 2D again,
before the accelerations can be evaluated at the particle's location.

3.3   Parallel Adaptivity

The PP calculation scales as the square of the number of local particles, which
becomes a problem as particle clusters develop. The AP³M code reduces the
amount of work by locating computationally expensive areas, namely these clusters, and replacing the chaining cells in these regions with refinements. These
refinements consists of another AP³M, but one with finer chaining- and FFT-
meshes and isolated boundary conditions. As the chaining cells are smaller they
will contain fewer particles and reduce the cost of the local calculation. This
process is repeated, as necessary, to add further levels of refinement - 'parent'
refinements can create 'child' refinements. The computation of the refinements
is parallelised using a task-farm (see Sect. 6).

The original serial code [1] places cubic refinements. Moreover, the refinement
placing routine was found to be intrinsically serial. A new parallelisable routine,
which places cuboid refinements is now employed.

4   Refinements

Consider a cuboid of parent chaining cells which will be excluded from the parent's PP work. The contribution to the acceleration on these particles, known as
the 'core' particles, is calculated within child refinements. In [1], any PP work
between the core and the remaining particles occurs at the parents level and if

the refinements abut, then the PP work at the common boundaries is performed at the parent level.

In HYDRA_MPI there are two types of refinement: 'traditional refinements', as in [1], and 'extended refinements'. For extended refinements, both the core particles and the particles which reside in the parent chaining cells that surround these core particles, known as buffer particles, are loaded. As with traditional refinements, all core particles are excluded from the parent's PP work, however, the forces on the core particles due to the buffer particles are now also excluded. Loading the buffer particles allows refinements to abut whilst excluding the expensive PP work at the parent level.

Under most circumstances, only the accelerations associated with the core particles are unloaded from the refinement's AP^{3}M. However, core particles for both the traditional and extended refinements can contain parental buffer particles (where some antecedent will have been an extended refinement). Therefore, for level one refinements, the 'unload region' is the same as the core region and for lower level refinements, the unload region is the intersection of the child core region and the unload region of the parent.

4.1   Refinement PP and PM work

Child refinements are placed immediately before their parent's PP and PM computations. Thus child refinements are started, and indeed could complete, before their parent has completed. This has implications for the unloading process (see Sects. 6.2).

During the refinement's PP work, the pair-wise force law is employed, even if one of the particles is not unloaded. This was found to be faster than avoiding force updates on buffer particles. Any excluded particles will be a child's core particles.

All the refinement particles contribute 'mass' to the refinement's FFT-mesh, however, only particles within the unload region evaluate the mesh contribution to their acceleration. The general convolution routine is now employed using isolated boundary conditions,

4.2   Selecting Refinement Regions

The parallel refinement-placing algorithm locates non-overlapping cuboid regions that form the core regions of the refinement. A brief description follows.

All refinements are placed based solely on the numbers of particles within the cyclic-block distribution of chaining cells. First we generate a per-chaining cell cost function that reflects the desirability of placing this cell in a refinement. Then, apply a threshold to the cost function values. All cells above a threshold are set to be refined away and are called 'target cells'. This is a varying threshold, determined using a simple search algorithm utilising a cost estimate returned by the refinement placing code. We now form unconnected clusters by tagging the target cells: this is achieved by a parallel 'flood fill'. A list of corner-point pairs, describing cuboid boxes containing each cluster, is then formed.

We now have the option to sub-divide some of these regions which are particularly long or where only a small fraction of the cells are actually target cells. Sub-dividing is performed by dividing up the longest dimension of the region and then shrinking the resulting sub-regions to fit the enclosed target cells. This process is repeated until there are no candidate regions left.

Lastly, we must clip the region boundary of child refinements against the core region of the parent, as any refinement may include parental buffer particles. Note that the parent core region does not necessarily correspond to chaining cell boundaries so we have to convert to real space coordinates at this point (see Sect. 5).

There are a number of user-defined parameters employed in refinement placing, i.e. the limit on the number of child refinements, the number of particles in refinements, number of processors, an upper limit on the size of the FFT-mesh, etc. As soon as it becomes obvious that the current threshold fails a constraint the placement is aborted and restarted with a new, higher, threshold value.

Lastly, all refinements are stored on a list of jobs, to be communicated around the task-farm (see Sect. 6). Each job consists of a single job `record`, which, essentially, is the list of parameters sufficient to describe a refinement, including, for instance, the pairwise force-softening parameters, mesh sizes and refinement boundaries.

5   A Global Integer Coordinate System

If we were to use Real-space parameters to designate the boundaries of the core region, we may find that rounding errors introduce a mis-match between the assignment to parent chaining cells and the child refinement boundaries. (Here, each refinement would convert particle coordinates from the parental coordinates. Particles would be assigned to chaining cells by an additional transformation to a coordinate system in yet another range.) Furthermore, two particles with distinct coordinates may map to the same location when the origin of a sub-mesh is moved.

This problem could be avoided by tagging particles with their refinement number when refinements are placed. However, we have followed a memory-saving methodology, where all refinements are loaded based on their position in the global coordinate system. When assigning a particle to chaining cells, the global coordinate system is converted to a fixed-point representation avoiding any mis-match due to different rounding errors.

6   Task-Farm

There are three main types of tasks, where each task is an AP³M cycle with isolated boundary conditions. There are tasks which require all of the processors, tasks which require only a single processor and tasks which require a fixed intermediate number of processors. These tasks are known as full-machine, single-PE and sub-machine tasks, respectively.

Tasks are currently limited to generate up to 200 further new tasks, where each new task requires the same number of processors or fewer. Thus, a sub-machine processor group can generate new sub-machine and single-PE groups, but it cannot generate further full-machine tasks.

The task-farm consists of a 'master' processor whilst all other processors are 'slave' processors - a conventional task-farm. The master holds a list of tasks to be done in the master_list. Each job is sent to the root slave processor and placed at the head of the slave_list. As the refinement is processed, any new jobs are stored in the slave_list. This list is then returned to the master after the refinements are placed and before the AP^{3}M cycle begins, thus reducing processor waiting time. All incoming job lists are appended to the master_list. Initially, both job lists are 200 elements long, however, the master_list is allocated more space if required - up to 1000 jobs.

The task-farm proceeds as follows. The base level refinement generates a list of tasks, which may consists of all three types. All of the full-machine tasks are performed first. Any new tasks are added to the master_list. Once no full-machine tasks remain, the machine is fragmented into disjoint processor groups. These groups constitute the task-farm, with of one master processor, a small number of sub-machine processor groups and the remaining processors are split into single-PE groups. Thus, for 16 processors, say, a typical configuration would be one master, two 4 processor groups and seven single-PE groups. The master processor then dispatches tasks to the relevant processor groups. All incoming jobs are appended to the master_list. Then, once no sub-machine tasks remain in the master_list, and it is guaranteed that no sub-machine groups will generate further sub-machine tasks, all the sub-machine groups are split into single-PE groups.

When the root slave processor receives a task, the root process loads a superset of the required particles then, once the refinement's AP^{3}M has completed, unloads the resultant, relevant acceleration contributions. This is described in more detail in the following section.

6.1 Loading Particles

Once a job has arrived, the root slave processor determines which base-level chaining cells overlap with the domain of the refinement. All particles within these base chaining cells are loaded into this root processor.

The particles are retrieved from their host processor using MPI_Get. Strictly speaking, only the mass and particle position need to be loaded. Loading particles naturally requires the use of MPI-2 single sided communications. The particles may belong to any of the processors in the global MPI communicator and the particle's host processor may be occupied with another refinement, therefore, the single-sided communications epoch: a shared MPI_Lock/MPI_Unlock, is employed. The particle mass and position are never updated during the task-farm so a shared lock is sufficient.

Once all particles from a particular processor are loaded, they are then sorted, according to the new chaining cells, and distributed across the slave processor group. All unnecessary particles are removed at this stage. This process is repeated for all the host processors which own relevant chaining cells. Once all the necessary particles have been sent to the relevant processor, the particles are then sorted and stored locally, relative to the new chaining cells.

6.2   Unloading Particles

Once the acceleration contributions have been calculated, the accelerations alone are accrued to the accelerations of the particles residing on their host processor. This is achieved using MPI_Accumulate.

In general, the particle distribution within a refinement does not match the base level distribution, thus we employ the MPI_Indexed_type which allows for the redistribution of particles on the fly.

As with the loading process, the host processors may be occupied with other refinements. Again we employ a shared MPI_Lock/MPI_Unlock for the communication epoch. The details of any new child refinements are added to the master_list as they are generated. This implies that a child refinement may complete before its parent and that contributions to the acceleration of a particular particle may come from several refinement levels - thus we require an exclusive lock rather than a shared lock.

7   Comments on the MPI-2 RMA Routines

As previously stated, the AP^{3}M part of HYDRA_MPI is based on the SHMEM code [5]. It was decided, during the early stages of code development, that the loading and unloading of refinements would require using the remote memory accessing, or RMA, routines of MPI-2, namely MPI_Put, MPI_Get, MPI_Accumulate and MPI_Fence and MPI_Lock/MPI_Unlock. This decision meant that the conversion of the SHMEM code to MPI (and MPI-2) was relatively straight-forward.

The global SHMEM barriers were either removed or replaced with barriers over subsets of processors. The barrier synchronised SHMEM calls were replaced by MPI_Put or MPI_Get calls within a MPI_Fence epoch. See [3] for further details. Indeed, many of the converted communications routines were re-used within refinements after some modifications. In principle, all of these communications routines could be replaced by conventional MPI-1.1 communications routines though this would take significantly longer than the conversion to MPI-2.

The loading and unloading processes, on the other hand, required the use of MPI_Lock/MPI_Unlock. However, not every parallel platform has these two routines within the locally available MPI libraries. This issue with portability has lead to an added feature which provides an alternative implementation of the load/unload operations. The 'Get' and 'Accumulate' operations are emulated by sending MPI-1.1 messages from the source processor to the target processor. An explicit routine, namely progress_fake_lock, must be called regularly by the target processor to poll for these messages and complete the operations.

Other than the loading and unloading operations a processor will only communicate with the master processor and other members of its own processor

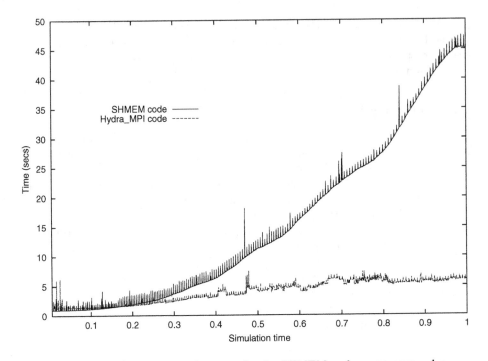

Fig. 2. Execution times per time step for the SHMEM and HYDRA_MPI codes

group. However, deadlock may occur if the target processor attempts to communicate with the source processor, waiting at a barrier, say, whilst the source is waiting for the target to respond to a 'Get' or 'Accumulate' operation. These deadlocks may therefore be avoided by ensuring that the communications calls in the task-farm master and the barriers at the end of the loading/unloading routines continue to call `progress_fake_lock` whilst waiting for other processors. In addition, once the barrier at the end of the loading/unloading routines is performed, communication within the processor group is known to be safe.

8   Conclusions

In the paper, we have briefly described an MPI version of the AP^3M algorithm. It is hoped, in the near future, to add an SPH module to a produce fast, portable, parallel version of HYDRA, i.e. a parallel AP^3M-SPH code.

During code development, the correctness of the code was always a high priority. To this end, an extensive and thorough regime of testing was undertaken, employing a varied number of cosmological simulations and numerical constructions. Furthermore, the code has been executed on a wide range of machines, namely the Cray-T3E, Sun E3500 SMP, a Compaq SMP, SGI Origin 2000 and Beowulf systems. The Beowulf systems included Sun systems, an SGI-1200 and a generic alpha cluster using the MPI libraries from Sun, SGI and LAM.

Naturally, the performance of HYDRA_MPI depends on the platform and the distribution of particles, however, as an example, both the SHMEM code, [5], and HYDRA_MPI were employed to simulate the universe from soon after the big bang till the present day for only 64^3 dark matter particles. The execution times per time step are presented in figure 2. Clustering commences around 0.25 time units into the simulation. Large spikes in the execution time are due to file I/O.

The code is now being used for production runs of dark-matter simulations by members of the Virgo Consortium [10].

Acknowledgements

The authors acknowledge the assistance of J-C Desplat, Adrian Jenkins, Tom MacFarland, Rob Thacker and Peter Thomas. Hugh Couchman acknowledges the support of NSERC (Canada) and the Canadian Institute for Advanced Research.

References

1. Couchman, H. M. P., Thomas, P. A. & Pearce, F. R.: Hydra: An Adaptive–Mesh Implementation of P^{3}M–SPH. Astrophys. J., **452** (1995) 797
2. Couchman, H. M. P.: Mesh-refined PPPM: a fast adaptive N-body algorithm. Astrophys. J., **368** (1991) L23
3. Desplat, J.C.: Porting SHMEM codes to MPI-2. EPCC-TR01-01, EPCC, (2001)
4. Hockney, R. W. & Eastwood, J. W.: Computer Simulation Using Particles, McGraw-Hill, (1981)
5. MacFarland, T, Couchman, H. M. P., Pearce, F. R., Pichlmeier, J.: A New Parallel P^{3}M Code for Very Large-Scale Cosmological Simulations. New Astronomy, **3** (1999) 687
6. Monaghan, J. J.: Smoothed Particle Hydrodynamics. Annu. Rev. Astron. Astrophys., **30** (1992) 543
7. Pearce, F. R. & Couchman, H. M. P.: Hydra: an adaptive parallel grid code. New Astronomy, **2** (1997) 411
8. Thacker, R. J., Couchman, H. M. P. & Pearce, F. R.: Simulating Galaxy Formation on SMP's. HPCS'98, Kluwer (1998) 71–82
9. Frido, M. & Johnson., S. G.: http://www.fftw.org
10. Frenk, C. S., et. al.: http://star-www.dur.ac.uk/ frazerp/virgo/virgo.html

Efficient Mapping for Message-Passing Applications Using the TTIG Model: A Case Study in Image Processing*

Concepció Roig[1], Ana Ripoll[2], Javier Borrás[2], and Emilio Luque[2]

[1] Universitat de Lleida, Dept. of CS
Jaume II 69, 25001 Lleida, Spain
roig@eup.udl.es
[2] Universitat Autònoma de Barcelona, Dept. of CS
08193 Bellaterra, Barcelona, Spain
a.ripoll@cc.uab.es, Javier.Borras@ntsig1.uab.es, e.luque@cc.uab.es

Abstract. In this paper we describe the development and performance of an image processing application with functional parallelism within the PVM framework. The temporal behaviour of the application is statically modelled with the new task graph model TTIG (Temporal Task Interaction Graph), that enhances classical models by capturing percentages of concurrency between adjacent tasks. We show how this information can be used in the mapping phase in order to obtain better assignments of tasks to processors. The effectiveness of the TTIG in allocation for the application under study is established through experimentation on a cluster of PCs.

1 Introduction

A fundamental problem in parallel/distributed programming is to find the most efficient mapping of a parallel program onto the processors of a parallel system, to optimize final execution time. This paper specifically addresses the development and the static mapping of an image processing application called BASIZ (Bright And Saturated Image Zones), implemented within the PVM framework. BASIZ exhibits functional parallelism, and it is composed of a set of communicating tasks (processes) performing different functions. The tasks are arranged in a pipeline structure that is fairly common in several application areas, including computer vision, image processing and signal processing [1].

The tasks in BASIZ communicate at different points inside them, and several situations of temporal behaviour of tasks exist in the application. The classical task graph models used in the literature for the purpose of mapping are not able to capture these differences in temporal behaviour of application tasks. Our goal in this work is to achieve an efficient mapping of BASIZ by modelling its behaviour with the new Temporal Task Interaction Graph (TTIG). The TTIG model captures the potential concurrency of adjacent tasks due to their mutual

---

* This work was supported by the CICYT under contract TIC98-0433.

Y. Cotronis and J. Dongarra (Eds.): Euro PVM/MPI 2001, LNCS 2131, pp. 370–377, 2001.

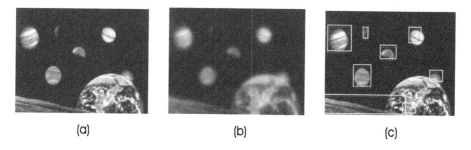

(a)   (b)   (c)

Fig. 1. (a) Original image. (b) Blured image. (c) Detected zones.

dependencies. In [2] the authors showed the effectiveness of using this model for a set of synthetic C+PVM programs that were randomly generated. In this work we show the effectiveness provided by the use of the TTIG model in the mapping phase of BASIZ. The execution results are compared with the use of a classical model and the PVM default allocation scheme in a PVM platform, providing the TTIG with significant improvements in all the cases.

This paper is organized as follows. Section 2 describes the BASIZ parallel application. The temporal behaviour of BASIZ is modelled with the new TTIG graph in section 3. Section 4 illustrates the execution results of BASIZ in a cluster of PCs using the different mapping strategies. Finally, section 5 outlines the main contributions.

2   The Parallel Image Processing Application

BASIZ is a parallel application that performs the detection of the zones with more brightness and colour intensity for a given image. The BASIZ application has been developed on the existing support of URT modular image processing [3], from which the application has been paralleled for the PVM framework. URT maintains a monolithic structure for levels, and each one provides the following level with the data required for its processing. An important use of BASIZ in the computer vision environment is to facilitate the identification of the most sensitive zones of an image to human vision, which is based on the application of a bluring process to the original image. This yields an image where the most coloured areas correspond to the zones that are most perceptible to human vision, meanwhile the less coloured areas correspond to the image background [4]. As an illustrative example, Figure 1 shows the blured image and the detected zones of an original image in colour.

The BASIZ application is composed of a set of 26 tasks (T0 to T25), implemented in C+PVM. The interconnection structure is shown in Figure 2, where each task is labeled with an identifying number and its process name.

The application tasks are organized in seven pipelined stages that allow the processing of several images concurrently. The functions carried out in each stage and the tasks composing each one are the following.

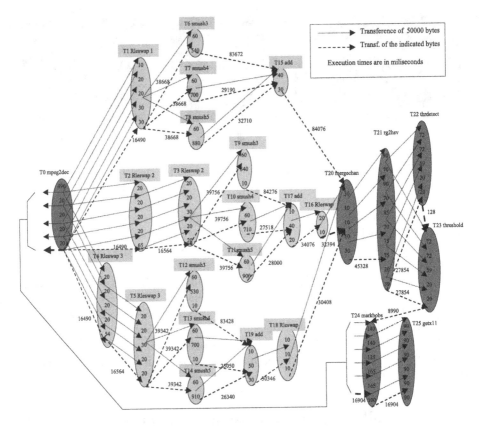

Fig. 2. Task structure of BASIZ application.

1. *Colour separation* (tasks T0 to T5). Separation of the three colour channels red, green and blue from the original image. This generates three images with only one colour each.

2. *Blur* (tasks T6 to T14). For each colour of the image, a bluring process is carried out with three distinct levels. Three, four and five level blured images are generated by applying a Gaussian filter.

3. *Adding* (tasks T15,T17,T19). For each colour, the pixels of the three blured images are added.

4. *Merging* (tasks T16,T18,T20). The three colour channels are merged into a single blured image, as the image of Figure 1(b) shows.

5. *Image conversion* (task T21). Conversion of the representation of the image from colour format RGB (Red, Green, Blue) to HSI (Hue, Saturation, Intensity).

6. *Threshold* (tasks T22,T23). The threshold for detecting the most sensitive zones is dynamically determined depending on the values of saturation and intensity for the global image. The result is a binary image where the pixels are codified with the values 255 or 0 depending on whether or not they belong to a sensitive zone respectively.

7. *Marking zones* (tasks T24,T25). Using the previous binary image, the current non-background areas are found, and the frames surrounding them are generated. These frames are applied to the original image as Figure 1(c) shows.

Each task is composed of a set of computation phases. These are a sequence of time periods during which the task is carrying out sequential computation of instruction sets, and they are expressed in milliseconds in the graph. Communication between neighbouring tasks is carried out with the *pvm_send* and *pvm_recv* primitives. These are represented in the graph with directed edges and have associated with them the amount of bytes involved in the transference. For instance, task T25 has six computation phases of 90 milliseconds each, and before each computation phase a *pvm_recv* primitive is executed from task T24.

The parameters included in the graph were derived by analysing the profile information from a set of executions. The processes that are applied to every image are always the same, and their execution time is not dependent on the input image. Therefore, this graph is representative of the stable computation and communication pattern of the application.

3 Modelling the Temporal Behaviour of BASIZ

As we described in the previous section, BASIZ is a message-passing application composed of a set of tasks performing different functions. A task is an indivisible unit of computation to be mapped on one processor. The mapping of tasks to processors is solved statically using a task graph that models application behaviour, where nodes represent the tasks and edges denote inter-task interactions. Weights are associated to nodes and edges to represent the estimated task computation times and communication volumes respectively. The classical options in the literature concerning parallel application modelling in a task graph are not adequate when dealing with applications where tasks communicate at any point within themselves as is the case of BASIZ. On the one hand, the Task Precedence Graph (TPG) is used to model applications whose tasks communicate only at the beginning and at the end. A task in a TPG graph cannot commence its execution until all its predecessors have already finished. As can be seen in Figure 2, this is not common in the task interaction pattern. For instance, task T3 does not need its predecessor T2 to have finished in order to start its execution. The other task graph model extensively used is the TIG (Task Interaction Graph). In the TIG, nothing is stated regarding the temporal dependencies between communicating tasks, and all tasks are considered to be simultaneously executable. This is a poor realistic assumption for the task structure of BASIZ since, while there are adjacent tasks that can advance their execution in parallel (as is the case of T24 and T25), there are other cases where there is a strict precedence relationship, and a sequential execution will be necessary (as is the case with T23 and T24).

To represent in a more realistic way the temporal behaviour of BASIZ, it was modelled using the new Temporal Task Interaction Graph (TTIG), proposed by the authors in [5]. This is an extension of the TIG model that includes a new concurrency measure, called degree of parallelism, for pairs of communicating tasks. The degree of parallelism is a normalized index belonging to the interval [0,1], and it represents the maximum percentage of parallel execution that two adjacent tasks can achieve. The degree of parallelism is obtained for each pair of adjacent tasks assuming that they are isolated from the rest of the graph and taking into account their mutual interaction. It is therefore calculated by considering their concurrent execution ability due to their mutual dependencies and without considering the time involved in communications. Extreme values for the degree of parallelism represent extreme situations in the possibility of concurrent execution. On the one hand, a degree of parallelism of 0 implies that a task has to be executed sequentially with its adjacent, as happens in the TPG model. On the other hand, a degree of parallelism equal to 1 states that, at maximum, a task can execute concurrently all the time with its adjacent, as is assumed in the TIG model. The detailed description for the calculation of the degree of parallelism is reported in [5]. In order to automatically generate the TTIG graph from the weights of computation and communication of a program, a tool has been designed and implemented, that is available in [6].

Figure 3 shows the TTIG graph that models the BASIZ application. It is composed of 26 nodes representing the application tasks, and each task has its global computation time associated (i.e. the sum of the computation phases inside). The arcs have two parameters associated: (a) the global amount of data to be transfered between the adjacent tasks and, (b) the degree of parallelism. The degree of parallelism is represented in the arc with respect to the task that has less computation time. For instance, the degree of parallelism 0.8 between tasks T24 and T25 indicates that task T25 will be able to parallelize with T24 a maximum of an 80% of its global computation time (this implies that tasks T24 and T25 will be able to concurrently execute a maximum of 540·0.8=432 milliseconds, due to their mutual dependencies).

By the use of the TTIG, all the situations related to maximum pair-wise concurrency of adjacent tasks can be represented in the graph. Therefore, the TTIG integrates the two classical models TPG and TIG, making it useful as a unified abstraction of program modelling for message-passing applications.

4   Execution of BASIZ for Different Mappings

In this section we present an experimental study of the performance provided by the parallel execution of BASIZ, with task allocations based on the new TTIG model, the classical TIG and the PVM default strategy. The experiments were carried out on a cluster of 10 Linux machines running PVM 3.4. Each machine was a Pentium II at 350 Mhz with 128 Mbytes of RAM. The interconnection network was a 10 Mbps Ethernet. The two following mapping strategies were used to statically allocate the tasks of BASIZ to the architecture processors:

(a) *TIG mapping*. In this case, the tasks were allocated using the TIG model. The TIG can be generated in a time complexity no worse than $O(rn^2)$ where

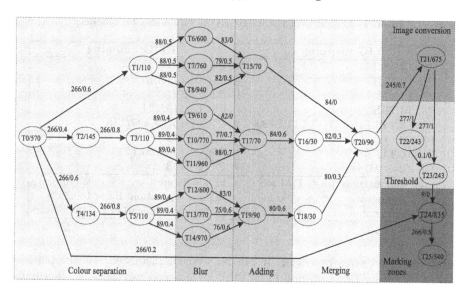

Fig. 3. TTIG graph for BASIZ application. Task computation times in milliseconds and communication volumes in Kbytes.

n is the number of tasks and r is the maximum number of computation phases for all the tasks. The heuristic is based on a two stage approach that first merges the tasks into as many clusters as the number of processors, and then assigns clusters to individual processors [7]. The merging stage is carried out with the goal of achieving load balancing and minimization of the communication volume among clusters.

(b) *TTIG mapping.* The mapping was based on the TTIG model of the application, which generation has an associated time complexity no worse than $O(r^2n^2)$. The heuristic evaluates the nodes of the graph in a breadth-first approach. It iteratively merges the more dependent tasks to the same cluster (i.e. the adjacent tasks that need more time when executed separately, according to their degree of parallelism and the communication volume), and the less dependent tasks are assigned to different processors by trying to keep load balancing among processors.

As this is a static approach, the task graph model is generated prior to execution. Thus, the differences in the execution time are due to the different task allocations provided by both mappings. Table 1 shows, as an illustrative example, the final assignment of BASIZ that was obtained for 8 processors using the two strategies based on the models TIG and TTIG.

It can be observed in the table that there are significant differences in the assignments of both mappings. On the one hand, the TIG mapping generates an allocation that equilibrates the computation better among processors, and minimizes the global communication volume to be transferred. On the other hand, the TTIG mapping generates an allocation where the more dependent tasks are

Table 1. Mappings of BASIZ based on TIG and TTIG models.

TIG mapping	TTIG mapping
P0: T2,T4,T14	P0: T3,T14,T22,T23
P1: T0,T1,T7	P1: T6,T12,T15,T19
P2: T5,T12,T13	P2: T7,T21
P3: T3,T9,T10	P3: T8,T16,T24
P4: T8,T15,T16,T17,T18,T19,T20	P4: T1,T5,T9,T17,T18
P5: T21,T22,T23	P5: T4,T11
P6: T6,T11	P6: T2,T10,T20,T25
P7: T24,T25	P7: T0,T13
min. computation: P5 = 1161 mS	min. computation: P4 = 930 mS
max. computation: P6 = 1560 mS	max. computation: P3 = 1805 mS
global communication = 4192 Kbytes	global communication = 5425 Kbytes

assigned to the same processor. It can be seen for instance, that the pairs of tasks (T6,T15), (T22,T23) and (T12,T19) are assigned together. This normally occurs in all assignments based on TTIG when the number of processors varies, while with TIG, these more dependent tasks are assigned jointly or separately, provided that they fulfil load balancing criteria as their aim.

Using the assignments obtained for a number of processors varying from 2 to 10, we carried out an experimental study by executing BASIZ application with 20 images. It was experimentally proved that this is the number of images that provided best throughput in the pipeline structure.

Figure 4 shows the average speedup in the execution time when the BASIZ application was executed using the TIG and TTIG mappings. Moreover, the speedup obtained when using the PVM default allocation scheme is reported [8]. It can be observed that in all the cases the TTIG mapping provides greater improvements in speedup. The reason is that this mapping takes into account potential task concurrency as an essential information in the decision allocation process and therefore provides more efficient assignments, although their computation and communication are not as balanced as with TIG mapping. Of the three, the PVM mapping provides the lowest speedup given that it is not based on the application structure.

As can be observed, the speedup attained on using 10 processors only increases with PVM mapping. In the other cases it remains constant or falls. This is due to the structure of dependencies between tasks, which allows an eight processor mapping to be enough for exploiting the application parallelism.

5   Conclusions

This paper reports on the development and performance of the BASIZ (Bright And Saturated Image Zones) parallel application within the PVM framework. The application exhibits functional parallelism and the tasks are arranged in a pipeline structure.

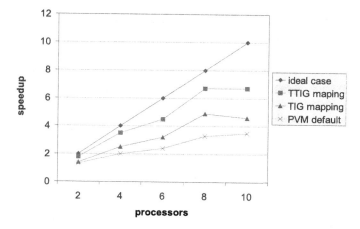

Fig. 4. Speedup for TTIG, TIG and PVM mappings.

The temporal behaviour of BASIZ was modelled statically with the new task graph approach TTIG (Temporal Task Interaction Graph), which includes task concurrency measures among tasks. The application was executed in a cluster of PCs, varying the number of processors, with the task assignments provided by the following strategies: (a) mapping based on the classical model TIG, (b) mapping based on the new model TTIG and (c) the PVM default allocation. The results showed that for all assignments the temporal information included in the TTIG had been essential to improving speedup in the final execution time.

References

1. Subhlok J. and Vondram G.: Optimal Use of Mixed Task and Data Parallelism for Pipelined Computations. J. of Parallel and Distr. Compt. 60. pp. 297-319. 2000.
2. Roig C., Ripoll A., Senar M.A., Guirado F. and Luque E.: Exploiting Knowledge of Temporal Behaviour in Parallel Programs for Improving Distributed Mapping. 6th. International Euro-Par conference. LNCS, vol:1900, pp. 262-271, 2000.
3. The Utah Raster Toolkit. Utah University http://www.utah.com.
4. Lindeberg T. Linear Scale-Space, a Geometry-Driven Diffusion in Computer Vision. Kluwer Academic Pub. 1994.
5. Roig C., Ripoll A., Senar M. A., Guirado F. and Luque E.: Modelling Message-Passing Programs for Static Mapping. IEEE Proc. 8th. Euromicro Workshop on Par. and Distr. Processing. pp. 229-236 Jan. 2000.
6. http://www.diei.udl.es/arees/atc. Computer Science Dep. Univ. de Lleida. Spain.
7. Senar M. A., Ripoll A., Cortés A. and Luque E.: Clustering and Reassignment-based Mapping Strategy for Message-Passing Architectures. (IPPS/SPDP 98) 415-421. IEEE CS Press USA, 1998.
8. Geist A., Beguelin A., Dongarra J., Jiang W., Manchek R. and Sunderam V.: PVM: Parallel Virtual Machine. A Users' Guide and Tutorial for Networked Parallel Computing. The MIT Press. Cambridge, Massachusets. 1994.

Text Searching on a Heterogeneous Cluster of Workstations

Panagiotis D. Michailidis and Konstantinos G. Margaritis

Parallel and Distributed Processing Laboratory
Department of Applied Informatics, University of Macedonia
156 Egnatia str., P.O. Box 1591, 54006, Thessaloniki, Greece
{panosm,kmarg}@uom.gr
http://macedonia.uom.gr/~{panosm,kmarg}

Abstract. In this paper we propose a high-performance flexible text searching implementation on a heterogeneous cluster of workstations using MPI message passing library. We test this parallel implementation and present experimental results for different text sizes and number of workstations.

1 Introduction

Text searching is a very important component of many problems, including text processing, information retrieval, pattern recognition and DNA sequencing. Especially with the introduction of search engines dealing with tremendous amount of textual information presented on the World Wide Web (WWW) as well as the research on DNA sequencing, this problem deserves special attention and any improvements to speed up the process will benefit these important applications.

The basic text searching problem can be defined as follows. Let a given alphabet (a finite sequence characters) Σ, a short pattern string $P=P[1]P[2]...P[m]$ of length m and a large text string $T=T[1]T[2]...T[n]$ of length n, where both the pattern and the text are sequences of characters from Σ, with $m \leq n$. The text searching problem consists of finding one or more generally all the exact occurrences of a pattern P in a text T. Survey and experimental results of well known algorithms for this text searching problem can be found in [3], [9], [11], [16].

The implementation of the text searching problem on a cluster of workstations or PCs [1] can provide the computing power required for the speed up the searching on large free text collections. In [8], [12] five sequential text searching algorithms were parallelised and tested on a homogeneous cluster giving very positive experimental results. In [13] a performance prediction model was proposed for static master-worker model on a homogeneous cluster. In [10] a parallel text searching implementation was presented for static master-worker model and results are reported for the Brute-Force text searching algorithm [11] on a heterogeneous cluster.

The contribution of this work is the implementation of a parallel flexible text searching algorithm using cluster computing technique. This algorithm realized

Y. Cotronis and J. Dongarra (Eds.): Euro PVM/MPI 2001, LNCS 2131, pp. 378–385, 2001.

the master-worker model with static allocation of texts. It is implemented in C in conjunction with the Message Passing Interface (MPI) library [6], [14], [15] which follows the SPMD (Single Program Multiple Data) model and run on a heterogeneous network of workstations.

The remainder of this paper is organized as follows: Section 2 briefly presents heterogeneous computing model and the metrics. Section 3 discusses the text partitioning strategy and the proposed parallel implementation. Section 4 gives the experimental results of the parallel implementation. Finally, Section 5 contains our conclusions and future research issues.

2   Heterogeneous Computing Model

A heterogeneous network (HN) can be abstracted as a connected graph HN(M,C), where

- M={M_1, M_2,...,M_p} is set of heterogeneous workstations (p is the number of workstations). The computation capacity of each workstation is determined by the power of its CPU, I/O and memory access speed.
- C is standard interconnection network for workstations, such as Fast Ethernet or an ATM network, where the communication links between any pair of the workstations have the same bandwidth.

Based on the above definition, if a network consists of a set of identical workstations, the system is homogeneous. Further, a heterogeneous network can be divided into two classes: a dedicated system where each workstation is dedicated to execute tasks of a parallel computation, and a non-dedicated system where each workstation executes its normal routines (also called owner workload), and only the idle CPU cycles are used to execute parallel tasks. In this paper we use a dedicated heterogeneous network of workstations.

2.1   Metrics

Metrics help to compare and characterize parallel computer systems. Metrics cited in this section are defined and published in previous paper [19]. They can be roughly divided into characterization metrics and performance metrics.

Characterization Metrics. To compute the power weight among workstations an intuitive metric is defined as follows:

$$W_i(A) = \frac{min\{T(A, M_j)\}}{T(A, M_i)} \tag{1}$$

where A is an application and T(A,M_i) is the execution time for computing A on workstation M_i. Formula 1 indicates that the power weight of a workstation refers to its computing speed relative to the fastest workstation in the network. The value of the power weight is less than or equal to 1.

To calculate the execution time of an application A, the speed, denoted by S_f of the fastest workstation executing basic operations of an application is measured by the following equation:

$$S_f = \frac{\Theta(A)}{t_A} \qquad (2)$$

where $\Theta(A)$ is a complexity function which gives the number of basic operations in a application A and t_A is the execution time of A on the fastest workstation in the network.

Using the speed of the fastest workstation, S_f, we can calculate the speeds of the other workstations in the system, denoted by S_i (i=1,...,p), using the computing power weight as follows:

$$S_i = S_f * W_i, i = 1, ..., p, and\, i \neq f \qquad (3)$$

where W_i is the computing power weight of M_i. So, by equation 3, the execution time of an application A across heterogeneous network HN, denoted by $T_{cpu}(A,HN)$, can be represented as

$$T_{cpu}(A, HN) = \frac{\Theta(A)}{\sum_{i=1}^{p} S_i} \qquad (4)$$

where $\sum_{i=1}^{p} S_i$ is the computing capacity of the heterogeneous network when p workstations are used. Here, T_{cpu} is considered the required CPU time for the application.

Performance Metrics. Speedup is used to quantify the performance gain from a parallel computation of an application A over its computation on a single machine on a heterogeneous network system. The speedup of a heterogeneous computation is given by:

$$SP(A) = \frac{min\{T(A, M_j)\}}{T(A, HN)} \qquad (5)$$

where T(A,HN) is the total execution time for application A on HN, and $T(A,M_j)$ is the execution time for A on workstation M_j, j=1,...,p.

Efficiency or utilization is a measure of the time percentage for which a machine is usefully employed in parallel computing. Therefore, the utilization of parallel computing of application A on a dedicated heterogeneous network is defined as follows:

$$E = \frac{SP(A)}{\sum_{j=1}^{p} W_j} \qquad (6)$$

The previous formula indicates that if the speedup is larger than $\sum_{j=1}^{p} W_j$, the system computing power, the computation presents a superlinear speedup in a dedicated heterogeneous network.

3   A Strategy for Static Load Balancing

To avoid the slowest workstations to determine the parallel execution time, the load should be distributed proportionally to the capacity of each workstation. The goal is to assign the same amount of time which may not correspond to the same amount of text.

A balanced distribution is achieved by a static load distribution made prior to the execution of the parallel operation. To achieve a good balanced distribution in the heterogeneous network, the relative amount of text assigned to each workstation should be proportional to its processing capacity compared to the entire network:

$$l_i = \frac{S_i}{\sum_{j=1}^{p} S_j} \tag{7}$$

In next subsection, we present the parallel algorithm that is based on previous strategy for static allocation of subtexts.

3.1   Master-Worker with Static Allocation of Subtexts

In order to present the algorithm, we make the following assumptions. First, the workstations have an identifier myid and are numbered from 0 to p-1, second the each subtext is stored in local disk (or file) of a workstation in name file format nametext.#myid in order to be separate and finally, the pattern is stored in main memory to all workstations. Figure 1 demonstrates a very simplified version of the static parallel implementation in C-like pseudo-code using MPI library. From the master and the worker sub-procedures it is clear that the application ends only when all the local text searching operations have been completed and their results have been collected by the master workstation. Further, it must be noted that the line 3 of the worker sub-procedure calls any sequential exact text searching algorithm. This entire program is constructed so that alternative sequential exact text searching algorithms can be substituted quite easily [9], [11]. In this paper we use the Shift-Or (in short, SO) flexible string matching procedure [2], [9], [11]. The main advantage of this algorithm is that very easy can support extended patterns, such as don't cares symbols, class of characters, complement of a character or a class, set of patterns and other extensions [18].

4   Experimental Results

In this section we present the experimental results for the performance of the parallel flexible string matching implementation, which is based on static master-worker model. This algorithm is implemented in ANSI C programming language [7] using the MPI library [6], [14], [15] for the point-to-point and collective communication operations. The target platform for our experimental study is a heterogeneous cluster connected with 100 Mb/s Fast Ethernet network. More specifically speaking, the cluster consists of 2 Pentium III workstations, based on 550 MHz with 384 MB RAM and 5 Pentium workstations, based on 100 MHz with

```
Main procedure
main()
{
    1. Initialize message passing routines;
    2. If (process==master) then call master_distribution();
    3. Call worker();
    4. If (process==master) then call master_collection();
    5. Exit message passing operations;
}

Master sub-procedure
master_distribution()
{
    1. Broadcast the pattern (P) to workers; (MPI_Bcast)
    2. Broadcast the name of text to workers; (MPI_Bcast)
}

Worker sub-procedure
worker()
{
    1. Receive the pattern (P) and the name of text from master;
       (MPI_Bcast)
    2. Open the file of the local subtext (T) and store the local
       subtext (T) in memory;
    3. Call matches=search(P,m,T,k) ; //using any sequential string
       matching approach
    4. Send the results (i.e. matches) to the master; (MPI_Reduce)
}

Master sub-procedure
master_collection()
{
    1. Receive the results (i.e. matches) from all workers and returns
       the total matches; (MPI_Reduce)
    2. Print the total number of occurrences of P in distributed T
       (i.e.total matches).
}
```

Fig. 1. Static master-worker parallel implementation

64 MB RAM. The MPI implementation used on the network is MPICH version 1.2. During all experiments, the cluster of workstations was dedicated. Finally, to get reliable performance results 10 executions occurred for each experiment and the reported values are the average ones.

The number of workstations, the pattern lengths and the text sizes, can influence the performance of the parallel string matching significantly and thus these parameters are varied in our experimental study.

Table 1. Experimental execution times (in seconds) for text size of 3MB and 27MB respectively using several pattern lengths

3MB					27MB			
p/m	5	10	20	30	p/m	5	10	20	30
1	0.1323	0.1322	0.1322	0.1323	1	1.0577	1.0569	1.0570	1.0566
2	0.0693	0.0694	0.0696	0.0694	2	0.5466	0.5462	0.5470	0.5454
3	0.0661	0.0663	0.0661	0.0661	3	0.5212	0.5217	0.5215	0.5214
4	0.0625	0.0623	0.0624	0.0627	4	0.4981	0.4983	0.4976	0.4982
5	0.0601	0.0600	0.0601	0.0604	5	0.4773	0.4765	0.4767	0.4766
6	0.0576	0.0576	0.0576	0.0583	6	0.4562	0.4569	0.4558	0.4565
7	0.0556	0.0556	0.0555	0.0561	7	0.4399	0.4390	0.4395	0.4394

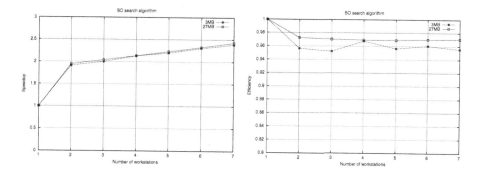

Fig. 2. Parallel text searching with respect to the number of workstations for several text sizes

Table 1 we show the execution times in seconds, for the SO string matching algorithm, the four pattern lengths, for two total English text sizes and for different number of workstations. The results of Table 1 are limited for pattern length smaller than the word size (m≤31) according to the SO algorithm [2]. Further, Figure 2 present the speedup and efficiency curves with respect to the number of workstations for English text of various sizes and for the SO string matching algorithm. It is important to note that the speedup and efficiency, which is plotted in Figure 2 is result of the average for four pattern lengths.

We can see a clear reduction in the computation time of the algorithm when we use the parallel algorithm. For example, with text size of 27MB and m=10, we reduce the computation time from 1.0569 seconds in the sequential version to 0.439 seconds in the parallel version using 7 workstations. In other words, we observe that for constant total text size there is an expected inverse relation between the parallel execution time and the number of workstations. As far as the pattern length we observe that there is not significant effect on the performance of the parallel implementation. Further, the static master-worker model achieves reasonable speedups and efficiency for all pattern lengths and workstations. We had an increasing speedup curves up to about 2.4 on 7 workstations which had the computing power of 2.5. Also, the decrease in the efficiency of text searching was less than 5%. Therefore, more time is spent in the text searching operation than communicating with the master workstation. Finally, from the speedup curves we observe that the parallel algorithm has significant degree of scalability when the number of workstations and the text size is increased.

5   Conclusions

We have proposed a parallel flexible text searching implementation for large free text databases on a cluster of heterogeneous workstations. The experimental results show that heterogeneous processing with load balancing can significantly reduce searching time compared to a homogeneous approach. In [10] a performance prediction model was presented for the proposed parallel implementation on a heterogeneous cluster. The experimental results of the SO flexible text searching algorithm incorporated in the proposed parallel scheme were confirmed with the theoretical values of the performance model [10].

Future work includes experiments for dynamic load balacing algorithms on a heterogeneous cluster of workstations.

References

1. T. Anderson, D. Culler and D. Patterson, A case for NOW (network of workstations), *IEEE Micro*, vol. 15, no. 1, pp. 54-64, 1995.
2. R. Baeza-Yates and G. Gonnet, A new approach to text searching, *Communications of the ACM*, vol. 35, no. 10, pp. 74-82, 1992.
3. M. Crochemore and W. Rytter, Text algorithms, Oxford University Press, 1994.
4. V. Donaldson, F. Berman and R. Paturi, Program speedup in a heterogeneous computing network, *Journal of Parallel and Distributed Computing*, vol. 21, no. 3, pp. 316-322, 1994.
5. I. Foster, Designing and building parallel programs: Concepts and tools for parallel software engineering, Addison-Wesley, 1995.
6. W. Gropp, E. Lusk and A. Skjellum, Using MPI: Portable parallel programming with the message passing interface, The MIT Press, Cambridge, Massachusetts, 1994.
7. B.W. Kernighan and D.M. Ritchie, The C Programming Language, Prentice Hall, Englewood Cliffs, NJ, 2nd edition, 1988.
8. P.D. Michailidis and K.G. Margaritis, Implementing string searching algorithms on a network of workstations using MPI, *to appear in Proc. of the 5th Hellenic European Conference on Computer Mathematics and its Applications*, 2001.
9. P.D. Michailidis and K.G. Margaritis, On-line string matching algorithms: Survey and experimental results, *International Journal of Computer Mathematics*, vol. 76, no. 4, pp. 411-434, 2001.
10. P.D. Michailidis and K.G. Margaritis, Parallel text searching application on a heterogeneous cluster of workstations, *to appear in Proc. of the 3rd International Workshop on High Performance Scientific and Engineering Computing with Applications*, 2001
11. P.D. Michailidis and K.G. Margaritis, String matching algorithms, Technical Report, Dept. of Applied Informatics, University of Macedonia, 1999 (in Greek).

12. P.D. Michailidis and K.G. Margaritis, String matching problem on a cluster of personal computers: Experimental results, *to appear in Proc. of the 15th International Conference Systems for Automation of Engineering and Research*, 2001.
13. P.D. Michailidis and K.G. Margaritis, String matching problem on a cluster of personal computers: Performance modeling, *to appear in Proc. of the 15th International Conference Systems for Automation of Engineering and Research*, 2001.
14. P.S. Pacheco, Parallel Programming with MPI, San Francisco, CA, Morgan Kaufmann, 1997.
15. M. Snir, S. Otto, S. Huss-Lederman, D.W. Walker and J. Dongarra, MPI: The complete reference, The MIT Press, Cambridge, Massachusetts, 1996.
16. G.A. Stephen, String searching algorithms, World Scientific Press, 1994.
17. B. Wilkinson and M. Allen, Parallel programming: Techniques and applications using networked workstations and parallel computers, Prentice Hall, 1999.
18. S. Wu and U. Manber, Fast text searching allowing errors, *Communications of the ACM*, vol. 35, no. 10, pp. 83-91, 1992.
19. Y. Yan, X. Zhang and Y. Song, An effective and practical performance prediction model for parallel computing on non-dedicated heterogeneous NOW, *Journal of Parallel and Distributed Computing*, vol. 38, no. 1, pp. 63-80, 1996.

Simulation of Forest Fire Propagation on Parallel & Distributed PVM Platforms

Josep Jorba, Tomàs Margalef, and Emilio Luque

Computer Architecture and Operating Systems (CAOS Group),
Departamento de Informatica, Universidad Autonoma de Barcelona,
08193 Bellaterrra, Spain
{josep.jorba,tomas.margalef,emilio.luque}@uab.es

Abstract. This paper summaries experiences designing and developing a parallel/distributed simulator for existing models of forest fire propagation using PVM. The methodology used to parallelise the application is described. Results obtained for the parallelisation on different platforms are presented: clusters of SUN workstations and PCs, and one parallel machine (Parsytec CC) with different internal node interconnections. The studies present several measurements of performance & scalability of the proposed solution for this forest fire application.

1   Introduction

Forest fire is one of the most critical environmental risks specially in southern Europe. In the literature there are several forest fire propagation models and it must be taken into account that the forest fire propagation is a very complex problem that involves several aspects that must be considered: a) Meteorological aspects: Temperature, wind, moisture; b) Vegetation features; and c) Topographical aspects: Terrain topography.

In a fire prevention phase use case, the simulation system can be used to simulate the evolution of the forest fire using different scenarios. The results would be very useful to decide which actions must be taken to minimize the fire risks and fire damages.

In a real emergency use case, these simulations require high computing capabilities because the decisions should be taken on real time in order to extinguish the fire as fast as possible and minimize the fire risks.

Parallelisation techniques are very useful to minimize the execution time of the applications and to provide fast results. The simulation of forest fire propagation models is a problem that offers certain geometrical and functional features that allow the parallelisation in a natural way. Moreover, the parallelisation incorporates additional benefits for testing the existing models: for example, allows to carry out a great number of simulations in order to test different scenarios or to analyse the sensibility of the propagation models to the input parameters in order to determine the precision required in field measures.

In section 2 the fire propagation models are described focusing in the particular one selected in our work. Section 3 describes the methodology applied to parallelise

Y. Cotronis and J. Dongarra (Eds.): Euro PVM/MPI 2001, LNCS 2131, pp. 386–392, 2001.

the selected model. Section 4 shows the implementations carried out and the results obtained. Finally, section 5 provides some conclusions.

2   Forest Fire Propagation Model

There are several models in the literature to describe the behaviour of forest fire propagation. First of all it must be pointed that the propagation models, in general, include two separate models: the global model and the local model. On one hand, the global model considers the fire line as a whole unit (geometrical unit) that evolves in time and space. On the other hand, the local models consider the small units (points, sections, arcs, cells, ...) that constitutes the fire line. These local models take into account the particular conditions (vegetation, wind, moisture, ...) of each unit and its neighborhood to calculate the evolution of each unit.

In the literature there are several approaches to solve the global models [1][3][5]. All these models can be generalized in a Global Fireline Propagation Model [1] that can be divided in the following steps:

1st) Subdivision of the fireline $\phi(t)$ into a partition of sections $\delta_i\phi(P_i,t)$.

2nd) Resolution of a certain Local Problem for each section $\delta_i\phi(P_i,t)$, giving as result a particular virtual fireline $\Phi_{v,i}(\Delta t)$.

3rd) Aggregation and coupling of the information inherent to the set $\{\Phi_{v,i}(\Delta t)\}$, providing the definition of $\Phi(t+\Delta t)$.

In this paper the model defined by André and Viegas in [2, 3] is used. The operational cycle of this model consists in a iterative calculation of the next position of the fireline considering the current fireline position (figure 1).

In this model, the global model allows the partitioning of the fireline into a set of sections [2,3]. Under these conditions the movement of the fireline can be considered as the separate movement of the different sections (taking into account its particular local conditions), and then it is possible to compose the fireline in the next time step by aggregating the particular movement of the different sections.

In order to calculate the movement of each section (by the local model) it is necessary to determine the propagation speed, which is calculated considering the direction normal to the particular section of the fireline. The computation of this speed (Rn) involves three separate factors:

a) Calculation of the speed in the maximum propagation direction (Rn0).
b) Calculation of the term that takes into account the difference between the normal and the maximum speed directions (fn).
c) Calculation of the effect of the curvature of the particular section (gn).

The speed in the normal direction is obtained by the multiplication of these three factors considering the vectors.

Some of these factors (figure 1) can be subdivided into several subcalculations:

a) Rn0, Factor in the direction of maximum speed: The calculation requires the local parameters: vegetation, topography and meteorology (wind). In order to do these calculations the Rothermel model [4, 5] is used.
b) fn, angular speed dependency: This factor considers the effect of the angle between the maximum propagation speed direction and the direction normal to

the section. However, this calculation is very complex due to the fact that the fireline usually has very strange and irregular forms. Therefore, the calculation is done based on regular forms already studied in the literature, such as the double ellipses described by Anderson [6].

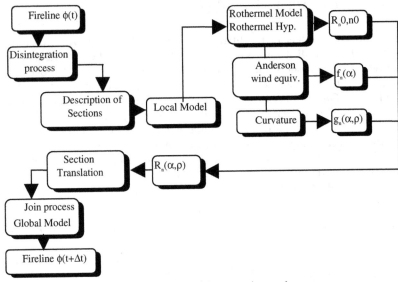

Fig. 1. Diagram of the operation cycle

As it has been shown this forest fire propagation simulation involves several processes that requires complex calculation. Therefore, the computing power required to solve this problem in an acceptable time is very high. In such situation, parallel/distributed computing should provide power computing to improve the simulation time and make such simulation feasible, and going to see, supply more attractive features for the applications.

3   Parallelisation of Propagation Models

The parallelisation of simulation models can be done in two main ways: data parallelism and functional parallelism [8]. At this moment, the implementation shows basically a data parallelism, using a typical master/slave paradigm.

In the data parallelism approach, the same algorithm can be applied to several sets of data independently. This means, in our case, that the movement of each section of the fireline can be calculated independently. In this case, we apply the same calculations to different sets of data (sections) to get the new fireline.

On the other hand, functional parallelism is obtained when different parts of an algorithm can be executed simultaneously. It implies that the algorithm can be divided into a set of tasks, which cooperate to solve the particular problem. Each task can be responsible of the calculation of a particular part of the simulation. Some of these tasks can be executed simultaneously and they can exchange data with other tasks (send and receive data from other tasks) of the simulation.

Determining the movement of one section of the fireline requires the calculation of the direction with the maximum local speed of propagation. This maximum speed is obtained by the multiplication of three factors: the speed in the direction normal to the section of the fireline, a second factor considering the orientation of the section and a third factor considering the radius of curvature of the section. The calculation of these three factors can be parallelised since they are no dependent. Another source of parallelism is the evaluation of the input parameters of the local model (wind, moisture) which can be done concurrently with the evaluation of the movement of the fireline. This distribution of the calculations among the processors of the system allows carrying on all these calculations in parallel. In the global model the fireline is composed of a set of independent sections.

These sections can be represented in a numerical form as an arc with a normal that points in the direction of the fireline movement. These arcs are interpolated from the points of the fireline. Therefore, the calculation of these sections (interpolation process) can be distributed among the resources of the parallel machine. It is assumed that the fireline is composed of N sections and the parallel system consists of M processing elements. Under these conditions, each processing element can calculate $K=N/M$ sections. This means that theoretically the speed-up that can be obtained is M with respect to the sequential model [9]. But the ratios of communications/computations limits this theoretical speed-up.

It is necessary to distribute the input data (sections of fireline) among the processors and after the calculation, the results must be collected. Moreover, it must be noted that although the sections of the fireline are independent, the extreme points of each section are shared with their neighbouring sections. This implies a clustering of neighboring sections for minimizing the communications of shared points. Each processing element, running the slave process, receives a fireline segment, composed of a group of continuous sections. The final results must be collected to generate the aggregated fireline, this process is carry out by the master process.

In our particular case, when one processor must calculate K sections, these sections should be consecutive because in this way the communications among different processors are reduced.

Due to the initial independence of the sections, the use of the local model to determine the movement of each section can also be done simultaneously on different sections. The calculation of the fn factor can also be done in parallel, due to the independence among the sections. Moreover, in this case there is a total independence among the calculations for each section.

In the previous discussion it has been explained that the use of parallel computing reduces the simulation time and this fact offers new attractive possibilities for the application of simulations:

a) Sensibility studies: Several simulations can be executed considering some variation in the input parameters in order to analyse the effect on the simulation results. These studies would provide which parameters are more critical for the forest fire propagation models.

b) Multiple scenario simulations: It would be possible to run several simulations considering different scenarios (wind conditions, moisture, temperature) to decide which actions must be taken in the real fire fight.

c) More complex models: Forest fire simulation requires a set of complementary models to consider the dynamic behaviour of the environment that affects the fire propagation. Parameters such as moisture or wind vary dynamically [9]. Wind is a very important factor that requires heavy models to solve the non linear equations involved. Such models are good candidates to be included in a simulation system based on parallel/distributed computing.

4   Implementation & Results

The main goal of all the implementations based on propagation models is to provide an integrated simulation system of forest fire propagation. The global and local models and the required environmental information must be integrated to obtain a simulation system that provides the space-time forest fire evolution.

The main components are the following ones:

a) Input information databases, concerning the physical environment, including: 1) Ignition point or current status of the fireline, 2) vegetation maps that include the characteristics of the vegetation of each region, 3) topographic information of the terrain where the fire is burning, and 4) meteorological information, usually the wind field.
b) Propagation models: The global and local models described in section 2.
c) Complementary models: These models include those parameters with a dynamic behaviour.

However, the current state of forest fire research does not allow to include all the components in a real system. There is active research in all these fields but there are not final results that can be included in the simulation systems. Therefore, the real current simulation systems has a simplified structure. However, it must be pointed that in the near future the research results will be introduced in these simulation systems.

In our work, considering the theoretical models discussed in section 2, several implementations have been developed on different machines. The first implementation was a sequential implementation that runs on a PC or SUN workstation. This sequential version was used to validate the results of the future parallel/distributed versions and also to compare the performance obtained.

The parallel/distributed version (discussed in section 3) was developed using PVM [11] libraries and there are two versions that run on different systems although they are absolutely equivalent from the design point of view:

a) A distributed version that runs on a cluster of workstations (or even PCs) connected by Ethernet. This solution offers the possibility of increasing the computing capabilities of the system at low cost.
b) A parallel version that runs on a parallel Parsytec CC with 8 processing nodes. In this case, the system includes an internal high-speed connection network that improves the interprocessor communications.

The distributed version exploits the capabilities of easily available systems although the communication speed is not very high and limits the speed up obtained.

Meanwhile, the parallel version tries to exploit the high capabilities of a real parallel system with a high speed communication network.

The experimental results show that the simulation time is reduced in both, parallel (CC) and distributed (cluster) implementations. It can be observed that the execution time is reduced when the number of processors is increased, but when we reach a certain number of processors the execution time is not significantly reduced. In this case the computation/communication ratio is too low. When the problem size is increased it is observed that when we increase the number of processors the execution time is also reduced.

Figure 2 summarizes the results obtained for a particular problem size (2048 points), showing the speedup obtained with different number of processors. It can be observed that the speedup is generally very low due to the low computation/communication ratio that implies that the processes requires many communications and performs few computations. This fact penalizes the parallel and distributed version in a significant way (in a small and medium size of the problem).

More precise models will be developed in the near future to simulate the evolution of the fireline in a more realistic way. In this models the wind field and wind prediction will be included, and also some complementary models related to the moisture contents of the vegetables, and so on. These new models will include much more heavy equations. Therefore, the computation/communication ratio of such models will be higher. This fact will allow to obtain better improvements by the application of parallelism. In any case it is clear that the geometry of the problem allow to apply parallel/distributed techniques in a quite easy way.

In our implementation, some simple load balancing techniques are used for controlling the value of that ratio. Defining a minimum threshold, based on a value of sections/processing element, we obtain a value of number of processing elements involved in the simulation. This threshold, is in this moment based on the results obtained in the experiments, and this value as fixed in 16 sections/processing element, but this value is platform dependent. The load balance involved in the execution is dynamic; in each simulation step, the number of sections are not equal to the previous. Each step could need a section interpolating process to define the fireline more precisely, due the grow of the fireline in each step, and the consequent loss of form definition. This affects the number of computing processors available for computation, in function of the mentioned static threshold.

5   Conclusions

The simulation of forest fire propagation, of their complex models should be fast in order to predict the fire behavior in advance and use this information to decide which actions should be taken to control fire propagation. These accurate models require high performance capabilities in order to provide the results in a satisfactory time. In some utilizations this speed are need for different reasons, for example the number of total simulations. In our experiences, we obtain favorable results in the parallelisation of existing models (incompletes in some aspects), for the scalability of these models, with a threshold of communication/computation rate, which demonstrates the models are easily parallelised, and is possible to run multiple simulations needed to test several different scenarios.

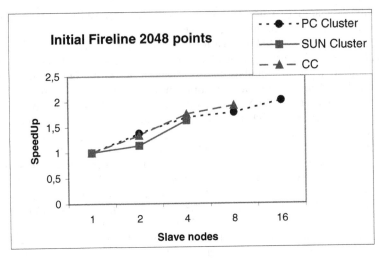

Fig. 2. SpeedUp of different versions

References

1. André, J.C.S, Viegas, D.X.: An Unifying Theory on the Propagation of the Fire Front of Surface Forest Fires. Proc. of the 3nd International Conference on Forest Fire Research. Coimbra, Portugal, (1998)
2. André, J.C.S. and Viegas, D.X., A Strategy to Model the Average Fireline Movement of a light-to-medium Intensity Surface Forest Fire, Proc. of the 2nd International Conference on Forest Fire Research, pp. 221-242. Coimbra, Portugal, (1994)
3. André, J.C.S., A theory on the propagation of surface forest fire fronts, PhD Dissertation (in portuguese), Universidade de Coimbra, Portugal, (1996)
4. Rothermel, R. C., A mathematical model for predicting fire spread in wildland fuels, USDA-FS, Ogden TU, Res. Pap. INT-115, (1972)
5. Rothermel, R. C., How to predict the spread and intensity of forest and range fires, USDA-FS, Tech. Rep. INT-143, (1983)
6. Anderson, H. E. Predicting wind-driven wildland fire size and shape, USDA-FS, Odegen TU, Res. INT-305, (1983)
7. Andrews, P.L. BEHAVE: Fire Behaviour Prediction and Fuel Modelling System – Burn Subsystem, Part I, USDA-FS, Ogden TU, Gen. Tech. Rep. INT-194 (1986)
8. Foster, I., Designing and Building Parallel Programs, Addison-Wesley, (1995)
9. Jorba, J.; Margalef, T.;Luque, E.;André, J.C.S. and Viegas, D.X., Application of Parallel Computing to the Simulation of Forest Fire Propagation, Proc. of the 3nd International Conference on Forest Fire Research. Coimbra, Portugal, (1998)
10. Lopes, A., Modelaçao numérica e experimental do escoamento turbulento tridimensional em topografia complexa: aplicaçao ao caso de um desfiladeiro, PhD Dissertation (in portuguesse), Universidade de Coimbra, Portugal, (1993)
11. Geist, Al; Beguelin, Adam; Dongarra, Jack; Jiang, W., Mancheck, R. and Sunderman, V., PVM : Parallel Virtual Machine – A User's Guide and Tutorial for Networked Parallel Computing, The MIT Press, Cambridge, (1994)

A Data and Task Parallel Image Processing Environment

Cristina Nicolescu and Pieter Jonker

Delft University of Technology
Faculty of Applied Physics
Lorentzweg 1, 2628CJ Delft, The Netherlands
cristina@ph.tn.tudelft.nl

Abstract. The paper presents a data and task parallel environment for parallelizing low-level image processing applications on distributed memory systems. Image processing operators are parallelized by data decomposition using algorithmic skeletons. At the application level we use task decomposition, based on the Image Application Task Graph. In this way, an image processing application can be parallelized both by data and task decomposition, and thus beter speed-ups can be obtained. The framework is implemented using C and MPI-Panda library and it can be easily ported to other distributed memory systems.

Keywords: data parallelism, task parallelism, skeletons, image processing

1 Introduction

In the recent years image processing became a very important application area for parallel and distributed computing. Many algorithms have been developed for parallelizing different image operators on different parallel architectures. Most of these parallel image processing algorithms are either architecture dependent, or specifically developed for particular applications and very difficult to be implemented by an usual image processing user without enough knowledge of parallel computing.

In this paper we present an approach for adding data and task parallelism to an image processing library using *algorithmic skeletons* [2] and the Image Application Task Graph (IATG).

The paper is organized as follows. Section 2 presents skeletons for parallel low-level image processing on a distributed-memory system. Section 3 describes the Image Application Task Graph used in the task parallel framework. The multi-baseline stereo vision application together with its data parallel code using skeletons versus sequential code and the speedup results for the data parallel approach versus the data and task parallel approach is presented in Section 4. Finally, concluding remarks are made in Section 5.

Y. Cotronis and J. Dongarra (Eds.): Euro PVM/MPI 2001, LNCS 2131, pp. 393–400, 2001.

2   Skeletons for Low-Level Image Processing

Skeletons are algorithmic abstractions which encapsulate different forms of parallelism, common to a series of applications. The aim is to obtain environments or languages that hide the parallelism from the user [2]. In this paper we develop algorithmic skeletons to create a parallel image processing environment ready to use for an easy implementation/developing of parallel image processing applications.

Low-level image processing operators use the values of the image pixels to modify the image in some way. They can be divided into *point operators, neighborhood operators* and *global operators* [1]. These operators can be further classified in monadic, dyadic and triadic operators, according to the number of images involved in the computation of the output image. Point, neighborhood and global image processing operators can be parallelized using the data parallel paradigm with a master-slave approach.

Based on the above observations we identify a number of skeletons for parallel processing of low-level image processing operators. They are called according to the type of the low-level operator and the number of images involved in the operation. Headers of a few skeletons are shown below. All the skeletons are implemented in C using MPI [14] as message passing library.

```
void ImagePointDist_1IO(unsigned int n,char *n1,void(*im_op)()
    // DCG skeleton for monadic point operators : 1 input/output image
void ImagePointDist_1I_1O(unsigned int n,char *n1,char *n2,void(*im_op)());
    // DCG skeleton for dyadic point operators : 1input image/1output image
void ImageWinDist_1IO(unsigned int n,char *n1,Window *win,void(*im_op)());
    // DCG skeleton for neighborhood operators
void ImageGlobalDist_1IO(unsigned int n,char *n1,void(*im_op)());
    // DCG skeleton for global operators
```

With each skeleton we associate a parameter which represents the task number n corresponding to that skeleton. This is used by the task parallel framework. Depending on the skeleton type, one or more identifiers of the images are given as parameters. The last argument is the point operator for processing the image(s). Depending on the operator type and the skeleton type, there might exist additional parameters necessary for the image operator.

3   The Task Parallel Framework

Recently, it has been shown that exploiting both task and data parallelism in a program to solve very large computational problems yields better speedups compared to either pure data parallelism or either pure task parallelism [5,11]. The main reason is that both task and data parallelism are relatively limited, and therefore using only one of them bounds the achievable performance.

In order to fully exploit the potential advantage of the mixed task and data parallelism, efficient support for task and data parallelism is a critical issue. This can be done not only at the compiler level [5,7,8,9], but also at the application

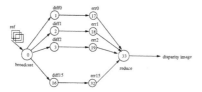

Fig. 1. IATG of the multibase-line stereo vision application

level and applications from the image processing field are very suitable for this technique. Mixed task and data parallel techniques use a directed acyclic graph, in the literature also called a Macro Dataflow Graph (MDG) [5], in which data parallel tasks are the nodes and the precedence relationships are the edges. For the purpose of our work we change the name of this graph to the Image Application Task Graph (IATG).

3.1   The Image Application Task Graph Model

A task parallel program can be modeled by a Macro Dataflow communication Graph [5], which is a directed acyclic graph $G = (V, E, w, c)$, where: V is the finite set of nodes which represents tasks (image processing operators), E is the set of directed edges which represent precedence constraints between tasks $e = (u, v) \in E$ if $u \prec v$, w is the weight function $w : V \to N^*$ which gives the weight (processing time) of each node (task)and c is the communication function $c : E \to N^*$ which gives the weight (communication time) of each edge.

An Image processing Application Task Graph (IATG) is, in fact, an MDG in which each node stands for an image processing operator and each edge stands for a precedence constraint between two adjacent operators, see Figure 1.

3.2   Processing Cost Model

A node in the IATG represents a processing task (an image processing operator applied via a skeleton) that runs non-preemptively on any number of processors. Each task is assumed to have a *computation cost*, denoted $T_{exec}(t, p)$, which is a function of the number of processors. The computation cost function of the task can be obtained either by *estimation* or by *profiling*.

For cost estimation we use Amdahl's law. According to it, the execution time of the task T_i is:

$$T_{i,exec}(t, p) = (\alpha + \frac{1 - \alpha}{p}) * \tau \quad (4)$$

where τ is the task's execution time on a single processor, p is the number of processors allocated to the node, and α is the fraction of the task that executes serially.

If we use profiling, the task's execution costs are either fitted to a function similar to the one described above (in the case that data is incomplete over the

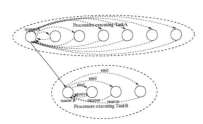

Fig. 2. Data redistribution between two tasks

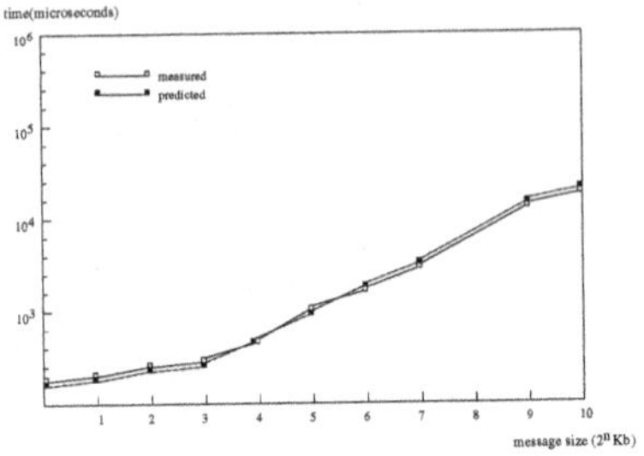

Fig. 3. Modelling the communication time

whole number of processors), or the profiled values can be used directly through a table.

3.3   Communication Cost Model

Data communication (redistribution) is essential for implementing an execution scheme which uses both data and task parallelism. Individual tasks are executed in a data parallel fashion on subsets of processors and the data dependences between tasks may necessitate not only changing the set of processors but also the distribution scheme.

We reduce the complexity of the problem first by allowing only one type of distribution scheme (row-stripe) and second by sending images only between two processors (the selected master processors from the two sets of processors), as shown in Figure 2.

An edge in the IATG corresponds to a precedence relationship and has associated a *communication cost*, denoted through $comm(T_i,T_j)$ which depends on the network characteristics (latency, bandwidth) and the amount of data to be transferred. Also in this case we can use either cost estimation or profiling to determine the communication time. It should be emphasized that there are two types of communication times. First, we have *internal communication time* which represents the time for internal transfer of data (distribution) between the processors allocated to a task. This quantity is part of the term α of the execution time associated to a node of the graph, see Formula 4. The internal communication time is actually the communication time of sending and receiving data (parts of images) between the master processor and the other processors allocated to a particular node. Secondly, we have *external communication time* which is the time of transferring data, i.e. images, between two processors. These two processors represent the master processors for the two associated image processing tasks (corresponding to the two adjacent graph nodes). This quantity is actually the *communication cost* of an edge of the graph.

Also in this case we can use either cost estimation or profiling to determine the communication time. In state-of-the-art of distributed memory systems the time to send a message containing L units of data from a processor to another processor can be modeled as:

$$T_{comm}(i,j) = t_s + L * t_b \quad (5)$$

where t_s,t_b are the startup and per byte cost for point-to-point communication and L is the length of the message, in bytes.

We run our experiments on a distributed memory system which consists of a cluster of Pentium Pro/200Mhz PCs with 64Mb RAM running Linux, and connected through Myrinet in a 3D-mesh topology, with dimension order routing [12]. As message passing library we have used MPI-Panda [14,15] which uses Panda [15] as a message passing layer. Figure 3 shows the performance of point-to-point communication operations and the predicted communication time. The reported time is the minimum time obtained over 20 executions of the same code. It is reasonable to select the minimum value because of the possible interference caused by other users' traffic in the network. From these measurements we perform a linear fitting and we extract the communication parameters t_s and t_b. In Figure 3 we see that the predicted communication time, based on the above formula, approximates very good the measured communication time.

3.4   IATG Cost Properties

A task with no input edges is called an *entry* task and a task with no output edges is called an *exit* task. Usually, the *entry* task is the node of the graph where all the input of data (initialization, reading of images from files) takes place. The *exit* task is the node of the graph which contains the output of the algorithm. The length of a path from the graph is the sum of the computation and communication costs of all nodes and edges belonging to the path. We define the *Critical Path* [5] (CP) as the longest path in the graph. If we have a graph with n nodes and t_i represent the finish time of node i associated with task t (time when task t finishes its execution), then the critical path is given by the formulas (6) and (7), where $PRED_i$ is the set of immediate predecessor nodes of node i, and p_i is the number of processors on which node (task) i is executed:

$$CP = t_n \quad (6)$$

$$t_i = \max_{p \in PRED_i} (t_p + T_{comm}(p,i)) + T_{i,exec}(t,p_i) \quad (7)$$

We define the *Average Area* [5] (A) of an IATG for a P processor system as in formula (8), where p_i is the number of processors allocated to task T_i.

$$A = \frac{1}{P} * \sum_{1}^{n} T_{i,exec} * p_i \quad (8)$$

The critical path represents the longest path in the IATG and the average area provides a measure of the processor-time area required by the IATG. Based on these two formulas, processors are allocated to tasks according to the results obtained by solving the following minimization problem:

$$\phi = min(max(A, CP)) \quad (9) \quad \text{subject to} \quad 1 \leq p_i \leq P \quad \forall i = 1, n$$

After solving the allocation problem, a scheduler is needed to schedule the tasks to obtain a minimum execution time. Multiple processor task scheduling is NP-complete and heuristics are used [5,6,7,11].

The intuition behind minimizing ϕ in equation (9) is that ϕ represents a theoretical lower bound on the time required to execute the image processing application corresponding to the IATG. The execution time of the application can neither be smaller than the critical path of the graph nor be less than the average area of the graph.

For solving the previous min-max problem a nonlinear solver SNOPT [13] can be used, available on the Internet. For solving the scheduling problem, one of the proposed scheduling algorithms [5,6,7,11] may be used.

4   Experiments

To evaluate the benefits of the proposed data parallel framework based on skeletons and also of the task parallel framework based on the IATG we compare the speed-ups obtained by applying only data parallelism to the application, with the speed-ups obtained with both data and task parallelism.

The multi-baseline stereo vision application uses an algorithm developed by Okutomi and Kanade [3] and described by Webb and al. [4], that gives greater accuracy in depth through the use of more than two cameras. Input consists of three $n \times n$ images acquired from three horizontally aligned, equally spaced cameras. One image is the *reference image*, the other two are named *match images*. For each of 16 disparities, $d = 0, .., 15$, the first match image is shifted by d pixels, the second image is shifted by $2d$ pixels. A *difference image* is formed by computing the sum of squared differences between the corresponding pixels of the reference image and the shifted match images. Next, an *error image* is formed by replacing each pixel in the difference image with the sum of the pixels in a surrounding 13×13 window. A *disparity image* is then formed by finding, for each pixel, the disparity that minimizes error. Finally, the depth of each pixel is displayed as a simple function of its disparity. Figure 1 presents the IATG of this application. It can be observed that the computations of the difference images can be executed in (a task) parallel (way). The same holds for the computation of the error images.

It can be observed that the computation of the difference images requires point operators, while the computation of the error images requires neighborhood operators. The computation of the disparity image requires also a point operator.

```
Input: ref, m1, m2 (the reference and the two match images)
for d=0,15
   Task T1,d: m1 shifted by d pixels
   Task T2,d: m2 shifted by 2*d pixels
```

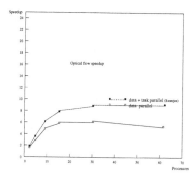

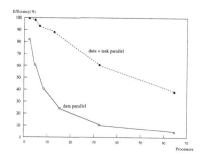

Fig. 4. Speed-up of the stereo vision algorithm

Fig. 5. Efficiency of the stereo vision algorithm

```
Task T3,d: diff = (ref-m1)*(ref-m1)+(ref-m2)*(ref-m2)
  Task T4,d: err(d) = sum diff[i,j]
Task T5: Disparity image = d which minimizes the err image
```

The results of the data parallel approach are compared with the results obtained using data and task parallelism on a distributed memory system which consists of a cluster of Pentium Pro/200Mhz PCs with 64Mb RAM running Linux [12], and connected through Myrinet in a 3D-mesh topology, with dimension order routing. In the task parallel framework we use a special mechanism to register the images on the processors where they are first created. Moreover, each skeleton has associated the task number to which it corresponds. We use 1, 2, 4, 8, 16, 32 and 64 processing nodes in the pool. The image size is 256×256. The code is written using C and MPI message passing library. In Figure 4 we show the speed-ups obtained for the data parallel approach versus the data and task parallel approach, for a 256×256 image size. The stereo vision application is an example of a regular and well balanced application, for which usually the schedulers [5,6,7,11] give the same results. The previous figure presents the results obtained using the scheduler described by Banerjee et al, see [5]. Figure 5 shows the efficiency of the same application using the two parallel paradigms. We can observe that the speed-ups become quickly saturated for the data-parallel approach while the speed-ups for the data and task parallel approach perform very good. In fact, we have pure task parallelism till 16 processors and data and task parallelism from 32 on. Using both data and task parallelism is more efficient than using only data parallelism.

5   Conclusions

We have presented an environment for data and task parallel image processing. The data parallel framework, based on algorithmic skeletons, is easy to use by any image processing user. The task parallel environment is based on the Image Application Task Graph and computing the communication and processing costs. If the IATG is a regular well balanced graph task parallelism can be applied without the need of these computations. We showed an example of using skeletons and the task parallel framework for the multi-baseline stereo vision application. The multi-baseline stereo vision is an example of an image processing application which contain parallel tasks, each of the tasks being a very simple image point or neighborhood operator. Using both data and task parallelism si more efficient than using only data parallelism. Our code for the data and task parallel environment, written using C and MPI-Panda library [14,15] can be easily ported to other parallel machines.

References

1. I.Pitas: Parallel Algorithms for Digital Image Processing, Computer Vision and Neural Networks, John Wiley&Sons, 1993.
2. M. Cole: "Algorithmic skeletons: structured management of parallel computations", Pitman/MIT Press, 1989.
3. M. Okutomi, and T. Kanade: A multiple-baseline stereo, in *IEEE Transactions on Pattern Analysis and Machine Intelligence*, 15(4):353-363, 1993.
4. J. Webb and al.: The CMU Task Parallel Program Suite, *Technical Report Carnegie Mellon University*, CMU-CS-94-131, 1994.
5. S. Ramaswamy, S. Sapatnekar and P. Banerjee: A framework for exploiting task and data parallelism on distributed memory multicomputers, in *IEEE transactions on parallel and distributed systems*, vol. 8, no. 11, November 1997.
6. T. Rauber, and G. Runger: Compiler support for task scheduling in hierarchical execution models, in *Journal of Systems Architecture*, vol.45:483-503, 1998.
7. J. Subhlok, and B. Yang: A new model for integrated nested task and data parallel programming, in *Proceedings of the Symposium on Parallel Algorithms and Architectures*, 1992.
8. T.I. Foster, and K.M. Chandy: Fortran M: A language for modular parallel programming, in *Journal of Parallel and Distributed Computing*, 26:24-35, 1995.
9. S.B. Hassen, H.E. Bal, and C.J. Jacobs: A task and data parallel programming language based on shared objects, in *ACM Transactions on Programming Languages and Systems*, 20(6):1131-1170, 1998.
10. R.L. Graham: Bounds on multiprocessing timing anomalies, in *SIAM Journal on Applied Mathematics*, 17(2):416-429, 1969.
11. A. Radulescu, C. Nicolescu, A. van Gemund and P.P. Jonker: CPR: Mixed Task and Data Parallel Scheduling for Distributed Systems, in *CDROM Proceedings of The 15th International Parallel & Distributed Symposium (IPDPS'2001)* , Best Paper Award, 2001.
12. The Distributed ASCI supercomputer (DAS) site, http://www.cs.vu.nl/das.
13. P.E. Gill, W. Murray, and M.A. Sanders: User's guide for snopt 5.3: A fortran package for large-scale nonlinear programming, *Technical Report SOL-98-1*, Stanford University, 1997.
14. M.Snir, S.Otto, S.Huss, D.Walker and J.Dongarra: "MPI - The Complete Reference, vol.1, The MPI Core", The MIT Press, 1998.
15. T.Ruhl, H, Bal, R. Bhoedjang, K. Langendoen and G. Benson: *Experience with a portability layer for implementing parallel programming systems*, Proceedings of International Conference on Parallel and Distributed Processing Techniques and Applications, pp. 1477-1488, Sunnyvale CA, 1996

Evaluating the DIPORSI Framework: Distributed Processing of Remotely Sensed Imagery*

J.A. Gallud[1], J. García-Consuegra[1], J.M. García[2], and L. Orozco[3]

[1] Departamento de Informática, Universidad de Castilla-La Mancha,
Campus Universitario, 02071 Albacete, Spain
jgallud@info-ab.uclm.es
[2] Departamento de Ingeniería y Tecnología de Computadores, Facultad de
Informática, Universidad de Murcia, Campus de Espinardo, 30080 Murcia, Spain
jmgarcia@ditec.um.es
[3] School of Information Technology and Engineering, University of Ottawa,
161 Louis Pasteur St, Ottawa, ON, K1N 6N5, CANADA
lbarbosa@uottawa.ca

Abstract. The recent advances in remote sensing are contributing to improve emerging applications like geographical information systems. One of the constraint in remotely-sensed images processing is the computational resources required to obtain an efficient and precise service. Parallel processing is applied in remote sensing in order to reduce spatial or temporal cost using the message passing paradigm. In this paper, we present our experiences in the design of a workbench, called DIPORSI, developed to provide a framework to perform the distributed processing of Landsat images using a cluster of workstations. We have focused in describing the results obtained in the implementation of the georeferring function on a biprocessor cluster. In addition, this work offers a set of reflections about the design of complex distributed software systems like the distributed geographic information systems are.

1 Introduction

Recent advances in remote sensing are contributing to improve the application domain of geographical information systems. Remote sensing involves the manipulation and interpretation of digital images which have been captured from remote sensors on board of satellite or aircraft systems. Such images collect information about the Earth's surface, which allow scientists to perform many environmental studies.

Remotely sensed image processing is an interesting application area for distributed computing techniques, which has become more and more attractive both for the spatial resolution increase of the new sensor generation, and for the

---

* This work has been partially supported by Spanish CICYT, under grants TIC 2000-1151-C07-03 and TIC 2000-1106-C02-02

Y. Cotronis and J. Dongarra (Eds.): Euro PVM/MPI 2001, LNCS 2131, pp. 401–409, 2001.

emerging use of Geographic Information Systems (GIS) in environment studies [12]. The large data volumes involved and the consequent processing bottleneck may indeed reduce their effective use in many real situations, and hence the need for exploring the possibility of splitting both the data and processing over several storing and computing units [14,1].

Remote sensed imagery processing involves several steps that begin when a image is acquired by the satellite, and continue with the image analysis and processing to obtain, finally, a variety of results. All the procedures involved in the image manipulation of such images may be categorized into one or more of the following four broad types of computer assisted operations: image rectification and restoration, image enhancement, image classification and data merging [11].

In this paper, we describe a distributed workbench designed to perform a considerable number of the former tasks by using a cluster of workstations composed by either Windows NT/Windows 2000 or Linux platforms which are connected by means of an ethernet network using the Message-Passing Interface standard (MPI). MPI provides an interface to design distributed applications that run on a parallel system [13].

DIPORSI is the name of our distributed workbench, and it is related with the distributed processing of remotely sensed imagery. DIPORSI has been designed to be one of the modules of a more ambitious software project which is a distributed geographic information system.

In addition, this paper shows how the long computation time required by remote sensing procedures can be reduced without using high-cost parallel machines. All the distributed applications are implemented by splitting the original images into several small ones, which are processed in each node parallelly.

These practical results have been obtained from the implementation of the Landsat imagery rectification operation on several platforms, as a computer intensive function useful to show the global behaviour of the system.

Another important consideration about the practical results is that they can help us to improve DIPORSI in order to implement real time GIS.

The following section explains the georeferring function. Section 3 outlines the related work. In section 4, we describe the structure of our distributed workbench. In section 5, the distributed version of the georeferring function is presented. Section 6 shows the results obtained. Finally, in section 7 the conclusions and future work are presented.

2   The Georeferring Process

In remotely sensed of images, a number of different errors appear distorting them, in such a way that they cannot be used as maps. In order to use a remotely sensed in a GIS, a rectify process must be carried out before images can be used in the GIS applications[11].

The process of georeferring a satellite image consists in the application of a set of mathematical operations on the original image to obtain a geometrically corrected image. The rectification function is the process carried out to correct both radiometric and geometric errors from a remotely sensed image due to different causes (the influence of the atmosphere, earth surface and so on). So, the purpose of geometric correction is to compensate for the distortions introduced by different factors (earth curvature, relief displacement, etc). The aim of this process is that the corrected image has the geometric integrity of a map [10].

This process is usually implemented using one of two possible procedures [8,11]. In the first one, the procedure uses the orbital model reproducing the physical conditions of the image acquisition. The second procedure applies the traditional polynomial correction algorithm with or without using a digital model terrain.

The most frequent cases are based on the second step related to random distortions. This kind of distortion is corrected by analyzing well-distributed ground control points occurring in an image. The GCP (ground control points) are relevant features of some known ground locations that can be accurately located on the digital imagery. Then, these values are submitted to a least-squares regression analysis to determine coefficients for two coordinate transformation equations that can be used to relate the geometrically correct coordinates and the distorted image coordinates. Once the coefficients for these equations are determined, the distorted image coordinates for any map position can be precisely estimated by using this expression:

$$X = f1(x, y)$$
$$Y = f2(x, y)$$

Where (X,Y) are the distorted image coordinates, (x,y) are the correct coordinates (map) and $f1$ and $f2$ are the transformation functions. The process is actually made inversely. An undistorted output matrix of empty map cells is defined, and then each cell is filled in with the correct coordinates, from the value of the gray level of the corresponding pixel, or pixels (depending on the method employed) of the distorted image. In a nutshell, the process is performed using the following operations:

1. The coordinates of each element in the undistorted image are transformed to determine their corresponding location in the original distorted image.
2. The intensivity value or digital number (DN) of the undistorted image is determined by using one of these usual methods: the nearest neighbor, bilinear interpolation and cubic convolution.

Therefore, it can be used three methods to obtain the digital number of each pixel. The nearest neighbor needs one only read for pixel. The bilinear interpolation uses four read operations to compute the digital number of each pixel. Finally, the cubic convolution method needs sixteen read operations to compute the DN of each pixel in the target image.

3   Related Work

There are a number of sensor features that must be considered before proceeding to perform the correction. So, firstly related work using the Themathic Mapper sensor aboard the Landsat satellite is presented, because this is the sensor we are interested in.

There are a few research work regarding the remotely sensed image geometric and radiometric correction process. An interesting paper covering the atmospheric correction of Landsat imagery can be mentioned [2]. In [4], we can find the radiometric and geometric SAR image rectification.

Despite the fact image rectification has not been enough treated in previous work, there is a explicit reference in many papers focused in image classification [1,14,6].

In [7], we can find the design of the MMIPPS system, a software package for multitemporal and multispectral image processing on parallel systems. The work offers a parallel approach to the classification process.

The purpose of many others research groups is the design of sophisticated remote sensing systems. In [15], the application of parallel paradigm to ecosystems monitoring can be found. A more ambitious system is the one designed by the U.S. Geological Survey to cover a variety of applications [9].

4   The DIPORSI Workbench

A first approach to the DIPORSI system was introduced in the development of a previous work [3]. As it was mentioned there, DIPORSI was configured as an execution module inside an image processing system.

The basic idea behind remotely sensed image processing is simple. The digital image is fed into a computer one pixel at a time, usually together with its neighbors. The computer is programmed to insert these data into an equation, or series of equations, and then store the results of the computation for each pixel. These results make up a new digital image that may be displayed or recorded in a raster format or may itself be further manipulated by additional programs [8].

Many remote sensing algorithms work in a parametric way, that is, the application performance is related to the initial parameters. The final process usually employs classifier algorithms and their behaviour are governed by a set of initial parameters. To get a classified image, the algorithm starts its computation with such inital values as yield the resultant image. This resultant image depends on both the initial parameters as well as the original multidimensional image. In many cases, the process must be repeated with other parameter values because the results are unacceptable. Thus, it is easy to understand the need to reduce both spatial and temporal costs when a trial-and-error process is employed.

Our goal was to implement an environment to perform remotely sensed image processing distributely where both the process and data can be distributed. So, we designed DIPORSI workbench, which stands for *DIstributed Processing Of Remotely Sensed Imagery.*

The final objective of our research team is to design a distributed geographic information system (D-GIS). One of the functions offered is the image processing module. In this module, a user can enter a set of complex image operations by means a script language.

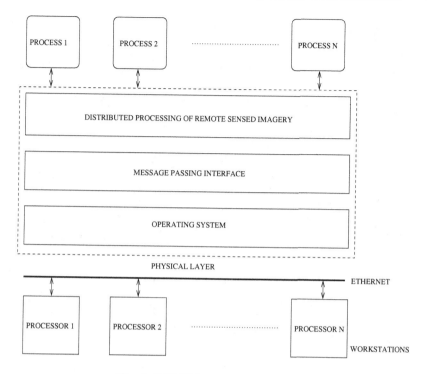

Fig. 1. DIPORSI functional diagram

In figure 1, the functional diagram of DIPORSI is exposed. DIPORSI appears as a layer between MPI functions and the code of each process. DIPORSI offers a set of functions and a message structure to the user for performing easily and distributively whatever remote sensing algorithm it chooses. DIPORSI runs in a batch way, and it needs a parametric file with the following information generates by the user:

- The algorithm, or the sequence of algorithms, to compute: i.e. the georeferring algorithm using the bilinear interpolation method.
- The workload allocation: how many nodes and how much information in each node must be defined.
- The number and the location of the data images.

DIPORSI offers a set of defined functions to generate special messages to interpretate them, a number of functions for managing files of different kinds such as text files, data files, raster images, etc.

Our workbench allows the user to work by distributing both spatially and temporally either of the remote sensing algorithms. That is, either user can make all the nodes perform the same computations on different data or the user can make each node runs different computation on the same image.

When a distributed approach of a particular algorithm is defined, its special features must be studied in order to improve the efficiency. DIPORSI follows a general scheme which is applied to all remote sensing algorithms. Such a scheme is structured into three steps:

1. The images (in our example, the distorted images) are broadcasted into the nodes.
2. The computations are performed following the particular algorithm in each node.
3. The resultant image is restored by the nodes usually on the master computer.

In order to confer a greater generality to our development, and with a view to further applications, two types of messages have been defined: control and data. The first ones are used to specify the activity to be carried out and the parameters necessary for the algorithm, as well as for the control of the MPI implementation. The second type of messages are used for the passing of data or image files, as well as pixel streams.

5   The Distributed Georeferring Process

In this section, the distributed algorithm is described and the platforms used to implement it are presented. So, we are going to show the scheme of the distributed algorithm.

```
compute_target_dimensions(max_rows,max_columns)
compute_distribution(nrows) distribute_control_data
[distribute_data]
 for b=1 to max_bands do
 for i=firstrow to nrows do
  for j=1 to max_columns do
    get_coordinates(i,j,x,y)
    nd=read_digital_number(b,x,y,method)
    write_pixel(b,i,j,nd)
  next j
 next i
next b
```

The *compute_distribution* function gets the number of rows each node has to compute. The *distribute_control_data* function is the method used by the master process to communicate each node the task to perform and the data to process. The *distribute_data* function appears between brackets because it is an optional function depending the data location (local or remote).

The first step is made by the master process, which opens a file that contains the information of the task to be solved. The root process is initiated with a parametric file, in which all the activities to be performed and their parameters are specified in a ordered way. The master process sends such information to all the nodes involved. Thus, each node knows what it must do (correction method) and how much information it must receive (number of bands -the resultant image to a given frequency-, resolution, the transformation functions, etc).

At this point, the master process sends the Landsat image to all the nodes. The special features of the georeference algorithm force us to send all the data to all the nodes. As explained above, the resultant image is a combination of rotations and translations of the original distorted image.

On the other hand, the MPI broadcast function allows us to distribute the image easily, but unfortunately at a high cost.

The second step is executed in each node, which acts on the region of the distorted image where the computations must perform the corrections. As soon as a row of the resultant image is computed, it is sent to the master process, so computation and communication are overlapped. In addition, when a node is involved in a communication, the others can carry out correction computations.

The second and third step run at the same time. The master process receives partial data from nodes, together with the information to restore the resultant image. This is made by using a special message with information about the location (x and y coordinates and the number of the band) of the received data.

6   Evaluation Environment and Results

A Landsat scene has 7 bands, 6 with 30x30mts and 1 with 120x120 mts of resolution respectively, with approximately 40MB each band, which explains the high value of the response time when geometric correction is computed in a single machine.

DIPORSI works by splitting the original distorted image into number-of-rows/number-of-nodes blocks in accordance with an uniform workload allocation, then each node computes a small submatrix and the computation time can be reduced and the overall performance of the algorithm can be improved.

There are different MPI implementations, most of them designed to run over Windows 2000, Unix or Linux workstations. We have used a MPI implementation on Windows NT named WinMPICH 0.9b version [5] and two known MPI implementations for Linux called LAM and MPICH.

We have tested different georeferring implementations of DIPORSI over different hardware and software platforms. In a previous work [3], we describe a detail of the results obtained from different platforms and different georeferring methods.

In this work we show the evaluation of DIPORSI in the next hardware platform: Pentium II 300MHz dual, 64MB and Linux RedHat and NFS over fast ethernet.

To the interpretation of the results, we must take into account the following considerations:

1. All data times show service or response time.
2. In order to measure the service time we have used the times function.
3. The geometric rectification of a three band scene means that we have to open three original images and three target ones. In the worst case, we have to open three 37MB original images and three over 50MB target images simultaneously.

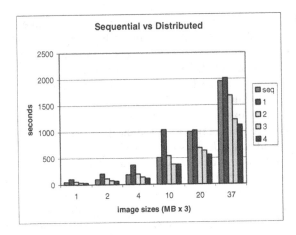

Fig. 2. Sequential vs distributed georeferring

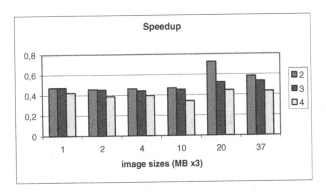

Fig. 3. Speedup of the data showed in the figure 2

4. We consider only the cubic convolution interpolation method.

Figure 2 shows a real view of the performance obtained when we compare the sequential version and the distributed ones. Figure 3 shows the speedup corresponding to the figure 2 execution. The benefits of using a cluster appears when large image sizes are used together with a intensive algorithm, as cubic convolution, is needed. We have tested that there is a lower advantage when nearest neighbor is selected.

7   Conclusions and Future Work

This paper describes a distributed workbench to implement the main remote sensing algorithms, using a cluster of workstations, called DIPORSI. The implementation shows the timing results when different image sizes are used, as well as the time gained with the parallel algorithm for georeferring. MPI proved useful for implementing distributed applications on low-cost platforms, which can contribute to designing efficient solutions in remote sensing.

Future work admits of several possibilities: adding dynamic workload allocation, the implementation of other algorithms to solve geometric correction, the implementation of radiometric correction, automatic detection of the ground control points, and other tasks intended to improve the communications between nodes.

References

1. D. A. Bader, *Parallel Algorithms for Image Enhancement and Segmentation by Region Growing with an Experimental Study* (In The Journal of Supercomputing. Vol 10, No 2, 1996) 141-168.
2. H. Fallah-Adl et al, *Fast Algorithms for Removing Atmospheric Effects from Sattellite Images* (In IEEE Computational Science and Engineering, Vol. 3, No. 2, Summer 1996).
3. J. A. Gallud, J. M. García y J. García-Consuegra, *Cluster Computing Using MPI and Windows NT to Solve the Processing of Remotely Sensed Imagery* (Recent Advances in Parallel Virtual Machine and Message Passing Interface. Lecture Notes in Computer Science 1697, Springer-Verlag, 1999) 442-449.
4. I. Glendinning, *Parallelisation of a Satellite Signal Processing Code - Strategies and Tools* (In ACPC '99. LNCS 1557, 1999) 388-397.
5. W. Gropp, E. Lusk, N. Doss, A. Skjellum, *A High-Performance, Portable Implementation of the MPI Message Passing Interface Standard* (Parallel Computing, Vol. 22, No. 6, 1996) 789-828.
6. F. M. Hoffman, W. W. Hargrove, *Multivariate Geographic Clustering Using a Beowulf-style Parallel Computer* (In Proceedings of the International Conference on Parallel and Distributed Processsing Techniques and Applications, PDPTA '99, Volume III, H. R. Arabnia, Ed. ISBN 1-892512-11-4, CSREA Press) 1292-1298.
7. J. Janoth et al, *MMIPPS - A Software Package for Multitemporal and Multispectral Image Processing on Parallel Systems* (In ACPC '99. LNCS 1557, 1999) 398-407.
8. T.M. Lillesand and R.W. Kiefer, *Remote Sensing and Image Interpretation 3rd Edition* (J.Wiley & Sons), 1994.
9. U.S. Geological Survey, *Earth Resources Observations Systems*, (MAGIC-II Project, 1997, http://edciss1.cr.usgs.gov).
10. B.L. Markham, *The Landsat Sensors' Spatial Responses* (IEEE Transactions on Geoscience and Remote Sensing, Vol. GE-23, No. 6, 1986) 864-875.
11. P.M. Mather, *Computer Processing of Remotely-Sensed Images* (John Wiley & Sons), 1987.
12. J.A. McCormick and R. Alter-Gartenberg and F.O. Huck, *Image Gathering and restoration: information and visual quality* (Journal of Optical Society of America, Vol 6, No. 7, 1989) 987-1005.
13. MPI Forum, *MPI:A message-passing interface standard* (International Journal of Supercomputer Applications 8(3/4), 1994).
14. J.C. Tilton, *A Recursive PVM Implementation of an Image Segmentation Algorithm with Performance Results Comparing the HIVE and the Cray T3E* (In Proceedings of the IEEE Symposium of Massively Parallel Processing, 1998).
15. C. J. Turner and G. J. Turner, *Adaptive Data Parallel Methods for Ecosystem Monitoring* (In Proceedings of the ACM Conference on Supercomputing '94, 1994) 281-291.

Scalable Unix Commands for Parallel Processors: A High-Performance Implementation[*]

Emil Ong, Ewing Lusk, and William Gropp

Mathematics and Computer Science Division
Argonne National Laboratory
Argonne, Illinois 60439 USA

Abstract. We describe a family of MPI applications we call the Parallel Unix Commands. These commands are natural parallel versions of common Unix user commands such as `ls`, `ps`, and `find`, together with a few similar commands particular to the parallel environment. We describe the design and implementation of these programs and present some performance results on a 256-node Linux cluster. The Parallel Unix Commands are open source and freely available.

1 Introduction

The oldest Unix commands (`ls`, `ps`, `find`, `grep`, etc.) are built into the fingers of experienced Unix users. Their usefulness has endured in the age of the GUI not only because of their simple, straightforward design but also because of the way they work together. Nearly all of them do I/O through `stdin` and `stdout`, which can be redirected from/to files or through pipes to other commands. Input and output are lines of text, facilitating interaction among the commands in a way that would be impossible if these commands were GUI based.

In this paper we describe an extension of this set of tools into the parallel environment. Many parallel environments, such as Beowulf clusters and networks of workstations, consist of a collection of individual machines, with at least partially distinct file systems, on which these commands are supported. A user may, however, want to consider the collection of machines as a single parallel computer, and yet still use these commands. Unfortunately, many common tasks, such as listing files in a directory or processes running on each machine, can take unacceptably long times in the parallel environment if performed sequentially, and can produce an inconveniently large amount of output.

A preliminary version of the specification of our Parallel Unix Commands appeared in [4]. New in this paper are a refinement of the specification based on experience, a high-performance implementation based on MPI for improved scalability, and measurements of performance on a 256-node Unix cluster.

---

[*] This work was supported by the Mathematical, Information, and Computational Sciences Division subprogram of the Office of Advanced Scientific Computing Research, U.S. Department of Energy, under Contract W-31-109-Eng-38.

Y. Cotronis and J. Dongarra (Eds.): Euro PVM/MPI 2001, LNCS 2131, pp. 410–418, 2001.

The tools described here might be useful in the construction of a cluster-management system, but this collection of user commands does not itself purport to *be* a cluster-management system, which needs more specialized commands and a more extensive degree of fault tolerance. Although nothing prevents these commands from being run by `root` or being integrated into cluster-management scripts, their primary anticipated use is the same as that of the classic Unix commands: interactive use by ordinary users to carry out their ordinary tasks.

2 Design

In this section we describe the general principles behind this design and then the specification of the tools in detail.

2.1 Goals

The goals for this set of tools are threefold:

- They should be familiar to Unix users. They should have easy-to-remember names (we chose `pt<unix-command-name>`) and take the same arguments as their traditional counterparts to the extent consistent with the other goals.
- They should interact well with other Unix tools by producing output that can be piped to other commands for further processing, facilitating the construction of specialized commands on the command line in the classic Unix tradition.
- They should run at interactive speeds, as do traditional Unix commands. Parallel process managers now exist that can start MPI programs quickly, offering the same experience of immediate interaction with the parallel machine, while providing information from numerous individual machines.

2.2 Specifying Hosts

All the commands use the same approach to specifying the collection of hosts on which the given command is to run. A host list can be given either explicitly, as in the blank-separated list `'donner dasher blitzen'`, or implicitly in the form of a pattern like `ccn%d@1-32,42,65-96`, which represents the list `ccn1,...,ccn32,ccn42,ccn65,...,ccn96`.

All of the commands described below have a hosts argument as an (optional) first argument. If the environment variable `PT_MACHINE_FILE` is set, then the list of hosts is read from the file named by the value of that variable. Otherwise the first argument of a command is one of the following:

`-all` all of the hosts on which the user is allowed to run,
`-m` the following argument is the name of a file containing the host names,
`-M` the following argument is an explicit or pattern-based list of machines.

Table 1. Parallel UNIX Commands

Command	Description
ptchgrp	Parallel chgrp
ptchmod	Parallel chmod
ptchown	Parallel chown
ptcp	Parallel cp
ptkillall	Parallel killall (Linux semantics)
ptln	Parallel ln
ptmv	Parallel mv
ptmkdir	Parallel mkdir
ptrm	Parallel rm
ptrmdir	Parallel rmdir
pttest[ao]	Parallel test

Command	Description
ptcat	Parallel cat
ptfind	Parallel find
ptls	Parallel ls
ptfps	Parallel process space find
ptdistrib	Distribute files to parallel jobs
ptexec	Execute jobs in parallel
ptpred	Parallel predicate

Thus

```
ptls -M "ccn%d-myr@129-256" -t /tmp/lusk
```

runs a parallel version of ls -t (see below) on the directory /tmp/lusk on nodes with names ccn129-myr,...,ccn256-myr.

2.3   The Commands

The Parallel Unix Commands are shown in Table 1. They are of three types: straightforward parallel versions of traditional commands with little or no output; parallel versions of traditional commands with specially formatted output; and new commands in the spirit of the traditional commands but particularly inspired by the parallel environment.

Parallel Versions of Traditional Commands. The first part of Table 1 lists the commands that are simply common Unix commands that are to be run on each host. The semantics for many of these is very natural – the corresponding uniprocessor version of any command is run on every node specified. For example, the command

```
ptrm -M "node%d@1-5" -rf old_files/
```

is equivalent to running

```
rm -rf old_files/
```

on node1, node2, node3, node4, and node5. The command line arguments to most of the commands have the same meaning as their uniprocessor counterparts.

The exceptions ptcp and ptmv deserve special mention; the semantics of parallel copy and move are not necessarily obvious. The commands presented here perform one-to-many copies by using MPI and compression; ptmv deletes the local files that were copied if the copy was successful. The command line arguments for `ptcp` and `ptmv` are identical to their uniprocessor counterparts with the exception of an option flag, `-o`. This flag allows the user to specify whether compression is used in the transfer of data. In the future the flags may be expanded to allow for other customizations. Handling of directories as either source or destination is handled as in the normal version of `cp` or `mv`.

Parallel `test` also deserves explanation. There are two versions of parallel `test`; both run `test` on all specified nodes, but `pttesta` logically ANDs the results of the tests, while `pttesto` logically ORs the results of the tests. By default, `pttest` is an alias for `pttesto`. This link allows the natural semantics of `pttest` to detect failure on any node.

Parallel Versions of Common UNIX Commands with Formatted Output. The second set of commands in Table 1 may produce a significant amount of output. In order to facilitate handling of this output, if the first argument to `ptfind`, `ptls`, or `ptcat` is `-h` (for "headers"), then the output from each host will be preceded by a line identifying the host. This is useful for piping into other commands such as `ptdisp` (see below). In the example

```
$ ptls -M "node%d@1-3" -h
[node1.domain.tld]
myfile1
[node2.domain.tld]
[node3.domain.tld]
myfile1
myfile2
```

the user has file `myfile1` on node1, no files in the current directory on node2, and the files `myfile1` and `myfile2` on node3. All other command line arguments to these commands have the same meaning as their uniprocessor counterparts.

To facilitate processing later in a pipeline by filters such as `grep`, we provide a filter that *spreads* the hostname across the lines of output, that is,

```
$ ptls -M "node%d@1-3" -h | ptspread
node1.domain.tld:  myfile1
node3.domain.tld:  myfile1
node3.domain.tld:  myfile2
```

New Parallel Commands. The third part of Table 1 lists commands that are in the spirit of the other commands but have no non-parallel counterpart.

Many of the uses of `ps` are similar to the uses of `ls`, such as determining the age of a process (respectively, a file) or owner of a process (respectively, a file). Since a Unix file system typically contains a large number of files, the Unix command `find`, with its famously awkward syntax, provides a way to search the file system for files with certain combinations of properties. On a single system, there are typically not so many processes running that they cannot be perused with `ps` piped to `grep`, but on a parallel system with even a moderate number of hosts, a `ptps` could produce thousands of lines of output. Therefore, we have proposed and implemented a counterpart to `find`, called `ptfps`, that searches the process space instead of the file space. In the Unix tradition we retain the syntax of `find`. Thus

```
ptfps -all -user lusk
```

will list all the processes belonging to user `lusk` on all the machines in a format similar to the output of `ps`, and

```
ptfps -all -user gropp -time 3600 -cmd ^mpd
```

will list all processes owned by `gropp`, executing a command beginning with `mpd`, that have been running for more than an hour. Many more filtering specifications and output formats are available; see the (long) `man` page for `ptfps` for details.

The command `ptdistrib` is effectively a scheduler for running a command on a set of files over specified nodes. For example, to compile all of the C files in the current directory over all nodes currently available, then fetch back all the resulting files, the user might use the following command:

```
ptdistrib -all -f 'cc -c {}' *.c
```

Here, the {} is replaced by the names of the files given, one by one. See the `man` page for more information.

The command `ptexec` simply executes a command on all nodes. To determine, for example, which hosts were available for running jobs, the user might run the following command:

```
ptexec -all hostname
```

No special formatting of output or return code checking is done.

The command `ptpred` runs a `test` on each specified node and outputs a 0 or 1 based on the result of the test. For example, to test for the existence of the file `myfile` on nodes node1, node2, and node3, the user might have the following session:

```
$ ptpred -M "node1 node2 node3" '-f myfile'
node1.domain.tld: 1
node2.domain.tld: 0
node3.domain.tld: 1
```

In this case, node1 and node3 have the file, but node2 does not. Note that `ptpred` prints the logical result of `test`, not the verbatim return value.

The output of `ptpred` can be customized:

```
$ ptpred -M "node1 node2 node3" '-f myfile' \
          'color black green' 'color black red'
node1.domain.tld: color black green
node2.domain.tld: color black red
node3.domain.tld: color black green
```

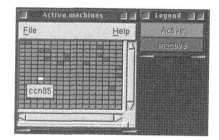

Fig. 1. Screenshots from `ptdisp`

This particular customization is useful as input to `ptdisp`, which is a general display tool for displaying information about large groups of machines. As an example, Figure 1 shows some screenshots produced by `ptdisp`.

The command `ptdisp` accepts special input from standard input of the form

```
<hostname>: <command> [arguments]
```

where `command` is one of `color`, `percentage`, `text`, or a number. The output corresponding to each host is assigned to one member of an array of button boxes.

As an example, one might produce the screenshot on the left in Figure 1 with the following command:

```
ptpred -all '-f myfile' 'color black white' \
                        'color white black' \
| ptdisp -c -t "Where myfile exists"
```

to find on which nodes a particular file is present. The command `ptdisp` can confer scalability on the output of other commands not part of this tool set by serving as the last step in any pipeline that prepares lines of input in the form it accepts. Since it reads perpetually, it can even serve as a crude graphical system monitor, showing active machines, as on the right side of Figure 1. The command to produce this display is given in Section 3. The number of button boxes in the display adapts to the input. When the cursor is placed over a box, the node name automatically appears, and clicking on a button box automatically starts an `xterm` with an `ssh` to that host if possible, for remote examination.

3   Examples

Here we demonstrate the flexibility of the command set by presenting a few examples of their use.

– To look for nonstandard configuration files:

```
ptcp -all mpd.cfg /tmp/stdconfig; \
```

```
ptexec -all -h diff /etc/mpd.cfg /tmp/stdconfig \
| ptspread
```

This shows differences between a standard file and the version on each node.
- To look at the load average on the parallel machine:

```
ptexec -all 'echo -n `hostname` ; uptime' | awk '{ print $1 \
": percentage " $(NF-1)*25 }' | sed -e 's/,//g' | ptdisp
```

The `percentage` command to `ptdisp` shows color-coded load averages in a compact form.
- To continuously monitor the state of the machine (nodes up or down)

```
(echo "$LEGEND$: Active black green Inactive black red"; \
while true; do (enumnodes -M 'ccn%d@1-256' \
| awk '{print $1 ": 0"}') ; sh ptping.sh 'ccn%d@1-256'; \
sleep 5; done) | ptdisp -t "Active machines" -c
```

We assume here that `ptping` pings all the nodes. This is admittedly ugly, but it illustrates the power of the Unix command line and the interoperability of Unix commands. The output of this command is what appears on the right side of Figure 1.
- To kill a runaway job

```
ptfps -all -user ong -time 10000 -kill SIGTERM
```

4   Implementation

The availability of parallel process managers, such as MPD [2], that provide pre-emption of existing long-running jobs and fast startup of MPI jobs, has made it possible to write these commands as MPI application programs. Each command parses its hostlist arguments and then starts an MPI program (with `mpirun` or `mpiexec`) on the appropriate set of hosts. It is assumed that the process manager and MPI implementation can manage `stdout` from the individual processes in the same way that MPD does, by routing them to the `stdout` of the `mpirun` process. The graphical output of `ptdisp` is provided by GTK+ (See http://www.gtk.org).

Using MPI lets us take advantage of the MPI collective operations for scalability in delivering input arguments and/or data and collecting results. Some of the specific uses of MPI collective operations are as follows.

- MPI_Bcast uses `ptcp` to move data to the target nodes.
- MPI_Reduce, with MPI_MIN as the reduction operation, is used in many commands for error checking.
- MPI_Reduce, with MPI_LOR or MPI_LAND as the reduction operation, is used in pttest.
- MPI_Gather is used in `ptdistrib` to collect data enabling dynamic reconfiguration of the list of nodes work is distributed to.

Table 2. Performance of some commands

Number of Machines	1	11	50	100	150	241
Time in seconds of a parallel copy of 10MB over Fast Ethernet	5.6	8.1	10.5	12.2	13.8	14.3
Time in seconds of a parallel execution of `hostname`	0.8	0.9	1.2	1.5	1.8	1.9

– Dynamically-created MPI communicators other than `MPI_COMM_WORLD` are
  used when the task is different on different nodes. An example of this situation occurs when the target specified in the `ptcp` command turns out to be
  a file on some nodes and a directory on others.

The implementation of `ptcp` is roughly that described in [5]. Parallelism is
achieved at three levels: writing the file to the local file systems on each host
is done in parallel; a scalable implementation of `MPI_Bcast` provides parallelism
in the sending of data; and the files are sent in blocks, providing pipeline parallelism. We also use compression to reduce the amount of data that must be
transferred over the network. Directory hierarchies are `tar`red as they are being
sent.

A user may have different user ids on different machines. Whether these
scalable Unix commands allow for this situation depends on the MPI implementation with which they are linked. In the case of MPICH [3], for example, it is
possible for a user to run a single MPI job on a set of machines where the user
has different user ids.

5   Performance

To justify the claims of scalability, we have carried out a small set of experiments
on Argonne's 256-node Chiba City cluster [1]. Execution times for simple commands are dominated by parallel process startup time. Commands that require
substantial data movement are dominated by the bandwidth of the communication links among the hosts and the algorithms used to move data. Timings for a
trivial parallel task and one involving data movement are shown in Table 2. Our
copy test copies a 10MB file that is randomly generated and does not compress
well. With text data the effective bandwidth would be even higher.

In Figure 2 we compare `ptpc` with two other mechanisms for copying a file
to the local file systems on other nodes. The simplest way to do this is to call
`rcp` or `scp` in a loop. Figure 2 shows how quickly this method becomes inferior
to more scalable approaches. The "chi_file" curve is for a sophisticated system
specifically developed for the Chiba City cluster [1]. This system, written in
Perl, takes advantage of the specific topology of the Chiba City network and the
way certain file systems are cross-mounted. The general, portable, MPI-based
approach used by `ptcp` performs better.

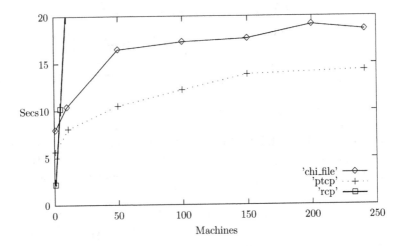

Fig. 2. Comparative Performance of ptcp

6   Conclusion

We have presented a design for an extension of the classical Unix tools to the parallel domain, together with a scalable implementation using MPI. The tools are available at http://www.mcs.anl.gov/mpi. The distribution contains all the necessary programs, complete source code, and man pages for all commands with much more detail than has been possible to present here. An MPI implementation is required; while any implementation should suffice, these commands have been most extensively tested with MPICH [3] and the MPD process manager [2]. The tools are portable and can be installed on parallel machines running Linux, FreeBSD, Solaris, IRIX, or AIX.

References

1. Chiba City home page. http://www.mcs.anl.gov/chiba.
2. R. Butler, W. Gropp, and E. Lusk. A scalable process-management environment for parallel programs. In Jack Dongarra, Peter Kacsuk, and Norbert Podhorszki, editors, *Recent Advances in Parallel Virutal Machine and Message Passing Interface*, number 1908 in Springer Lecture Notes in Computer Science, pages 168–175, September 2000.
3. William Gropp and Ewing Lusk. MPICH. World Wide Web.
   ftp://info.mcs.anl.gov/pub/mpi.
4. William Gropp and Ewing Lusk. Scalable Unix tools on parallel processors. In *Proceedings of the Scalable High-Performance Computing Conference*, pages 56–62. IEEE Computer Society Press, 1994.
5. William Gropp, Ewing Lusk, and Rajeev Thakur. *Using MPI-2: Advanced Features of the Message-Passing Interface*. MIT Press, Cambridge, MA, 1999.

Low-Cost Parallel Text Retrieval
Using PC-Cluster

A. Rungsawang, A. Laohakanniyom, and M. Lertprasertkune

Massive Information & Knowledge Engineering
Department of Computer Engineering, Faculty of Engineering
Kasetsart University, Bangkok, Thailand
{fenganr,b4105118,b4105102}@ku.ac.th

Abstract. We present a parallel vector space based text retrieval prototype implemented on a low-cost PC cluster running Linux operating system, using the PVM message passing library. We also embed the inverted file structure into our proposed prototype for fast retrieval. From several experiments derived from the standard TREC-9 collection, this prototype can index up to 500,000 web pages per hour using a simple x86 machine. We also obtain 5.4 seconds query response time on searching in the one and a half million TREC-9 web pages, using 2 machines.

1 Introduction

Indexing and retrieval of web documents gathered from the Internet are among the challenging problems in high-performance computing issue. Due to its intrinsic nature, heterogeneous web data spans over many computers and platforms interconnected with unpredefined network topology and bandwidth. Its content is changed dynamically, while its volume grows at the exponential rate, till many terabytes now. This triggers the need for efficient tools and powerful computer hardware to manage, retrieve and filter out unrelevant information from the web databases. We here focus on text retrieval although there are techniques to search for other non-textual web data, such as images, mp3, etc., they cannot be applied well on a large scale [1].

There are much research to adopt parallel processing techniques to large-scale text retrieval. Initial implementations were based on overlap-encoding signatures [14,8,2], frame sliced partitioning on signature files [6], parallel inverted file structures [15,13], the vector space model [11,3]. Most of these attempts had been made on the real environment of the supercomputer, such as the Connection Machine [14,15,13], GCel3/512 Parasytec machine [3], high performance Transputer network [2].

Here we propose a low-cost large-scale text retrieval prototype that can index a portion of web documents as a full-text database using PC-cluster running Linux operating system. The prototype is based on the vector space model, implemented using the PVM message passing library [5]. Our prototype can index up to 500,000 pages, around 2 gigabytes of web data, per hour, using a

Y. Cotronis and J. Dongarra (Eds.): Euro PVM/MPI 2001, LNCS 2131, pp. 419–426, 2001.

simple x86 machine. We also include inverted file method into our prototype so as to be able to perform fast document retrieval, and obtain on average 5.4 seconds query response time on searching in TREC-9 small web track collection using 2 simple x86 machines.

We organize this paper in the following way. We first describe the vector space retrieval model and the inverted file structure, and continue with the proposed design of our prototype (both sequential and parallel version) on PC-cluster. Then, we present some experimental results using a standard TREC-9 web collection. Finally, we conclude this paper.

2 Vector Space Model and Inverted File

The vector space retrieval model has been used as the basis for many ranking retrieval experiments, especially the SMART system experimented under the direction of Prof. Salton and his associates [10]. The model assumes that a given textual database is composed of n unique terms. A document d_i can be represented by a vector $(w_{i1}, w_{i2}, w_{i3}, .., w_{in})$, where w_{ij} represents a weight given to term j appearing to document i. Simple binary weight gives the value of w_{ij} equal to 1 if term j presents in document i, and 0 otherwise. The weight w_{ij} can be represented by the frequency of a term, or a specified term-weight, in the given document. This term weight usually provides substantial improvement of retrieval effectiveness [12,7]. A query can be represented in the same manner. To determine which document best matches the query, a simple dot product, or a cosine similarity measure, of the query vector and each document vector is made to rank the resulting documents.

Computing a large number of dot product or cosine similarity operations in a large document collection like web data is a time consuming process, the inverted file structure has been proposed to perform the similarity computation between a given query and only documents that terms in query appear [1]. An inverted file is a list of terms, with each term having links to the documents containing that term. The use of an inverted file structure improves searching time efficiency by several order of magnitude, a necessity for very large collections. The penalty paid for this efficiency is the additional time need to build, more space to keep the inverted file, and a need to update that file as data in the collection changes.

3 Design of Retrieval Prototype

3.1 Sequential WORM

We call our sequential text retrieval prototype, "WORM". WORM is composed of 4 main parts: (see Fig. 1). We exclude the parser that is needed to preprocess the web data into WORM's input format. This parser is responsible for filtering out all HTML tags, non-textual data such as images, keeping some necessary URL links, extracting only portions of text required by user to be a part of a document to be indexed, converting document to WORM format.

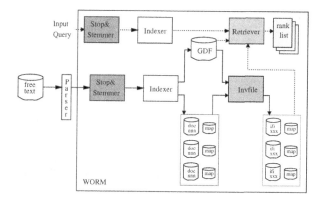

Fig. 1. Sequential WORM retrieval prototype.

WORM's first part, called "Stop&Stemmer", is responsible for removing words that are common in English language, such as "a", "and", "the", by using a specific stopword list [1]. It also fuses and replaces remaining words in each document with stems, general terms for the process of matching morphological term variants, using the Lovin's or the Porter's algorithm as presented in [4].

The resulting stems is then input into WORM's second part, called "Indexer". Indexer reads input stream of documents' stems, records their term-frequencies, updates the document frequency table, and build document vectors using term-frequency weight [7]. The resulting document vectors is called "doc.nnn", to respect SMART weighting convention [12,7], and the main document frequency table is called "GDF" (Global Document Frequency). Since the Linux V2.4.0's file system cannot support a file's size more than 2 gigabytes, WORM's Indexer has been designed to create several chunks of doc.nnn files, of which each is less than 2 gigabyte barrier, while data in each chunk can be navigated by using additional file called "doc.nnn.map".

WORM's third part, called "Invfile", is responsible for building inverted file structure of the whole document collection. It first reads the GDF data and build the main dictionary in memory. It then reads each doc.nnn chunk, statistically weights each document vector according to the SMART convention (e.g. atc, ntc, atn weights), and creates the corresponding inverted file structure, called "ifi.xxx" file, where "xxx" represents the SMART weighting scheme the user gives as indexing parameter. Each ifi.xxx file has its associating map written by WORM, for easy navigation later, in "ifi.xxx.map" file.

WORM's last part, called "Retriever", is responsible for retrieving relevant documents to answer an user query. Retriever first reads the main GDF data, receives input free-text query from the user, removes common English words and substitutes query words by corresponding stems, weights and creates query vector in the same way as that of document. It then reads each ifi.xxx file chunk and computes document scores, using dot product function, and finally re-sorts the document scores and outputs the first k documents, where k is the number

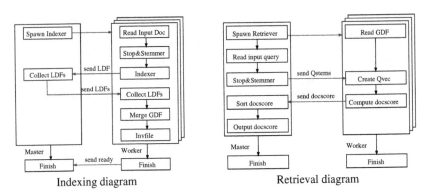

Indexing diagram   Retrieval diagram

Fig. 2. Parallel WORMS.

of documents a user would like to see as his answer in the collection that best match the requested query.

3.2   Parallel WORMS

We call our parallel WORM prototype, "WORMS". WORMS has been devised using PVM message passing library, and master-workers programming model [5]. One WORMS' indexing farm consists of a master process working in cooperation with several worker processes (see Fig. 2). WORMS works as follows: The master first spawns several indexing workers. Each worker then reads input documents, passes them to the Stop&Stemmer, starts the Indexer to update the LDF (Local Document Frequency) table and creates the corresponding document vectors. When there is no more document to index, worker sends its LDF to the master and waits for receiving the LDF of the other machines from the master.

In our design, we need that all document scores computed from each worker can be easily comparable, the master then multicasts each receiving LDF to all workers, except the one that owns the underlying LDF. After receiving all other LDFs, the worker starts its LDF merging process to get its own GDF table using an additional routine called "Merger", and creates the corresponding inverted file structure for all documents it possesses. When finishing its inverted file creation, worker send signal to the master, and terminates itself from the indexing farm.

During retrieval phase, the master first spawns several retrieval workers. Each worker then reads its own GDF table, and waits for input query from the master. The master then reads user input query, passes it to the Stop&Stemmer, multicasts query stems to all workers, and waits for returning document scores. After receiving query stems, the worker creates the query vector, using the same method as that of document vector, computes document scores using its inverted files, sends document scores back to master, and terminates itself from the retrieval farm. The master finally sorts all document scores, and outputs top k documents that best match the input query.

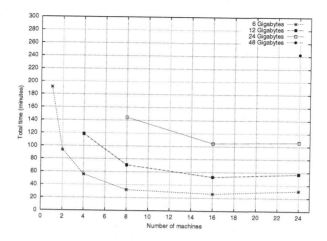

Fig. 3. Average total time needed by WORMS indexing process.

4 Experimental Results

We run several experiments using a cluster of 24 x86 PC based machines. Each machine is equipped with an Athlon 700 MHz CPU, 256 megabytes of SDRAM, and a 10 gigabytes IDE harddisk, interconnected with a simple fast Ethernet network. We choose the small web track TREC-9 documents [9] as our test collection. This TREC-9 collection is composed of 1.6 million, 10 gigabytes of web pages. We also use the TREC-9 standard query set, number 451-500, as virtual user queries to test the Retriever of our prototype.

We first use a special purposed parser to parse the TREC-9 documents, i.e. excluding non-textual data, HTML tags, converting them into WORMS format. The resulting collection, called "6G", is composed of 1,673,490 documents, totally 6 gigabytes. We duplicate the 6G to build more virtual collections of 12, 24 and 48 gigabytes, called "12G", "24G" and "48G", respectively.

4.1 Indexing Experiments

Indexing experiments cumulate the average time needed for reading WORMS's documents, passing them to Stop&Stemmer, creating document vectors using Indexer, merging the LDFs to get the GDF table using Merger, and building the inverted file using Invfile. Fig. 3 depicts the total time needed to index the 6G, 12G, 24G, and 48G document collections. Due to the harddisk space limitation of each machine in our PC cluster, we cannot have timing data for 12G collection using 1 and 2 machines, 24G collection using 1, 2 and 4 machines, and can only have timing data of 48G collection using 24 machines. From Fig. 3, the average total time needed to index the whole 1.6 million documents of the 6G collection is 191 minutes using 1 machine. This means that WORMS can index around 0.5 million web pages per hour.

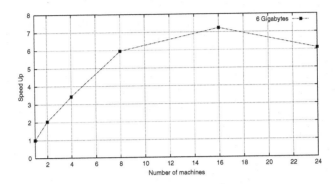

Fig. 4. WORMS speedup curve derived from the 6G collection.

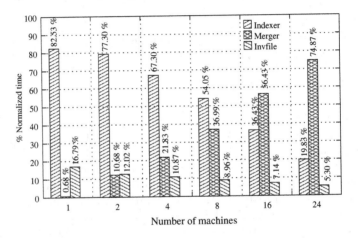

Fig. 5. WORMS normalized indexing time derived from the 6G collection.

Fig. 4 depicts the speedup curve of the 6G collection. Note that speedup decreases drastically when more machines have been added to participate in the indexing farm. The negative result of the 8, 16, and 24 machines seen from both Fig. 3 and 4 astounds us very much, since WORMS needs more time to index the same set of documents. Figure 5 which plots the normalization time needed for each parts of WORMS (i.e. the Indexer, the Merger, and the Invfile, respectively) for the 6G collection, explains all this doubt. When there are more machines, WORMS spends much time for Merger to merge all LDFs data. It spends up to 36.99%, 56.43% and 74.87% of the total time needed for the whole indexing process when using 8, 16 and 24 machines, respectively.

4.2   Retrieval Experiments

Retrieval experiments gather the average time required for reading input queries, passing them to Stop&Stemmer, creating query vectors, computing and resorting

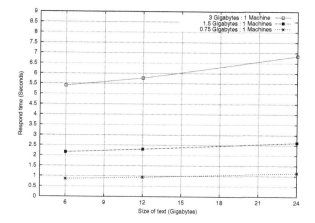

Fig. 6. WORMS average query response time.

document scores. Fig. 6 reports the average query response time that WORMS needs to search for an user query. In these experiments, we add more machines when searching in larger virtual collections. For example, the curve called "3 Gigabytes: 1 Machine" is obtained from the average time to employ 2, 4 and 8 machines to search 50 TREC-9 queries in 6G, 12G and 24G collections, respectively (the average total time for 2 machines needed to search 50 TREC-9 queries in 6G collection is 270 seconds, but we plot the curves using the total time divided by 50). From these experiments, we can see that WORMS Retriever is able to give almost uniform query response time when larger web collection is assigned to search, and more machines are added to the searching farm.

5   Conclusion

In this paper, we propose an efficient vector space based parallel text retrieval prototype, embedded with the inverted file structure. The parallelization of the sequential WORM into new WORMS prototype, using the PVM library and a low-cost cluster of x86 PC based machines, is the contribution key that distinguishes our effort from previous works. Using the small web track TREC-9 collection as the test database, WORMS can index up to 500,000 web pages, or around 2 gigabytes per hour per machine.

We have learned many things from our experiments. The negative result when using more machines during indexing is paid for LDF merging part. When there are more web data to index and more machines participating in indexing farm, WORMS spends much time to merge all LDF data computed from all indexing machines. However, we obtain nearly uniform query response time when adding more machines in addition with more documents in WORMS retrieving farm.

In our future work, we believe that well redesigned of the current LDF merging part is important. In addition, a query response time less than that we have obtained from WORMS is needed in current searching Internet environment. We are looking forward to doing more experiments using the large TREC-9 web track, when we can possess more disk space in each machine in our cluster.

Acknowledgement

We would like to thank all MIKE staffs, especially Jirus, Paricha, for their great help. We also thank to the department of computer engineering, the faculty of engineering for the support of computing facilities. This work is granted by KURDI, Kasetsart University, Bangkok, Thailand.

References

1. R. Baeza-Yates and B. Ribeiro-Neto, editors. Modern Information Retrieval. Addison-Wesley, 1999.
2. J. Cringean, R. England, G. Manson, and P. Willett. Parallel Text Searching In Serial Files using a Processor Farm. In *Proceedings of the ACM SIGIR*, 1990.
3. P. Efraimidis, C. Glymidakis, and B. Mamalis. Parallel Text Retrieval on a High Performance Supercomputer using the Vector Space Model. In *Proceedings of the ACM SIGIR*, 1995.
4. W.B. Frakes and R. Baeza-Yates, editors. Information Retrieval: Data Structures & Algorithms. Prentice Hall, 1992.
5. A. Geist, A. Beguelin, J. Dongarra, W. Jiang, R. Manchek, and V. Sunderam. PVM: Parallel Virtual Machine–A Users' Guide and Tutorial for Networked Parallel Computing. MIT Press, 1994.
6. F. Grandi, P. Tiberio, and P. Zezula. Frame Sliced Partitioned Parallel Signature Files. In *Proceedings of the ACM SIGIR*, 1992.
7. J.H. Lee. Combining Multiple Evidence from Different Properties of Weighting Schemes. In *Proceedings of the ACM SIGIR*, 1995.
8. C. Pogue and P. Willett. Use of Text Signatures for Document Retrieval in a Highly Parallel Environment. *Parallel Computing*, 4, 1987.
9. TREC-9 publications. See http://trec.nist.gov, TREC web site.
10. G. Salton, editor. The SMART Retrieval System, Experiments in Automatic Document Processing. Prentice-Hall, 1971.
11. G. Salton and C. Buckley. Parallel Text Search Methods. *Communications of the ACM*, 31(2), 1988.
12. G. Salton and C. Buckley. Term-Weighting Approaches in Automatic Text Retrieval. *Info. Processing and Management*, 24(5), 1988.
13. C. Stanfill. Partitioned Posting Files: A Parallel Inverted File Structure for Information Retrieval. In *Proceedings of the ACM SIGIR*, 1990.
14. C. Stanfill and B. Kahle. Parallel Free Text Search on the Connection Machine System. *Communications of the ACM*, 29(12), 1986.
15. C. Stanfill, R. Thai, and D. Waltz. A Parallel Indexed Algorithm for Information Retrieval. In *Proceeding of the ACM SIGIR*, 1989.

Parallelization of Finite Element Package by MPI Library

Felicja Okulicka-Dłużewska

Faculty of Mathematics and Information Science
Warsaw University of Technology, Pl. Politechniki 1, 00-661 Warsaw
Poland
okulicka@mini.pw.edu.pl

Abstract. The parallelization of the sequential finite element package is presented. A professional finite element code Hydro-Geo developed at Warsaw University of Technology is used for modelling, calculation and analysis of the elasto-plastic behaviour of the geotechnical constructions. The parallel versions for shared memory and distributed memory machines are developed and implemented on supercomputers Sun 10000E at Warsaw University of Technology and on Sun 65000 and Cray 3TE at Edinburg Parallel Computing Centre. The Cray Fortran compiler directives and the Message Passing Interface library were used for the parallelization of source code

Key words: finite element method, parallel algorithms, supercomputer

1 Introduction

In this paper we will introduce the parallelization of Finite Element Method algorithms. The structure FEM algorithm is very appropriate for parallelization.

The FEM package Hydro-Geo oriented at hydro and geotechnical problems is presented. The program is developed at Warsaw University of Technology and next extended to allow the parallel calculations. The sequential version of the program is the starting point for the development of the parallel versions step by step. The package is composed of the three main programs: the preprocessor for mesh generation and preparing the data, processor for main mechanical calculation and graphical post processor. These programs are running under the management shell. In the paper the parallel version of the processor in which the package numerical algorithm is implemented is described.

Section 2 contains the short description of numerical procedure used in the finite element modelling of the elasto-plastic behaviour of the constructions in geomechanics and implemented in Hydro-Geo package. The section is recalled after [3]. In section 3 the parallel machines, on which the MPI version of the package was developed and tested are presented. The structure of the algorithm implemented in the sequential package processor is described in section 4. In section 5 the algorithms for parallel processes created using MPI on base on the sequential version is described. The examples used for tests with some computing results are presented in section 6.

The distributed version were implemented and tested due the support of the European Community - Access to Research Infrastructure action of the improving Human potential programme (contract No HPRi-1999-CT00026).

Y. Cotronis and J. Dongarra (Eds.): Euro PVM/MPI 2001, LNCS 2131, pp. 427–436, 2001.

2   Numerical Procedure

The virtual work principle, continuity equation with boundary conditions is the starting points for numerical formulation. The finite element method is applied to solve initial boundary value problems. Several procedures stemming from elasto-plastic modelling can be coupled with the time stepping algorithm during the consolidation process. The elasto-plastic soil behaviour is modelled by means of visco-plastic theory (Perzyna, 1966). The finite element formulation for the elasto-plastic consolidation combines overlapping numerical processes.

The elasto pseudo-viscoplastic algorithm for numerical modelling of elasto-plastic behaviour is used after Zienkiewicz and Cormeau (1974). The stability of the time marching scheme was proved by Cormeau (1975). The pseudo-viscous algorithm developed in finite element computer code Hydro-Geo is successfully applied to solve a number of boundary value problems, Dłużewska (1993). The visco-plastic procedure was extended to cover the geometrically non-linear problems by Kanchi et al (1978) and also developed for large strains in consolidation, Dłużewska(1997). The pseudo-viscous procedure is adopted herein for modelling elasto-plastic behaviour in consolidation. In the procedure two times appear, the first (t) is the real time of consolidation and the second time (t) is only a parameter of the pseudo-relaxation process.

The global set of equations for the consolidation process is derived as follows

$$\begin{bmatrix} \mathbf{K_T} & \mathbf{L} \\ \mathbf{L}^T & -(\mathbf{S}+\theta\Delta t\ \mathbf{H}^i) \end{bmatrix}\begin{bmatrix} \Delta \mathbf{u}^i \\ \Delta \mathbf{p}^i \end{bmatrix}=\begin{bmatrix} \mathbf{0} & \mathbf{0} \\ \mathbf{0} & -\Delta t\mathbf{H}^i \end{bmatrix}\begin{bmatrix} \mathbf{u}^i \\ \mathbf{p}^i \end{bmatrix}+\begin{bmatrix} \Delta \mathbf{F}^i \\ \Delta \mathbf{q} \end{bmatrix} \tag{1}$$

where $\mathbf{K_T}$ is the tangent stiffness array, considering large strains effects, $\mathbf{L}$ is the coupling array, $\mathbf{S}$ is the array responsible for the compressibility of the fluid, $\mathbf{H}$ is the flow array, $\mathbf{u}$ are the nodal displacements, $\mathbf{p}$ are the nodal excesses of the pore pressure, $\Delta\mathbf{F}^i$ is the load nodal vector defined below

$$\Delta\mathbf{F}^i=\Delta\mathbf{F_L} + \Delta\mathbf{R}_{I}^{\ i}+\Delta\mathbf{R}_{II}^{\ i} \tag{2}$$

$\Delta\mathbf{F_L}$ is the load increment, $\Delta\mathbf{R}_I^i$ is the vector of nodal forces due to pseudo-visco iteration, $\Delta\mathbf{R}_{II}^i$ is the unbalanced nodal vector due to geometrical nonlinearity. $\Delta\mathbf{R}_I^i$ takes the following form

$$\Delta\mathbf{R}_I^i = \int\limits_{\substack{(i-1)\mathrm{V} \\ t+\Delta t}} \mathbf{B}_{(i-I)}^{T} \mathbf{D}({}_{t+\Delta t}^{t+\Delta t}\Delta\varepsilon_i^{\ vp})_{(i-I)}^{t+\Delta t} \, dv \tag{3}$$

and is defined in the current configuration of the body. The subscripts indicate the configuration of the body, and superscripts indicate time when the value is defined (notation after Bathe (1982)). $\Delta\mathbf{R}_I^i$ stands for the nodal vector which results from the relaxation of the stresses. For each time step the iterative procedure is engaged to solve the material non-linear problem. The i-th indicates steps of iterations. Both local and global criterions for terminating the iterative process are used. The iterations are continued until the calculated stresses are acceptable close to the yield surface, F < Tolerance at all checked points, where F is the value of the yield function. At the same time the global criterion for this procedure is defined at the final configuration of the body. The global criterion takes its roots from the conjugated variables in the virtual work principle, where the Cauchy stress tensor is coupled with the linear part of the Almansi strain tensor. For two-phase medium, the unbalanced nodal vector $\Delta \mathbf{R}_{II}^{i}$ is calculated every iterative pseudo-time step.

$$\Delta \mathbf{R}_{II}^{i} = \mathbf{P} - \int\limits_{\substack{(i-1) \\ t+\Delta t}}^{V} \mathbf{B}_{(i-1)}^{T} ({}_{t+\Delta t}^{t+\Delta t}\sigma^{\prime\,(i-1)} + \mathbf{m}_{t+\Delta t}^{t+\Delta t} p^{(i-1)}) \, {}_{(i-1)}^{t+\Delta t} dv \qquad (4)$$

The square norm on the unbalanced nodal forces is used as the global criterion of equilibrium. The iterative process is continued until both criterions are fulfilled.

3   Parallel Machines

During execution parallel programs create some number of threads or processes, which are running concurrently. The way of designing of parallel programs depends on the architecture of parallel machine where it is executed. Generally we divide the parallel programs in two groups: programs designed and implemented for shared memory machines and programs for distributed environment. Parallel programs for shared memory supercomputers create during execution the set of threads, which have access to shared variables. In programs running on distributed memory machines each thread or process created during the execution has own private memory invisible for others

In COI PW (Computing Centre of Warsaw University of Technology) there is the shared memory multiprocessor Sun 10000E. It can be characterised by: symmetric architecture - 12 processors, 768 MB of RAM, 100 GB of disk memory with the possibility of extension, Solaris 2.5 operating system. The Cray FORTRAN, which enables the automatic parallelization of the sources, is used for implementation of our shared memory parallel Hydro-Geo [7].

The Hydro-Geo distributed version is created using Message Passing Interface (MPI) and tested in Edinburg Parallel Computing Centre on Sun 6500 with 18 processors: 400Mhz UltraSPARC-II, 18 Gbyte shared memory, 108 Gbyte disc space and Peak performance equal 14.4Gflops.

The MPI version can be executed on clusters. The supercomputers are very expensive. Clusters became lately more and more popular because they can be built really cheap, comparing the cost of parallel machines.

4   Sequential Hydro-Geo Processor Algorithm

We are deal with the parallelization of the main part of the package - processor. In the processor the main numerical part of the finite element calculation is implemented. The program is the most time consuming part of the package. It is really worth to be done in parallel to take advantages from the speed up.

The parallel versions are built on base of the sequential one. The structure of the package is not changed. In the first step all auxiliary files for keeping data during the calculation process are cancelled. All data are kept into the memory. We reduce the number of read/write on/from the disk operations. That allows us to parallelize the main loops for calculating the local values and the local matrixes for single elements.

The Hydro-Geo processor algorithm can be written as follows:

```
Start
Data reading, initial computations
For each stage of the construction and each increment of load do
    For each element do
        Calculate the local stiffness matrix, coupling
        matrix, flow matrix
        Calculate the initial stresses
    End do
    Calculate the global set of fully coupled system
    First part of the solver (forward substitution)
    For each plastic point do
        Second part of the solver (backward substitution)
        For each element do
            Calculation of strains and stresses
        End do
    End do
End do
Stop
```

The calculation consists of three parts: first - calculation of local values (ocal stiffness matrix, coupling matrix, flow matrix, the initial stresses), second - calculation of the global set of equations and third - calculation of plasticity in which part the loop for element values is contained. Both loops for element values can be executed in parallel.

5   Distributed Version of Hydro-Geo

For distributed memory machines the number of processes is created during the execution of the parallel program. The processes are distinguished by own unique names called ranks, which are integer numbers. They have their own private variables. Each process has to have copies of all variables needed for calculations. All results calculated by one process and needed by another should be send or broadcast. In our approach only one process reads data. It will be called "master". Others obtain the data by broadcast from master, make calculations using received data and send the results needed for solving global set of equation and printing back to master. They are called "slaves".

The calculation is done concurrently in such a way that the loops, which calculate the local values for each element, are divided between processes. Each process calculates local values connected with single element (the local stiffness matrix, coupling matrix, flow matrix, initial stresses) for own private subset of elements. The subsets are determined at the beginning and remain fixed during whole calculation.

All processes know which subset of elements belongs to each process. Each process keeps data connected with elements for his private subset only. It is a kind of domain decomposition. When the local matrixes are calculated, they are sent to master, which calculates the global matrix for fully coupled system and solves the set of linear equations. The result is broadcast to all processes to allow them to continue the calculation.

For master process - process number 0 the algorithm can be written as follows:

```
Start
Data reading, initial computations
Sharing the computation - determine the subsets of
    elements calculating by separate processes
Broadcast read data to other processes
For  each stage of the construction and each increment of load  do
    For  my subset of the set of elements  do
        Calculate the local stiffness matrix, coupling
        matrix, flow matrix
        Calculate the initial stresses
    End do
    Gather the local stiffness matrixes from all
    processes
    Calculate the global set of fully coupled system
    For  each plastic point  do
        Solve the set of equations
        Broadcast the solution to all processes
        For  my subset of the set of element  do
            Calculation of strains and stresses
        End do
    End do
    Receive the results from all processes
    Print the calculated values
    Barrier()
End do
Stop
```

Processes with ranks greater than 0 receive the read data, initialize the data connected with their subsets of elements, calculate the values connected with their private subsets of elements and send local matrixes to the master. The algorithm for other processes it means for processes with numbers different than 0 is as follows:

```
Start
Receive of data from process 0
For  each stage of the construction and each increment of load  do
    For  my subset of the set of element  do
        Calculate the local stiffness matrix, coupling
        matrix, flow matrix
        Calculate the initial stresses
```

```
End do
Send the local matrixes to process number 0
```
For *each plastic point* do
```
        Receive the solution from the process number 0
```
        For *my subset of the set of element* do
```
                Calculation of strains and stresses
        End do
    End do
    Send the results to process 0
    Barrier()
End do
Stop
```

Some synchronization points are added to secure the proper communication and exchange of data. The implementation is made using Message Passing Interface library for communication and synchronization between processes.

6   Engineering Problems

6.1   Settlement of Tall Building

The settlement of the tall building by means of finite element approach in parallel version is modeled, calculated and analyzed. The main results which can be also presented in graphical way due to the post-processor are: strain, stresses, displacement and plastic points The mesh is presented on Fig 1. The calculated deformation is shown on Fig. 2.

6.2.   Embankment Rising

To study the influence of the large deformation description rising of the embankment on peat is modelled.

The material parameters are listed in the Table 1, where E - Young modulus, v - Poisson's ratio, c - cohesion, ϕ - friction angle, k_0 - initial permeability, n – initial porosity, A - coefficient of the pore closing.

The Coulomb-Mohr yield criterion and the non-associated flow rule is used with dilatancy angle equals zero. The permeable boundary below the peat layer is assumed. The layer is 10m thick. The embankment slope is 1:2. The embankment is built in four stages.

At the beginning the initial stresses are introduced into subsoil, $K_0 = 0.75$. The first stage of the embankment is risen up to the height of 2.0m, the second up to 4.0m, the third up to 6.0m and the fourth up to 7.0m.

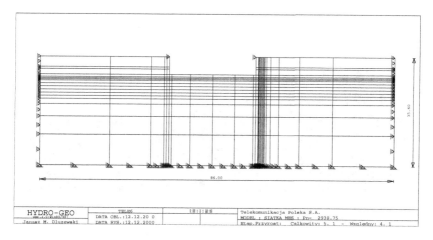

Fig. 1. The mesh for calculated problem

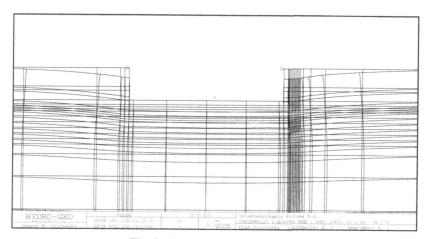

Fig. 2. The calculated deformation

Table 1. Material parameters

E	ν	γ	c	φ	k_0	n	A
MPa	-		kN/m³	kPa	°	m/day	-
Peat	0.2	.37	13.0	10.	18	0.0005	0.75
Sand	10.0	0.30	18.5	0.	33	100-	

The mesh contains 1879 nodes. The example of the calculating results - the pressure are presented on Figure 3.

The three different times are compared: real wall time, processor calculation time – user time from the point of view of system and system time i.e. time for synchronization , management of disk and memory access.

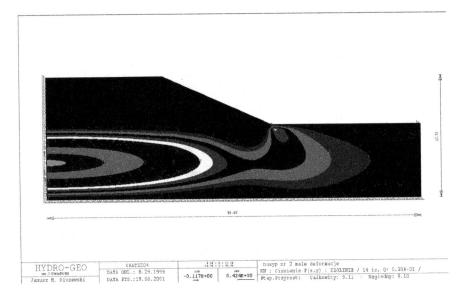

Fig. 3. Excess of the pore pressure

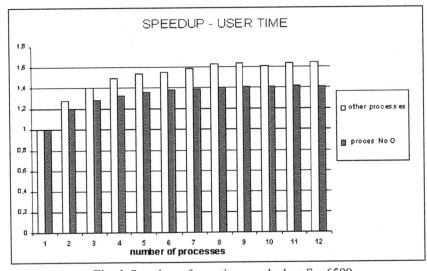

Fig. 4. Speed up of user time reached on Sun6500

The calculations were made for elastic model on Sun6500. The reached speedup is presented on Figure 4. The maximum speed-up for user time is 1,6. By Amdhal law it is bounded by slow procedure for solving the set of linear equations, which is a main part of calculations. The next step of the development of the package is to apply parallel quick procedure to solve the set of linear equations.

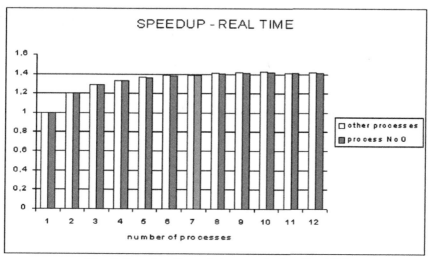

Fig. 5. Speed up for real time reached on Sun6500

7   Conclusions

Parallel programming opens new possibility for modelling and calculating engineering problems. The shared memory programming is easier but distributed programs can be run not only on very expensive supercomputers but on clusters as well.

References

1. B.S.Andersen,   F.Gustavson,   A.Karaivanov,   J.Wasniewski,   P.Y.Yalomov, LAWRA: Linear Algebra with Recursive Algorithms, *Proceedings of the Third International conference on Parallel Processing and Applied Mathematics*, Czestochowa 1999
2. K.M.Chandy, J.Misra, *Parallel Program Des*ign, Addison-Wesley 1989
3. J. M. Dluzewski, *HYDRO-GEO – finite element code for geotechnics, hydrotechnics and environmental engineering*, (in Polish), Warsaw University of Technology, Warsaw, 1997
4. J.M.   Dluzewski,   P.Popielski,   K.Sternik,   M.Gryczmanski,   Modelling   of consolidation in soft soils,in Numerical models in Geomechanics", Preceedings of the seventh International Symphosium Numerical Methods in Geomecjanics, NUMOG VII Gratz, Austria,1-3 September 1999, pp 569-572
5. *MPI: A Message Passing Interface Standard*, Message Passing Interface Forum, University of Tennessee, Knoxville, Tennessee, 1995
6. F. Okulicka, High-Performance Computing in Geomechanics by a Parallel Finite Element Approach, in Applied Parallel Computing, Proceedings of the 5th International Workshop, PARA 2000}, Bergen, Norway, June 2000, Lecture Notes in Computer Science 1947, pp 391-398

7. F. Okulicka, "Block parallel solvers for coupled geotechnical problems",10th International Conference on Computer Methods and Advances in Geomechanics, January 7 - 12, 2001, Tucson, Arizona USA- vol 1, A.A.Balkema,Rotterdam, Brookfield, 2001, pp861-866

8. B.H.V.Topping, A.I.Khan, *Parallel Finite Element Computations*, Saxe-Coburg Publications, Edinburg, 1996

9. J.G.G. van de Vorst, The Formal Development of a Parallel Program Performing LU-Decomposition, *Acta Informatica 26* ,1988, 1-17.

Author Index

Lecture Notes in Computer Science

For information about Vols. 1–2104
please contact your bookseller or Springer-Verlag